Tropical Agroforestry

Alain Atangana • Damase Khasa
Scott Chang • Ann Degrande

Tropical Agroforestry

Springer

Alain Atangana
Renewable Resources
University of Alberta
Edmonton
Canada

Institute of Integrative
and Systems Biology
Université Laval
2208 Pavillon C.-E. Marchand
Quebec G1V 0A6
Canada

Damase Khasa
Forest and Wood Sciences
Université Laval
Quebec
Canada

Scott Chang
Renewable Resources
University of Alberta
Edmonton
Canada

Ann Degrande
West and Central Africa Regional Program
World Agroforestry Centre
Yaounde
Cameroon

ISBN 978-94-007-7722-4 ISBN 978-94-007-7723-1 (eBook)
DOI 10.1007/978-94-007-7723-1
Springer Dordrecht Heidelberg London New York

Library of Congress Control Number: 2013951527

Printed on acid-free paper

Springer is part of Springer Science+Business Media (www.springer.com)

Preface

Agroforestry, the deliberate introduction or retention of trees on farmlands, is widely recognized as a sustainable land-use management practice, including for tropical landscapes that are threatened as a result of non-sustainable land-use practices. There is a need to promote environmentally friendly land management practices and tropical ecosystems that provide local people with their everyday needs for food, cash, shelter and medicine, among others. Agroforestry is a land-use practice that has come of age. Agroforestry techniques have been practiced traditionally worldwide for millennia. Since the recognition of agroforestry as a discipline of agricultural science, numerous institutions have been dedicated to agroforestry research either on a global scale (e.g., the World Agroforestry Centre, formerly known as the International Centre for Research in Agroforestry, ICRAF) or regionally. In addition, agroforestry curricula have been developed for undergraduate and graduate trainings in many universities. Agroforestry practices have been particularly popular in the tropics.

Despite rapid developments in agroforestry practices and improvements in agroforestry theory, textbooks on tropical agroforestry are lacking. The authoritative textbook on agroforestry by Nair (1993) was published 20 years ago, and that was before the advent of tree domestication, an important agroforestry practice today. In addition, many other research activities and emerging issues, such as agroforestry for integrated pest management, biofuel production, carbon sequestration, mitigation of climate change and REDD+(reducing emissions from deforestation and forest degradation, including conservation and sustainable management of forests and the enhancement of forest carbon stocks) mechanism, have become prominent in the agroforestry agenda of recent years. Therefore, there is an urgent need to develop and make available up-to-date educational material on tropical agroforestry for teaching agroforestry to students in agroforestry programs in general, and to students in tropical regions in particular. This textbook strives to provide up-to-date information on tropical agroforestry and, thus, to provide educational material specific to the tropical context.

This textbook is intended for agroforestry students, teachers and practitioners. This textbook is divided into five main parts. Part I describes the tropical biomes and the traditional land-use systems practiced in the tropics. It also highlights the negative impact of non-sustainable land-use systems on land and forest resources. This background is followed by an introduction to agroforestry, including the rationale, history and definition of agroforestry, and the description of major agroforestry systems that are found in the humid and semiarid tropics. Agroforestry tree domestication in the tropics constitutes the fourth and last chapter of the first part of the textbook. In Part II, the benefits and services of agroforestry systems, including tree-crop interactions, nitrogen fixation and mycorrhizal associations, soil conservation, carbon sequestration, biodiversity and integrated pest management in agroforestry are discussed. Research methods in agroforestry, including diagnosing methods, experimental design and on-farm research are covered in Part III. Part IV deals with economic and cultural considerations in agroforestry. The last section, Part V, provides an outlook on agroforestry in the 21st century. Lastly, this part covers biofuel production, phytoremediation, carbon markets, and modeling in tropical agroforestry.

As this is the first edition of the textbook, errors and omissions are unavoidable. The authors would greatly appreciate feedback from readers, instructors, and students who use this textbook for their agroforestry classes. Suggestions and comments can be sent to any of the four authors: alainatangana@yahoo.com (ARA), Damase.Khasa@ibis.ulaval.ca (DPK), Scott.Chang@ualberta.ca (SXC) and a.degrande@cgiar.org (AD).

The authors

Acknowledgments

The completion of this textbook was mainly supported by the International Office of the Department of Foreign Affairs, Trade and Development Canada (formerly the Canadian International Development Agency (CIDA)) through a project on training and management of natural resources in the Congo Basin (FOGRN). The CIDA-funded project was a collaborative effort led by Université Laval with the aim of strengthening forestry and agroforestry training in three countries most representative of the Congo Basin: Democratic Republic of Congo (DRC), Cameroon and Gabon. Project partners were the University of Alberta, Centre d'enseignement et de recherche en foresterie (CERFO), Oxfam-Québec (Canada), and eight institutions in the Congo Basin: University of Dschang and École nationale des eaux et forêts-ENEF (Cameroon); École nationale des eaux et forêts-ENEF (Gabon); the Institut supérieur d'études agronomiques (ISÉA) de Tshéla, Institut supérieur agro-véterinaire (ISAV) Kimwenza, Université de Kinshasa, and the École régionale d'aménagement intégré des forêts et territoires tropicaux-ERAIFT (DRC); and the Réseau des institutions de formation forestière et environnementale d'Afrique centrale-RIFFEAC (Cameroon). We are indebted to CIDA for funding this project. We also acknowledge the Natural Sciences and Engineering Research Council of Canada (NSERC) for partially supporting this work through funding to the University of Alberta (SXC) and Université Laval (DPK).

Many people provided photos for this textbook. We thank Prof. Roger Leakey, Drs. Serge Bobo Kadiri (University of Dschang), Jules Bayala (ICRAF), and Pierre-Marie Mapongmetsem (University of Ngaoundéré, Cameroon) for their support. We also thank Dr. Bill Parsons (Centre for Forest Research, University of Sherbrooke, Canada) and Ms. Kether Hayden for providing language editing of earlier drafts of this textbook. Dr. Estelle Campagnac (IBIS, Université Laval) provided useful comments on an earlier draft of the eightieth chapter of this textbook. Dr. Julie Godbout provided drawings for this textbook.

The authors

Table of Contents

Part V Outlook on Tropical Agroforestry

List of Figures

List of Tables

Part I
Tropical Biomes, Land Use Issues and Introduction to Agroforestry Systems

Chapter 1
Tropical Biomes: Their Classification, Description and Importance

Abstract Tropical biome classifications are mainly based on climate, the structure and function of plant communities, and soil type, with climate being the most frequently used criterion. The most common climate classification systems are the Köppen-Geiger system, which is composed of 5 groups, and Holdridge Life Form classifications, which consist of 38 classes. The Köppen-Geiger system classification is calculated from long term averages of temperature and precipitation at annual, seasonal and monthly time-scales to delineate climatic zones, whereas the Holdridge system uses rainfall and temperature as the main determinants of vegetation type in a given location. Biome classification is also based on soil nutrient status and function of the system. These classification systems have allowed researchers to describe the major biomes that are encountered the tropics, including the Amazon Basin, the Congo Basin, the Borneo-Mekong Basin, and Oceania. Tropical biomes include forests, savannas, mosaics of forest-crop and forest-savanna, woodlands and other plant formations. Tropical savannas include savanna woodlands, savanna parkland, savanna grassland, low tree and scrub savanna, and scrub communities. Tropical forests include mangroves, dense evergreen forests, semi-deciduous, transitional, gallery and fresh swamp forests. In mountainous areas around the equator, tropical cloud forests occur. These dense evergreen forests are located at elevations between 2000 and 3500 m in humid, marine, and equatorial conditions. Tropical forests are significant carbon sinks; they also harbor biodiversity hotspots, and provide agricultural land for people living around or inside these forests. Forest products contribute significantly to tax revenues and the gross domestic products of tropical countries.

1.1 Tropical biomes: Classification and Description

The tropical climate zone covers about 30% of the globe and encompasses several terrestrial biomes, which are groups of distinctive plant and animal communities that are adapted to environmental conditions in these particular geographic locations. While several biome classification systems exist, most are based on climate, the structure and function of the constituent plant communities, and soil type. Within the tropics, these biomes include rainforests, dry forests, savannas, marshes, and wetlands, depending upon their respective moisture regimes. Tropical biomes

A. Atangana et al., *Tropical Agroforestry,* DOI 10.1007/978-94-007-7723-1_1,

Fig. 1.1 A rainforest rich in Caesalpinioideae and Ulmaceae along the Munaya River (Korup) in Cameroon, Africa (Source: Dr. Serge Bobo K. (Univ. of Dschang, Cameroon))

typically extend from the equator to latitudes of 20 °N and S, and include rainforest and tropical dry forest. These systems have biotemperatures (Holdridge 1947) of 24 °C and greater, depending upon the elevational gradient. From 20 to 30 °N and S latitudes, dry forest and grassland extend into subtropical regions. This section describes the most frequently used biome classification systems and tropical biomes.

1.1.1 Classification of Biomes Based on Climate

Climate is the most common criterion used for biome classification. There are at least six classification systems that are based on climate (Köppen-Geiger, Whittaker, Holdridge, Walter, Bailey, and World Wide Fund for Nature). The most commonly used systems are those that were originally described by Köppen (1936) and Holdridge (Holdridge 1947).

1.1.1.1 The Köppen-Geiger Classification System

This classification system is grounded in the concept that indigenous vegetation is the best expression of the climate. It uses annual and monthly mean temperatures, and the seasonality of rainfall to distinguish different biome categories. In this system, vegetation distribution is used to delineate climatic zones. Five groups are included in this classification, as was updated by Peel et al. (2007):

Group A: Tropical/mega thermal

This group includes tropical rainforests, and is characterized by a tropical monsoon climate with heavy rainfalls, and savannah climate or tropical moist and dry climate. The amount of precipitation in the driest month is equal to or greater than 60 mm (Peel et al. 2007), and the temperature of the coldest month is greater or equal to 18 °C (Peel et al. 2007).

Tropical rain forests (Fig. 1.1) are mostly found in Central and West Africa, Amazonia, Central America, Mexico, Madagascar, Southeast Asia, and in many islands

Fig. 1.2 An example of a savanna ecosystem in northern Cameroon, with sweet detar (Detarium microcarpum Guill. & Perr. Fabaceae) as the dominant tree species (Source: P. M. Mapongmetsem, University of Ngaoundéré, Cameroon)

of the Pacific, Caribbean, and Indian Oceans (Peel et al. 2007). Vegetation of the rainforests includes semi-deciduous forests, which are found inland, and evergreen forests, which usually grow along coasts and have heavy rainfall. In semi-deciduous forests of Cameroon (Central Africa), mean annual rainfall is about 1962 mm (Letouzey 1985). The highest annual rainfalls, which can be up to 11000 mm, occur at Debunscha, Cameroon, on the coast of the Atlantic Ocean. The area would be classified as evergreen moist forest and has a monsoon climate. While the pattern of rainfall in semi-deciduous forest is bimodal, that of an evergreen forest is unimodal. Mean annual temperature in rainforests can vary between 23 and 25 °C. Relative humidity is usually between 73 and 84 %, but can rise on some mornings to 100 % in the coastal zone.

Savanna biomes (Fig. 1.2) have distinct wet and dry seasons. During the wet season, rivers flow and plant growth is lush. In the dry season, which is cooler than the wet season, some rivers and streams dry up and many plants shrivel, leading to migration of animals in search of new grazing areas. Temperatures vary between 20 and 25 °C in the winter (dry season), and between 25 and 30 °C in the wet season. Annual rainfall is between 100 and 400 mm.

Group B: Arid

Fig. 1.3 Semiarid zone in northern Cameroon with black plum or Vitex doniana Sweet (Verbenaceae) (Source: P. M. Mapongmetsem, University of Ngaoundéré, Cameroon)

Arid climates, where mean annual precipitation ten-fold lower than the precipitation threshold (amount of rainfall for a given duration that generates a critical discharge in a given river cross-section) and rainfall is lower than potential evapotranspiration, are found in desert areas or in the tropics. The climate is characterized by annual temperatures that are lower than those encountered at similar elevations elsewhere. These conditions are intermediate between humid and desert climates and tend to support scrubby vegetation, which is dominated by grasses (Fig. 1.3). Peel et al. (2007), in their updated world map of the Köppen-Geiger classification, included desert (mean annual precipitation lower than 5 times the precipitation threshold) and steppe (mean annual precipitation greater than or equal to 5 times the precipitation threshold) in this section. Steppes are divided into two categories: 'hot' (mean annual temperature greater than or equal to 18 °C) and 'cold' (mean annual temperature lower than 18 °C).

Group C: Temperate

In temperate climates, the average temperature of the hottest month is greater than 10 °C, and varies between 0 and 18 °C in the coldest month (Peel et al. 2007). Temperate climates can be divided into five categories: hot summer, warm summer, cold summer, dry summer, and dry winter. Hot summer, warm summer and cold summer climates are determined by temperature, and such climates do not have a dry season. Dry summer climates occur when the amount of precipitation of the driest month in the summer is less than 40 mm, and less than 1/3 of the precipitation of the wettest month in the winter (Peel et al. 2007). Dry winter climates are those characterized by precipitation in the driest month of winter being less than 1/10 of the precipitation of the wettest month in the summer. Temperate climates include Mediterranean, humid subtropical and oceanic climates, together with temperate climates with dry winters, and subpolar oceanic climates.

Group D: Cold

An area has a cold climate when the temperature of the hottest month is greater than 10 °C and that of the coldest month is less than or equal to 0 °C. Like areas of the temperate zone (Group C), this group can also be divided into five categories: dry summer, dry winter, hot summer, cold summer, and warm summer climates. The cold climate group can occur in continental climates with hot summers, hemi-boreal climates, continental subarctic or boreal climates, and continental subarctic climates with extremely severe winters.

Group E: Polar

The mean temperature of the hottest month in a polar climate is less than 10 °C. This group includes tundra polar, where the mean temperature of the hottest month is above 0 °C, and cold polar, where the mean temperature of the hottest month is less than equal to 0 °C (Peel et al. 2007).

1.1.1.2 The Holdridge Life Zones System

Four properties are included (three properties in Holdridge's original 1947 formulation) in the Holdridge Life Zones system for classifying global vegetation (Holdridge 1967):

- Total annual precipitation (evaluated on a logarithmic scale);
- Potential evapotranspiration (PET);
- Evapotranspiration potential (ETP) ratio (mean annual PET divided by mean total annual precipitation);
- Mean annual bio-temperature (i.e., average temperature above freezing, after eliminating below-freezing temperatures, together with those exceeding 30 °C; Holdridge 1947).

In this system, rainfall and bio-temperature are the main determinants of the vegetation type in a given location. Latitude, elevation and humidity are used to establish a life zone function (Fig. 1.4). The Holdridge classification system includes 38 classes that range from polar desert to tropical rainforests, and those latter classes include tropical desert, tropical desert scrub, tropical thorn woodlands, tropical very dry forests, tropical dry forests, tropical wet forests, tropical moist forests, and tropical rainforests. Tropical rainforests are also known as lowland equatorial evergreen rainforests; these forests receive more than 2000 mm of annual precipitation, and are found in the Amazon basin, Congo basin, Indonesia and New Guinea. Tropical moist forests include moist deciduous forests, montane rainforests and flooded forests. Montane rainforests can be divided into premontane and montane cloud forests which are found on the Andes, Mount Cameroon, Monteverde Costa Rica, Ruwenzori' Mountains of Uganda and the Democratic Republic of Congo. Tropical moist, wet and rainforests are found in Africa, Asia, South and Central America, Australia, and the Pacific Islands (Oceania biogeographic realm). They are characterized by high rainfall and mean annual temperature ranging between 21 and 45 °C. Soils are mostly Oxisols (Ferralsols, FAO/UNESCO classification). Tropical rainforests are also named after the dominant species occurring in these forests (e.g., dense evergreen Ulmaceous and Caesalpiniaceous forests; lowland mixed dipterocarp forests) or seasonal events (e.g., tropical seasonal rainforests).

Tropical thorn woodland, tropical dry and very dry forests, or tropical desert scrub in this classification, surround tropical wet, moist and rainforests. The dry forests are found in northeastern Australia, Myanmar (Burma) and Thailand, India, the Sahel region and the southern part of Africa, and Central and South America. These

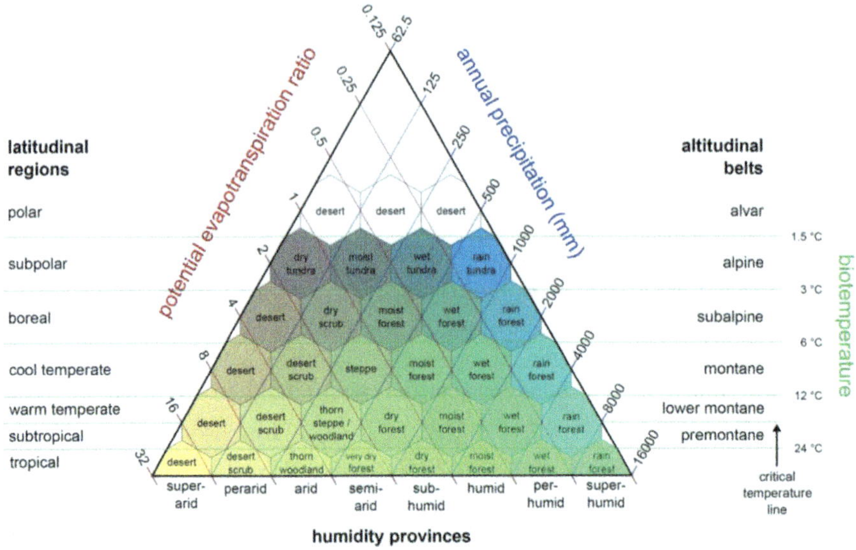

Fig. 1.4 The Holdridge life zone classification scheme (Source: http://en.wikipedia.org/wiki/Holdridge_life_zones)

ecosystems, which are also known as savannas and dry forests, constitute most of the semiarid tropic biomes. Savanna biomes are also classified as savanna woodland, savanna parkland, savanna grassland, low tree and shrub, thicket, and scrub savannas (Murphy et al. 1986). Tropical savannas are found in Australia, India, Africa (Sahel and southern Africa), Central and South America, and in the Caribbean. Those in Latin America include the Brazilian and Paraguayan "Cerrado" (Portuguese or Spanish for "closed") and the Venezuelan-Colombian "Llanos" (Spanish for the "plains"). The Cerrado is characterized by wooded savannas, park savannas, forest savannas, savanna wetlands, and gramineous-woody savannas, whereas the Llanos include savannas and flooded grasslands. In South America, semiarid tropical biomes also include "Caatinga" (dry forest, from the Tupi word meaning "white forest") and "Chaco" (a mosaic of dry woodlands and savannas, from the Quecha word meaning "hunting land"). Chaco constitutes the second largest biome in South America after the Amazon, with an area of 1200 km^2 extending into Argentina, Bolivia and Paraguay.

1.1.2 Other Classification Systems of Biomes

1.1.2.1 Forest Classification According to Plant Species

Forests are classified according to the 'temperament' of their constituent species and successional phase. Vegetation type at a given location depends on the climate, the physico-chemical properties of the soil, topography, elevation, and site history.

The functions of all combined species yield a community function. Thus, community classification implies a functional classification. Plant communities are often named after a predominant or group of species (e.g., Caesalpiniaceous and Ulmaceous evergreen forests in Cameroon).

The temperament of a species involves its growth and mode of reproduction. Oldeman and van Dijk (1991) have described two temperaments in tropical trees that occur within closed-canopy forests, which they refer to as 'fighters' and 'gamblers'. Fighters produce only a few seedlings, which grow slowly and can survive shady conditions. The life expectancy of these seedlings is high, resulting in higher rates of spread despite the low number of seeds that the species produces. In contrast, gamblers produce many seedlings and scatter them widely. The seeds cannot survive in shady or overgrown areas and need light entering through gaps in the canopy to grow rapidly. Gaps are often created by the fall of large-canopy trees. Gambler reproductive success depends on the few seeds that end up in safe sites, and which will be able to flourish.

Plant species that invade recently abandoned lands or pastures, or perturbations that destroy the structure of mature forests, can initiate forest succession. Simply put, forest succession is the natural replacement of plant species in an area, resulting from the creation of new habitat or the disturbance of the original habitat. Disturbances can be caused by logging, agricultural practices, or treefall that is incurred through windthrow. Pioneer species enter and dominate the newly formed habitat, and are slowly replaced by more stable communities until a new equilibrium is reached. Pioneer tree species (e.g., *Musanga cecropioides* R. Br. & Tedlie, Urticaceae; *Vernonia arborea* Buch.-Ham ex. Buch –Ham, Asteraceae; *Vernonia amygdalina* Delile, Asteraceae; *Trema orientalis* (L.) Blume, Cannabaceae) are generally fast-growing but have wood with a low specific-gravity (0.29; Kpikpi 1992).

1.1.2.2 Forest classification Based on Soil Nutrient Status

Soil fertility can be used for forest ecosystem classification (Jordan 1985). Soils in tropical rainforests generally have moderate to low fertility levels (Table 1.1; Sanchez 1981), despite having high humus contents. Accumulation of iron and aluminum oxides and the concomitant loss of silica and base cations characterize laterization, a process of intense weathering that results in the red-colored Oxisols (7th approximation, Ferralsols, UNESCO/FAO) typical of tropical rainforests. Oxisols, together with and Ultisols (Acrisols, FAO), which are strongly leached soils with less than 35 % base saturation, constitute the main soil orders in the Congo basin. In the forest lowlands of Amazonia, the decreased weathering observed in a sequence of Spodosols, Oxisols, Ultisols and Alfisols represents a gradient of increasing fertility (Moran et al. 2000). However, the gradient of soil fertility in tropical forest also depends on the pattern of nutrient cycling (Vitousek and Sanford 1986). On the basis of soil fertility levels, tropical moist forests have been classified as (i) productive forests supported by moderate fertile soils (ii) forests on oxisols/ultisols that are rich in nitrogen, and which cycle small quantities of phosphorus and calcium, and

Table 1.1 Proportional extent of major soils of the tropics. Sites with mean temperature >22 °C, annual precipitation >1500 mm, and a dry season less than 4 months year^{-1} are included (Sanchez 1981)

Soil and soil fertility class	Area (10^6 ha)	Percentage of area
Moderately lo very low		
Oxisols	525	35.3
Ultisols	413	27.7
Moderately fertile		
Alfisols	53	3.6
Tropepts	94	6.3
Andepts	12	0.8
Mollisols	7	0.5
Fluvents	50	0.3
Vertisols	5	0.3
Other	2	0.1
Very low		
Spodosols	19	1.3
Psamments	90	6.0
Variable		
Aquepts	120	8.1
Low		
Lithic (shallow)	72	4.8
Histosols	27	1.8
Total	1489	

(iii) montane tropical forests that are low in nitrogen (Vitousek and Sanford 1986). Jordan (1985), in contrast, used three gradients to classify tropical forests. The first gradient correlated with the state of weathering of the soil in evergreen forests, and allowed the distinction to be made between oligotrophic forests and eutrophic forests on highly weathered soils (oxisols, spodosols; Jordan 1985). The second gradient revealed altitudinal changes in ecosystem, whereas the third gradient was used to classify moisture-stressed forest ecosystems (Jordan 1985).

Oligotrophic forests are adapted to very low fertility soils, while eutrophic forests establish in relatively fertile soils. Net primary productivity is low in oligotrophic forests, and much higher in eutrophic forests. Eutrophic forests have greater diversity, fewer aboveground roots, lower root-to-shoot ratios, broader leaves, greater nutrient concentrations in leaves, and higher leaf litter decomposition rates.

1.1.2.3 Forest Classification Based on Function

Function is the most important criterion for forest classification among various users of forest products. Forests can be classified as productive, cultural, community and conservative in their function. Production forests are mostly intended for wood production (through logging), whereas cultural forests are intended for ritual or ceremonial services. Community forests belong to peasants, who develop and apply a management plan to resources found in these forests. Natural parks and forest reserves serve as conservation areas.

1.1.2.4 Classification of Semiarid Tropical Biomes

Semiarid tropical biomes consist mainly of dry woodlands and savannas. Savannas were classified on the basis of their physiognomy and floristic composition (Cole 1986). This classification includes savanna woodlands, savanna parkland, savanna grassland, low tree and shrub savanna and thickets, and scrub communities. Cole (1986) reported clear distinctions among the five categories of savanna vegetation from one another both physiognomically and floristically in Africa and Australia. According to this author, savanna woodlands (Miombo forests) are characterized by *Brachystegia* (Caesalpinioideae), *Isoberlinia* (Fabaceae), and *Julbernadia* (Fabaceae) woodlands, whereas savanna parklands are dominated by *Acacia* (Fabaceae), *Terminalia* (Combretaceae), *Piliostigma* (Caesalpinioideae), and *Combretum* grasslands. *Hyparrhenia*, *Themeda*, *Setaria* and *Echinochloa* grasslands occur in savanna grasslands, whereas *Chrysopogon*, *Aristida* and *Cenchrus* grasslands dominate low and tree shrub savannas. Baruch (2005) classified savannas on the basis of floristic composition, community physiognomy, species richness and plant cover. Bridgewater et al. (2002) described six savannas types in Belize on the basis of vegetation, including grassland and scrub savanna, pine-palmetto savanna, palmetto thicket, savanna orchard, woodland and pine ridge, and oak thicket.

Dry woodlands are classified on the basis of functional, structural and successional traits of the vegetation (Murphy and Lugo 1986). Functional traits include net primary productivity, foliage persistence and reproductive phenology, whereas structural traits include the number of tree species and canopy strata, complexity index (i.e., the product of number of species, basal area (m^2 0.1 ha^{-1}), maximum tree height (m), and number of stems 0.1 ha^{-1} times 10^{-3} in a 0.1 ha plot; Holdridge 1967; Holdridge et al. 1971), leaf area index, ground vegetation cover and basal area of trees. Successional traits include resistance to disturbance, resilience, vegetation cover and longevity of the soil seed bank (Murphy and Lugo 1986). The same authors also classified dry forests on the basis of life zone and forest utilization.

In Africa, semiarid biomes surround the Congo basin, and consist of the Sahel (immediately south of the Sahara desert), Sudanian savanna (consisting of the Niger, Lake Chad and Middle Nile Basins, which are three physiographically distinct regions), Guinean and Congolian forest-savanna mosaics, the Serengeti (savanna plains found in Kenya and Tanzania), grasslands, bushlands, thickets, Miombo (Swahili for *Brachystegia* spp.) and Mopane (*Colophospermum mopane* (Kirk ex Benth) Kirk ex J.Léon., Caesalpinioideae) woodlands, Kalahari woodlands, bushveld (in southern Africa, straddling the Republic of South Africa, Botswana and Zimbabwe), and forest-savanna mosaics of the Lake Victoria Basin (in Central and East Africa). In Australia and Southeast Asia, semiarid biomes mainly consist of tropical savannas and grasslands, while cerrado (eastern Brazil), llanos (Columbia and Venezuela), gran chaco (eastern Bolivia, Paraguay, northern Argentina, and parts of southern Brazil), and campos montane savannas and tropical savannas (eastern Brazil) are found in Latin America.

1.1.3 Main Tropical Humid and Semiarid Biomes

Tropical biomes are encompassed, for the most part, by the Tropic of Cancer (23°26' N) and the Tropic of Capricorn (23°26' S). Tropical rainforests are surrounded by semiarid biomes, and are found in the Congo basin, Madagascar, Central and South America, and Southeast Asia and Australia.

1.1.3.1 Congo Basin Rain Forests

Tropical forests that are found in the Congo basin can be classified as 1) mangroves, 2) dense evergreen forests, which are found along the Atlantic Coast of Angola, Gabon, Equatorial Guinea, Cameroon, Congo, Democratic Republic of Congo (DRC), and Nigeria, 3) semi-deciduous forests that are found in north-east Gabon, southeast Cameroon, southwest Central African Republic (CAR), north Congo, and part of Congo basin in DRC, 4) transition forests that are found in the eastern part of coastal evergreen formations in the north and south DRC evergreen forest, 5) gallery forests that are found along rivers such as the Sanaga, and 6) fresh swamp forests that are located around the Congo River (Letouzey 1985).

While providing detailed descriptions of each of the above systems is beyond the scope of this textbook, we provide some brief description of some of the important forest types below. For example, evergreen forests are characterized by high rainfall, which can average as much as 2,000–3,000 mm per year, and are often mixed with mangroves. The forests are located along the coast of the Atlantic coast of Africa. The most common species of evergreens belong to the clade Fabideae (Angiosperm Phylogeny Group III), such as bitter bark or cherry Mahogany (*Sacoglottis gabonensis* (Baill.) Urb.; Humiriaceae), which is associated with Okoumé or Gaboon Mahogany (*Aucoumea klaineana* Pierre, Burseraceae; in Gabon) and dwarf red ironwood or false shea (*Lophira lanceolata* Van Tiegh. Ex Keay, Ochnaceae; in Cameroon). In these forests, mean annual temperature varies between 26 and 27 °C, and relative humidity is very high, often exceeding 90 % (Letouzey 1985).

Tropical mangroves consist of trees and shrubs that grow most often on saline coastal sediments. Trees of the genus *Rhizophora* dominate mangrove forests. Dense evergreen forests are found near the equator, and are characterized by high annual rainfall (more than 2,000 mm). Most often, in mountainous areas around the equator, tropical montane cloud forests, which are dense evergreen forests located between 2000 and 3,500 m above sea level, and sometimes above 1,000 m (Hawai'i) in the humid, marine and equatorial conditions, can occur. Tropical montane cloud forests differ from lowland evergreen forests in terms of their reduced tree height and increased trunk size. Mean annual rainfall is very high (it can reach 11,000 mm, as in the case at Debunscha, on the slopes of Mount Cameroon) and high humidity.

Semi-deciduous forests are less humid. Mean annual rainfall can vary between 1,200 and 1,600 mm, and relative humidity usually sits between 73 and 85 %. The average annual temperature is between 23 and 25 °C. The flora is a combination of evergreen and deciduous species (in the dry season). Obeche (*Triplochiton scleroxylon* K. Schum, Malvaceae) and *Terminalia superba* Engl. & Diels (Combretaceae) are among the most common species.

Transition forests, which are located between dense evergreen and semi-deciduous forests, are characterized by abundant evergreens and a few deciduous species. Swamp forests are less dense than upland forests. Medium-sized trees in swamp forests develop stilts, and the most abundant species are *Oubanguia africana* Baill. (Scytopetalaceae (APG: Lecythidaceae)), and bubinga or kevazingo (*Guibourtia demeusii* Benn., Fabaceae). Gallery forests are forests forming corridors along riverbanks. These forests are common in tropical Africa, Latin America and Southeast Asia (e.g., along the Mekong River). Fresh swamp forests are forests that are inundated with fresh water in flood plains. They are common in the Amazon and in tropical Africa.

On the basis of vegetation cover and geography, tropical forests can also be classified as dense or lowland dense forests, sub-montane forests that are found between 900 m and 1,500 m above sea level, montane forests that are found over 1,500 m above sea level, swamp forests, mangroves, mosaic of forest-crop and forest-savanna, woodlands and other plant formations (Letouzey 1985). Dense forests occupied approximately 162 million ha over a 4048,470 km^2 area of Cameroon, Central African Republic, Congo, Democratic Republic of Congo, Equatorial Guinea, and Gabon (de Wasseige et al. 2012), representing the largest forest type in the region. Lowland humid forests occupy 88% (142,183,413 ha) of the land in the dense forest zone of the Congo Basin (de Wasseige et al. 2012).

1.1.3.2 Central and South American Rain Forests

Rainforests in Central and South America extend from southern Mexico to Bolivia. These forests mostly consist of moist deciduous and semi-evergreen seasonal forests, which are found in southern Mexico, some islands in the Caribbean, Panama, Costa Rica, Nicaragua, Honduras, and Belize, and lowland equatorial forests mostly that are found in the Amazon Basin (Brazil, Bolivia, Peru, Ecuador, Colombia, Venezuela, Guyana, and Suriname). These forests are also found in coastal Brazil, and northern and western South America (from Peru to Venezuela). The climate is warm and humid, with mean annual shade temperature of 23 °C, and mean annual rainfall of 2,300–3,200 mm in Upper Amazon. Soils are mostly Oxisols on plateaus, Inceptisols and Ultisols on slopes, with Andean alluvium. Within this region, the dominant tree species are members of the Arecaceae, Moraceae, Myristicaceae, and Violaceae (Pitman et al. 2000), together with the Dipterocarpaceae and flowering legumes. However, ter Steege et al. (2006) reported that the most abundant tree families in seven of the nine countries of the Amazon Basin and the Guiana shield are Fabaceae, Sapotaceae, Lecythidaceae, Moraceae, Burseraceae, Chrysobalanaceae, Malvaceae, Euphorbiaceae, Lauraceae and Myristicaceae, with the Fabaceae accounting for one-quarter of all large trees. These authors also identified congruence between gradients in tree composition and function and gradients in soil fertility and dry season length in Amazonian forests. Campbell et al. (1986) and Fujisaka et al. (1998) also reported that the most abundant tree families in Brazilian

Amazon forests was the Fabaceae, with Poaceae, Caesalpinioideae, Euphorbiaceae, Palmae, Lechythidaceae, Moraceae, Bombacaceae and Sterculiaceae being other abundant families in the region. The most abundant forest species were *Psychotria* sp. (Rubiaceae), *Protium apiculatum* Swart (Burseraceae) and jutahy (*Dialium guianense* (Aubl.), Fabaceae; Fujisaka et al. 1998). The most abundant forest tree species in the Brazilian Amazon are *Orbignya* (Arecaceae), Brazil nut (*Bertholletia excelsa* H. & B., Lecythidaceae), and *Theobroma speciosum* Willd. Ex Spreng. (Sterculiaceae; Campbell et al. 1986; Fujisaka et al. 1998).

1.1.3.3 Southeast Asian Tropical Rain Forests

South East Asian tropical rainforests are located between 20° N and 16° S latitude, and longitudes 95 to 105° E. These forests cover 12 countries, namely Brunei, Cambodia, China, Indonesia, India, Laos, Malaysia, Myanmar, the Philippines, Singapore, Thailand and Vietnam. The climate associated with those forests is hot and humid year-round, and the temperature of the coldest month averages at least 18 °C, whereas daily temperature varies between 10 and 25 °C, and the letter code used under Köppen's classification for tropical rainforest climate is Af. The forest canopy is dominated by dipterocarp species (Ashton 1988; Hamann et al. 1999), which is typical of wet tropical evergreen forests that are found across Southeast Asia. Dipterocarps are commonly mixed with Myristicaceae (the nutmeg family) or Fabaceae. Other species dominating forest canopy in Southeast Asian rainforests are Lauraceae, Sapotaceae, Burseraceae and Melastomaceae (Hamann et al. 1999). Lowland evergreen forests, which are also known as lowland mixed-dipterocarp forests in Southeast Asia, can also be found in Southwest India, i.e., the South Western Ghats rainforests, which are characterized by the abundance of *Dipterocarpus indicus* Bedd. (Dipterocarpaceae), *Kingiodendron pinnatum* (DC.) Harms (Fabaceae), *Humboldtia brunonis* Wall. (Fabaceae-Caesalpiniodeae), *Vateria indica* Linn (Dipterocarpaceae) and *Myristica dactyloides* Gaertn. (Myristicaceae; Pascal 1984; Pélissier 1998). These forests are located in the foothills of Ghats, between 500 and 600 m elevation. The climate is hot and humid, and the annual rainfall is above 5,000 mm, whereas the annual mean temperature is 22.5 °C.

Seasonal tropical rainforests are also found in Southeast Asia. For instance, these forests can occur in the valleys or foothills of Southwest China, at altitudes ranging between 500 and 900 m asl. These forests are characterized by three formations, with the *Terminalia myriocarpa* Van Heurck and Mull.Arg. (Combretaceae) and *Pometia tomentosa* (BI.) Teysm. Et Binn. (Sapindaceae) formations occupying the largest area (Zhang and Cao 1995; Zhu et al. 2006).

1.1.3.4 Australian Savannas

Savannas are defined as landscapes with a continuous grass layer and scattered trees (Scholes and Archer 1997). In Australia, savanna vegetation ranges from pure grassland to dense woodlands and can be classified on the basis of plant available moisture and nutrients. Australian savannas can exist as *Eucalyptus* savannas, savannas of *Acacia aneura* (F.Muell. ex. Benth.) and various mixtures of these species with grasses (Walker and Landridge 1997). Mott et al. (1985) and McKeon et al. (1990) have subdivided Australian savannas into numerous types. 1) The monsoon grasslands are dominated by many native graminoids that are also widely spread throughout Australia, Africa, southern Asia, and Oceania, and which include kangaroo grass (*Themeda triandra* Forssk.), firegrass (*Schizachyrium fragile* (R.Br.) A.Camus), *Heteropogon* spp., and *Sorghum* spp., together with golden beard grass (*Chrysopogon fallax* S.T.Blake), which is endemic to the Australasian realm. 2) Tropical tall grass is dominated by narrow-leaved ironbark (*Eucalyptus crebra* F.Muell.), white gum (E. *alba* Reinw. ex Blume), and bloodwood (*Corymbia dichromophloia* (F.Muell.) K.D.Hill & L.A.S.Johnson = *Eucalyptus dichromophloia* F.Muell.) in the overstory, while the understory consists mainly of tanglehead (*Heteropogon contortus* (L.) P. Beauv. ex Roem. & Schult.), *Themeda* and species of beardgrass or bluestem (*Bothrichloa*). 3) Sub-tropical tallgrass, mildgrass on clay, which results from clearing of *Acacia harpophylla* forests; dominant grasses are silky blue grass (*Dichantium sericeum* (R.Br.) A. Camus), Australian bluestem (*Bothrichloa bladhii* (Retz) S.T. Blake), and species of windmill grass (*Chloris*), Queensland blue grass (dominated by *D. sericeum*), blue grass-browntop (dominated by *D. fecundum* S.T. Blake, silky browntop (*Eulalia fulva* (R.Br.) Kuntze), *Cyprus* spp. and *Sorghum* spp.), mildgrass (dominated by *Eucalyptus populnea* F. Muell. or *Eucalyptus microneura* Maiden & Blakely, with the understory consisting of *Aristida* spp., *Bothriochloa bladhii* (Retz.) S.T. Blake, *B. decipiens* Kuntze and *Chloris* spp.), *Acacia* grasslands (the understory is dominated by *Digitaria* spp., *Monochather paradoxus* (R.Br.) Steud., *Eriane* spp.), Gidgee pastures, Tussock grasslands (dominated by *Astrebla*, *Iseilema* and *Dactyloctenium* species), channel pastures (dominated by Chenopods, *Trigonella* spp.) and Hummock grasslands (dominated by *Spinifex* spp.). Eucalypt species are abundant in northern Australia savannas. Other commoner tree species of northern Australian savannas include *Erythrophleum chlorostachys* (F.Muell.) Baillon, *Xanthostemon paradoxus* F.Muell., *Terminalia ferdinandiana* Exell, *Planchonia careya* (F.Muell.) R.Knuth, *Cochlospermum fraseri* Planch. and *Casuarina* spp.

Australian savanna ecosystems are used as pastures. Mean annual rainfall of Australian savannas ranges from 160 to 1,204 mm, and soil textures range from sands or loams to clays (Walker and Landridge 1997; Ludwig et al. 1999). The composition and structure of Australian savannas are strongly influenced by rainfall and soil texture. Indeed, Williams et al. (1990) found that woody species richness, height and diameter decline with decreasing rainfall and increasing soil clay content.

1.1.3.5 African Savannas

African savannas are found on the edges of tropical African rainforests between tropical forests and desert biomes. African savannas consist of northern African savannas (Sahel and Sudanian savannas) above the Equator, savannas below the

Equator, and east African savannas (i.e., Serengeti plains and Masai Mara) east of the Congo Basin. The Sahel is a wide band of semi-arid ecoregions consisting of savannas, grasslands, steppes, and thorn shrublands. Sahel ranges from Senegal and Mauritania on the Atlantic coast to Sudan and Eritrea on the Red Sea coast, and lies between the Sahara to the north and the wooded Sudanian savannas to the south. The most abundant woody species in northern African savannas are kinkeliba or *Combretum micranthum* G. Don (Combretaceae), *Grewia bicolor* Juss. (Malvaceae), *Pterocarpus lucens* Lepr. ex Guill. & Perrott. (Papilionoideae), and *Anogeissus leicoarpus* (DC.) Guill. & Perr. (Combretaceae) (Couteron and Kokou 1997). Other species found in northern African savannas are *Acacia* spp., *Boscia senegalensis* (Pers.) Lam. Ex Poiret (Capparaceae), *B. salicifolia* Oliv. (Capparaceae), *Adansonia digitata* L. (Malvaceae), *Commiphora Africana* (A.Rich.) Engl. (Burseraceae), *Combretum glutinosum* Perr. Ex DC. (Combretaceae), *Combretum nigricans* Lepr. Ex Guill. & Perr. (Combretaceae), *Grewia flavescens* L. (Malvaceae), *Guiera senegalensis* J.F. Gmel (Combretaceae) and *Dalbergia melanoxylon* Guill. & Perr. (Fabaceae; Couteron and Kouakou 1997). Other indigenous species of the Sahel are Marula tree (*Sclerocarya birrea* (A. Rich.) Hochst.; Anacardiaceae), shittah tree (*Acacia seyal* Del.; Fabaceae) and tamarind (*Tamarindus indica* L. Fabaceae-Caesalpinioideae). These species are also naturally distributed across east African savannas.

The most abundant woody species in east African savannas are *Acacia tortilis* Hayne (Fabaceae), *Balanites aegyptiaca* (L.) Delile (Zygophyllaceae), *Combretum* spp., *Commiphora* spp. (Burseraceae, the myrrh family), *Cordia* spp. (Boraginaceae), *D. melanoxylon*, *Grewia* spp., *Maerua triphylla* A. Rich. (Capparaceae) and *Markhamia* spp. (Bignoniaceae) (van de Vijver et al. 1999). Other Acacia species are also found in east African savannas, including *A. brevispica* Harms., *A. drepanolobium* Harms ex B.Y. Sjöstedt, *A. etbaica* Schweinf., *A. mellifera* (Vahl) Benth. and *A. senegal* (L.) Willd., Fabaceae). There is a continuum of savanna ecosystems from East Africa to southern Africa; therefore, the tree species that are found in these two ecoregions are nearly the same. For instance, *A. tortilis*, *A. digitata*, *Burkea africana* Hook. (Fabaceae), *Discrostachys cinerea* Wight et Arn. (Fabaceae), *Grewia flavescens* Juss. (Malvaceae), *Ochna pulchra* Hook. (Ochnaceae), *Euclea nataliensis* A.DC. (Ebenaceae), *T. indica*, and *S. birrea* are abundant in southern African savanna ecosystems. However, species distribution in southern African savannas is influenced by soil fertility. Indeed, the Combretaceae and Caesalpiniaceae dominate sandy nutrient-poor soils, whereas Mimosoidae are abundant on basalt or dolerite nutrient-rich soils (Scholes 1990).

Southern African savannas consist of central African savannas (forest-savanna mosaics and Miombo woodlands extending from the Congo, Democratic Republic of Congo and Tanzania to Namibia, Zambia, and Botswana) that are characterized by broadleaf and thornless trees, and south African savannas (*Baikiaea* (Fabaceae), Miombo and Mopane woodlands, bushveld, and coastal-flooded savannas extending from Namibia to Swaziland and Mozambique) characterized by fine-leafed and broad-leafed trees, and creeping grasses such as *Chrysopogon* and *Themedia* species.

1.1.3.6 Central American and Caribbean Islands Savannas

Savannas occur with a patchy or discontinuous distribution across Mexico, Belize, Honduras, Nicaragua, Panama and Caribbean Islands (Huber 1987). Fabaceae and Pinaceae are dominant woody plant families in Central American savannas (Hughes and Styles 1984; Kellman 1979). Beard (1944) described four types of savannas in Central and South America on the basis of their vegetational composition: open savannas (pure grass stands consisting of *Andropogon, Cymbopogon* or *Sporobolus*), orchard savannas (grassland with scattered bushes), pine savannas (grassland with pines) and palm savannas (grassland with occasional fan-palms). Common shrubs are chaparro or *Curatella americana* L. (Dilleniaceae), *Byrsonima* spp., sucupira or *Bowdichia virgilioides* Kunth (Papilonoideae), and cashew or *Anacardium occidentale* L. (Anacardiaceae), whereas palms include *Copernicia, Acoelorrhaphe wrightii* (Griseb. & H.Wendl.) H.Wendl. ex Becc. or Everglades palm, and *Chrysophila* (e.g., Give-and-take or *C. argentea* Bartlett, Arecaceae; Beard 1944). *Hyparrhenia* grasses are also common in the understory of Central American savannas.

Common savanna woody species in Central America also include *Acacia deamii* (Britton & Rose) Standl., *A. farnesiana* (L.) Willd. Syn. *Vachellia farnesiana* (L.) Willd., *A. pennatula* (Schltdl. & Cham.) Benth., *Albizia guachapele* (Kunth) Dugand, *Apoplanesia paniculata* C.Presl,, *Ateleia herbert-smithii* Pittier, *Caesalpinia coriaria* (Jacq.) Willd., *C. eriostachys* Benth., *C. velutina* (Britt. & Rose) Standl,, *Gliricidia sepium* (Jacq.) Kunth ex Walp., *Leucanea diversifolia* Benth., *L. leucocephala* (Lam.) de Wit., *L. shannoni* Donn. Smith, *Prosopis juliflora* (Sw.) DC., which belong to the Fabaceae, and *Simarouba glauca* DC. (Simaroubaceae), and *Crescentia alata* Kunth (Bignoniaceae). Pines (*Pinus caribea* and *P. oocarpa*) are dominant woody species in the savannas of Belize, whereas *Acacia farnesiana* is common in Costa Rica (Kellman 1979). Other common broadleaved species in Central American savannas are *Byrsonima crassifolia* (L.) Kunth, *Clethra hondurensis* Britton, *Quercus shippii* Standl., *Q. oleoides* Schltdl. & Cham. and the shrub *Miconia albicans* Ruiz & Pavón (Kellman 1979).

1.1.3.7 South American Savannas

Savanna ecosystems in South America are also known as *Llanos* (Colombian-Venezuelan llanos, Magdalena river valley llanos, Llanos de Mojos in Bolivia) and *Cerrados* (Campos cerrado and campos do Humaita in Brazil; Huber 1987). The basic floristic composition of Neotropical savannas includes herbs (Asteraceae, Cyperaceae and Poaceae) and shrub or low trees belonging to the families Dilleniaceae (*Curatella americana*), Malpighiaceae (*Byrsobima crassifolia* (L.) Kunth), and Flacourtiaceae (wild-coffee or *Casearia sylvestris* Sw.; Huber 1987). The common herbs belong to the Poaceae (*Axonopus aureus*, *Leptocaryphium lanatum* and *Trachypogon plumosus*), Cyperaceae (*Bulbostylis capillaris* and *Rhynchospora barbata*), and Asteraceae (*Eupatorium amygdalinum* and *Orthopappus angustifolius*; Huber 1987).

The Brazilian cerrado is located at the edge of the Amazonian forest, in Central Brazil, and covers about 2 million km^2 (Ratter et al. 1997). Average annual rainfall ranges between 800 and 2,000 mm, with average annual temperature ranging from 18 to 28 °C. Soils are mostly Oxisols, with low pH (reviewed by Ratter et al. 1997). Brazilian cerrado is rich in endemic woody species, as it harbors more than 500 species of trees and large shrubs (Ratter et al. 1996). The ground layer is dominated by Fabaceae, Compositae, Myrtaceae and Rubiaceae, whereas the most abundant woody species are plants in the Leguminosae, Mapighiaceae, Myrtaceae, Melastomataceae and Rubiaceae families (reviewed by Ratter et al. 1997). Grasses such as *Brachiaria*, *Hyparrhenia rufa* (Nees) Stapf and *Panicum maximum* are planted in pasture systems (Ratter et al. 1997).

Llanos are found on highly leached soils, and dominant trees are sclerophyllous evergreens (Medina and Silva 1990). Tree/grass ratios vary with water availability, and two main ecosystems characterize these savannas (Medina and Silva 1990). These ecosystems consist of (i) dense tree-savanna on clay-sandy and acid soils with *Curatella americana* L. (Dilleniaceae) and *Bowdichia virgilioides* Kunth (Fabaceae) as dominant woody species, and the ground layer dominated by *Thrasya petrosa* (Trin.) Chase (Poaceae), *Trachypogon plumosus* (Humb. & Bonpl.) Nees (Poaceae) and *Axonopus purpusii* (Mez) Chase (Poaceae); (ii) *Trachypogon*-savanna dominated by *Trachypogon plumosus*, with scattered trees of *Curatella americana* and *Byrsonima cirassifolia* (Medina and Silva 1990). Blydenstein (1967) described three distinct savannas types in the Columbian Llanos on the basis of grass composition. These ecosystems consist of the *Melinis minutiflora* P.Beauv. (Poaceae) savanna occurring on fine-textured soils or on high alluvial terraces, *Trachypogon ligularis* Nees-*Paspalum carinatum* (J.Presl) K.Schum & Hollrung savanna on the dunes of aeolian plains or on coarse-textured soils, and *P. carinatum* savanna on dissected high plains and on eroded soils (Bladystein 1967).

1.2 The Importance of Tropical Forest and Semiarid Ecosystems

Tropical forests are significant carbon sinks. By taking up atmospheric carbon dioxide (Philips et al. 1998; Lewis et al. 2009), they play an important role in reducing greenhouse effects and mitigating climate change (Justice et al. 2001). Tropical forests account for 34 % of terrestrial gross primary production, more than that of tropical savannahs (24 %; Beer et al. 2010). For example, the annual gross primary productivity of an undisturbed rainforest of Amazonia was estimated to be 24 tonnes C ha^{-1} $year^{-1}$ (Lloyd et al. 1995). Terrestrial gross primary production (GPP) is the largest global CO_2 flux driving several ecosystem functions. Using eddy covariance flux data and various diagnostic models, Beer et al. (2010) estimated the flux at 123 ± 8 petagrams of carbon per year (Pg C $year^{-1}$, Table 1.2).

Table 1.2 Terrestrial gross primary production (GPP) for biomes of the world (Beer et al. 2010)

Biome	GPP (Pg C year $^{-1}$)	GPP=2NPP* (Pg C year $^{-1}$)
Tropical forests	40.8	43.8
Temperate forests	9.9	16.2
Boreal forests	8.3	5.2
Tropical savannahs and grasslands	31.3	29.8
Temperate grasslands and shrublands	8.5	14
Deserts	6.4	7
Tundra	1.6	1
Croplands	14.8	8.2
Total	121.7	125.2

*NPP = Net Primary Productivity = Photosynthesis Rate – Plant Respiration Rate = NEE-soil respiration where NEE is the net ecosystem exchange, i.e. the difference between photosynthesis and ecosystem respiration (plant and microbial respiration)

Tropical forests and savannahs account for 34 and 24% of the global terrestrial GPP, respectively. Terrestrial gross primary production over 40% of the vegetated land is associated with precipitation. Based on the C4 plant distribution, more than 20% of terrestrial GPP originated from C4 vegetation. Also, understanding the relationships between climate and carbon exchange by terrestrial ecosystems is critical to predicting future levels of atmospheric carbon dioxide because of the potential accelerating effects of positive climate–carbon cycle feedbacks (Yi et al. 2010).

Tropical savannas also can play important roles in carbon (C) sequestration (Chen et al. 2003). Indeed, Chen et al. (2003) estimated that the total C stock of a tropical savanna in Northern Australia to be 204±53 tonnes C ha $^{-1}$. Also, tropical forests are hotspots of biodiversity (Mittermeier et al. 1998), and provide agricultural land for people living around or within these forests. For example, in the Peruvian Amazon, peasant charcoal production provides significant cash income for local people (Coomes and Burt 2001). In Africa, farmers in the Congo Basin depend heavily on agriculture and forest products such as *Dacryodes edulis* fruits, *Irvingia gabonensis* kernels, and *Gnetum africanum* leaves for food, medicine, shelter, and income. Further, forest products contribute to tax revenues and the gross domestic product. In the Congo Basin, the forest sector accounts for about 10.0, 6.0, 5.0, 4.0, and 1.0% of the gross domestic products of Cameroon, Gabon, Congo, Central African Republic, and the Democratic Republic of Congo, respectively (de Wasseige et al. 2012). Forestry also generates as much as 62.1 million € in tax revenue annually for the nation of Cameroon (de Wasseige et al. 2012).

Savannas also provide a livelihood for the peoples living in these regions. In the Malian savannas, the fat that is extracted from the nuts of *Vitellaria paradoxa* or shea butter, has represented the major source of income for women living in rural areas (Becker 2001). Savanna grasses are used for pastoralism, which is widespread land use system in tropical savannas and one that contributes significantly to the local economy.

Bibliography

Ashton PS (1988) Dipterocarp biology as a window to the understanding of tropical forest structure. Ann Rev Ecol Sys 19:347–370

Baruch Z (2005) Vegetation-enrichment relationships and classification of the seasonal Savannas in Venezuela Flora 200(1):49–64

Beard JS (1944) Climax vegetation in Tropical America. Ecology 25(2):127–158

Becker LC (2001) Seeing green Mali's wood: Colonial legacy, forest use and local control. Ann assoc Am Geog 9(1):504–526

Beer C, Reichstein M, Tomelleri E, Ciais P, Jung M, Carvalhais N, Rödenbeck C, Altaf Arain M, Baldocchi D, Bonan GB, Bondeau A, Cescatti A, Lasslop G, Lindroth A, Lomas M, Luyssaert S, Margolis H, Oleson KW, Roupsard O, Veenendaal E, Viovy N, Williams C, Woodward FI, Papale D (2010) Terrestrial gross carbon dioxide uptake: global distribution and covariation with climate. Science 329(5993):834–838

Bladystein J (1967) Tropical Savanna vegetation of the Llanos of Colombia. Ecology 48(1):1–15

Bridgewater S, Ibanez A, Ratter JA, Fyrley P (2002) Vegetation classification and floristics of the Savannas and associated wetlands of the Rio Bravo conservation and management area, Belize. Edinb J Bot 59:421–442

Campbell DG, Daly DC, Prance GT, Maciel UN (1986) Quantitative ecological inventory of terra firme and Várzea tropical forest on the Rio Xingu, Brazilian Amazon. Brittonia 38(4):369–393

Chen X, Hutley LB, Eamus D (2003) Carbon balance of a tropical savanna of Northern Australia. Oecologia 137:405–416

Cole MM (1986) The Savannas—Biogeography and Geobotany. Academic Press, Florida, pp 438

Coomes OT, Burt GJ (2001) Peasant charcoal production in the Peruvian Amazon: rainforest use and economic reliance. For Ecol Manag 140(1):39–50

Couteron P, Kokou K (1997) Woody spatial patterns in a semi-arid savanna of Burkina Faso, West Africa. Plant Ecol 132:211–227

de Wasseige C, de Marcken P, Bayol N., Hiol Hiol F, Mayaux Ph, Desclée B, Nasi R, Billand A, Defourny P, Eba'a AR, eds (2012) Les Forêts du Bassin du Congo—Etat des Forêts 2010. Office des publications de l'Union Européenne. Luxembourg, pp–276. ISBN 978-92-79-22717-2, doi:10.2788/48830

Fujisaka S, Escobar G, Veneklaas E (1998) Plant community diversity relative to human land uses in an Amazon forest colony. Biodivers Conserv 7:41–57

Hamann A, Barbon EB, Curio E, Madulid DA (1999) A botanical inventory of a submontane tropical rainforest on Negros Island, Philippines. Biodivers Conserv 8:1017–1031

Holdridge LR (1947) Determination of world plant formations from simple climatic data. Science 105:367–368

Holdridge LR (1967) Life zone ecology. Tropical Science Center, Costa Rica, pp 206

Holdridge LR, Grenke WC, Hatheway WH, Liang T, Tosi JA Jr (1971) Forest environments in tropical life zones. Pergamon, New York, pp 747

Huber O (1987) Neotropical savannas: Their flora and vegetation. Trends in Ecol Evol 2(3):67–71.

Hughes CE, Styles BT (1984) Exploration and seed collection of multiple purpose dry zone trees of Central America. Intern Tree Crop J 3(1):1–31

Jordan CF (1985) Nutrient cycling in tropical forest ecosystems: principles and their application in management and conservation. Wiley, Chichester, pp 189

Justice C, Wilkie D, Zhang Q, Brunner J, Donoghue C (2001) Central African forests, carbon and climate change. Clim Resear 17:229–246

Kellman M (1979) Soil enrichment by neotropical savanna trees. J Ecol 67:565–577

Köppen W (1936) Das geographisca System der Klimate. In: Köppen W, Geiger G (eds) Handbuch der Klimatologie 1. C. Gebr, Borntraeger. pp 1–44

Kpikpi WM (1992) Rating *Musanga cecropioides* and *Delonix regia* as papermaking hardwoods. Fiber supply—Tappi J. pp 73–75

Letouzey R (1985) Notice de la carte phytogeographique du Cameroun au 1/500000. 5-B I. Bibliographie et index des noms scientifiques. Institut de la Carte Internationale de la Vegetation, Toulouse, France, pp 143–240
Lewis SL, Lopez-Gonzalez G, Sonke B, et al. (2009) Increasing carbon storage in intact African tropical forests. Nature 457:1003–1007
Lloyd J, Grace J, Miranda AC, Meir P, Wong SC, Miranda HS, Wright IR, Gash JHC, McIntyre J (1995) A simple calibrated model of Amazon rainforest productivity based on biochemical properties. Plant Cell Environ 18(10):1129–1145
Ludwig JA, Tongway DJ, Eager RW, Williams RJ, Cook GD (1999) Fine-scale vegetation patches decline in size with increasing rainfall in Australian Savannas. Landscape Ecol 14:557–566
McKeon GM, Day KA, Howden SM, Mott JJ, Orr DM, Scattini WJ, Weston EJ (1990) Northern Australian Savannas: management for pastoral production. J Biog 17:355–372
Medina E, Silva JF (1990) Savannas of Northern South America: A steady state regulated by water-fire interactions on a background of low nutrient availability. J Biog 17(a/5):403–413
Mittermeier RA, Myers N, Thomsen JB, Da Fonseca GAB, Olivieri S (1998) Biodiversity hotspots and major tropical wilderness areas: areas to setting conservation priorities. Conser Biol 12(3):516–520
Moran EF, Brondizio ES, da Tucker JM, Silva-Forsberg MC, McCracken S, Falesi I (2000) Effects of soil fertility and land-use on forest succession in Amazonia. For Ecol Manag 139(1–3):93–108
Mott JJ, Williams J, Andrew MH, Gillison AN (1985) Australian savanna ecosystems. Ecology and management of the world's Savannas. In: Tothill JC, Mott JJ (eds) Australian Academy of Science, Canberra, ACT, Australia, pp 56–82
Murphy PG, Ariel E, Lugo AE (1986) Ecology of tropical dry forests. Ann rev Ecol Sys 17:67–88
Oldeman RAA, van Dijk J (1991) Diagnosis of the temperament of tropical rain forest trees: Rain forest regeneration and management. Man Biosph Ser 6:119–136
Pascal JP (1984) Les forêts denses humides sempervirentes des Ghâts occidentaux de l'Inde: écologie, structure, floristique, succession. Travaux de la Section Scientifique et Technique, Tome XX, Institut Français de Pondichéry, Inde, pp 318
Peel MC, Finlayson BL, McMahon TA (2007) Updated world map of the Köppen-Geiger climate classification. Hydrol Earth Sys Earth Sci 11:1633–1644
Pélissier R (1998) Tree spatial patterns in three contrasting plots of a southern Indian tropical moist evergreen forest. J Trop Ecol 14:1–16
Philips OL, Malhi Y, Higuchi N et al (1998) Changes in the carbon balance of tropical forests: evidence from long-term plots. Science 282:439–442
Pitman NCA, Terborgh JW, Silman MR, Nunez VP, Neill DA, Ceron CE, Palacios WA, Aulestia M (2000) Dominance and distribution of tree species in upper Amazonian terra firme forests. Ecology 82(8):2101–2117
ter Steege H, Pitman NCA, Phillips OL, Chave J, Sabatier D, Duque A, Molino J-F, Prévost M-F, Spichiger R, Castellanos H, von Hildebrant P, Vásquez R (2006) Continental-scale patterns of canopy tree composition and function across Amazonia. Nature 443:444–447
Ratter JA, Bridgewater S, Atkinson R, Ribeiro JF (1996) Analysis of the floristic composition of the Brazilian cerrado vegetation II: comparison of the woody vegetation of 98 areas. Edinb J Bot 53:153–180
Ratter JA, Ribeiro JF, Bridgewater S (1997) The Brazilian cerrado vegetation and threats to its biodiversity. Ann Bot 80:223–230
Scholes RJ (1990) The influence of soil fertility on the ecology of Southern African dry savannas. J Biogeogr 17(4–5):415–419
Sanchez PA (1981) Soils of the humid tropics. In: Col. William and Mary, Department of Anthropology, Williamsburg Va (ed) Blowing in the wind: deforestation and low-range implications. pp 347–410
Scholes RJ, Archer SR (1997) Tree-grass interactions in savannas. Ann Rev Ecol Sys 28:517–544

van de Vijver CADM, Foley CA, Olff H (1999) Changes in the woody component of an East African savanna during 25 years. J Trop Ecol 15(5):545–564

Vitousek PM, Sanford RL Jr (1986) Nutrient cycling in moist tropical forest. Annl Rev Ecol Sys 17:137–167

Walker BH, Landridge JL (1994) Predicting savannah vegetation structure on the basis of plant available moisture (PAM) and plant available nutrients (PAN): a case study from Australia. J Biogeogr 24:813–825

Yi C, Ricciuto D, Li R, Wolbeck J et al. (2010) Climate control of terrestrial carbon exchange across biomes and continents. Environ Res Lett 5 034007(10 pp)

Zhang J, Cao M (1995) Tropical forest vegetation of Xishuangbanna, SW China, and its secondary changes, with special references to some problems in local nature conservation. Biodivers Conserv 73:229–238

Zhu H, Cao M, Hu H (2006) Geological history, flora and vegetation of Xishuangbanna, southern Yunnan. Biotropica 38:310–317

Chapter 2
Major Land Use Issues in the Tropics, and the History of Agroforestry

Abstract Deforestation is the main land use issue facing tropical regions. Forest cover in Central Africa has been reduced from 248 538 000 ha in 1990 to 236 070 000 ha in 2005, and in West Africa, from 8 865 000 in 1990 to 7 437 200 ha in 2005. About 83% of forest losses in Africa occurred from 1990 to 2000, mostly due to slash-and-burn practices that were employed to clear the land for agricultural uses. Similarly, 65% of forest loss in Asia from 1990 to 2005 resulted from land use changes to agriculture. Twenty-three percent of this loss could be directly attributed to intensification of slash-and-burn agriculture, while 13% was attributable to direct land use changes on small-size farms. In Latin America and the Caribbean, forest areas have been reduced from 923 807 000 ha in 1990 to 859 925 000 ha in 2005. The majority (47%) of this loss was due to forestland conversion into large farms. In Brazil, the conversion of forest area to pastureland significantly reduced forest cover. Slash-and-burn agriculture, chemical inputs and extensive grazing are harmful to forest soils and biodiversity. The introduction of trees and/or livestock in agricultural plots was advocated to overcome the unsustainable use of natural resources and reduce poverty in the tropics. The World Agroforestry Centre (known as the International Centre for Research in Agroforestry, ICRAF, before 2002) was created to develop and promote agroforestry practices in the tropics and worldwide.

2.1 Introduction

Conservation of tropical forests is a global priority. Population growth and the poverty of local populations have led to increased pressures on natural resources in the tropics. Tropical forests provide land for agriculture, fuelwood, bushmeat, fruits and nuts, and other non-timber forest products (NTFPs) to local people for their everyday needs for food and cash. Non-timber forest products are not only consumed by farmers; these products are traded regionally and internationally, and provide substantial cash for stakeholders involved in the market chain. For these reasons, forest resources are increasingly exploited in the tropics. This chapter addresses the issue of sustainable management of tropical ecosystems, as well as alternatives to current unsustainable uses of these ecosystems. Agroforestry is one such promising alternative, and its history and rationale are discussed in this chapter.

A. Atangana et al., *Tropical Agroforestry*, DOI 10.1007/978-94-007-7723-1_2,

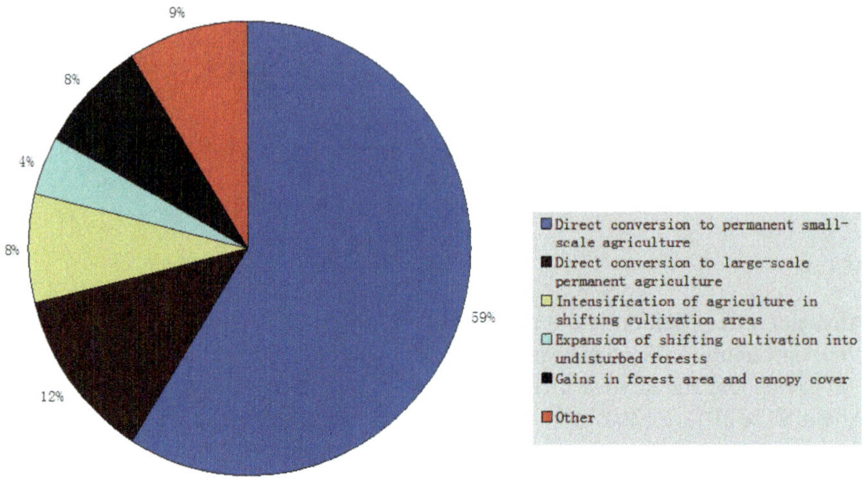

Fig. 2.1 Direct causes of forest area changes in tropical African countries between 1990 and 2000 (FAO 2001)

2.1.1 Impacts of Traditional Natural Resource Use on Tropical Ecosystems

2.1.1.1 Deforestation

Deforestation is the long-term or permanent removal of forest cover, whether it be naturally or anthropogenically and conversion to a non-forested land use. In contrast, forest degradation implies changes within the forest which negatively affect the structure or function of the stand or site, thereby lowering the capacity to supply products and/or services (Puustjärvi and Markku 2002). Forest cover in Central Africa has been reduced from 248 538 000 ha in 1990 to 236 070 000 ha in 2005, and in West Africa, from 8 865 000 to 7 437 200 ha over the same 15-year period (FAO 2009). This change represents a 26 812 000 ha loss, or about an 8 %, reduction in forest cover. In total, Africa has registered forest losses of 63 949 000 ha from 1990 to 2005 (FAO 2009), with 83 % of this loss occurring from 1990 to 2000, as a result of changes in land use to agriculture (Fig. 2.1). Intensification of slash-and-burn agriculture practices in agricultural areas accounted for 8 % of forest cover losses, while an additional 4 % was the result of slash-and-burn practices in wild forests. Smallholders have reduced forest cover to the greatest degree, with 59 % of lost forest cover having been converted to small-sized farms. In Southeast Asia, forest area has been reduced from 245 600000 to 203 887 000 ha between 1990 and 2005. Oceania's forest cover dropped from 212 514 000 to 206 254 000 ha within the same time period (FAO 2009). The annual rate of net deforestation in the Congo Basin is estimated to be 0.09 % between 1990 and 2000, with of net degradation of 0.05 %. Between 2000 and 2005, annual net deforestation in the Congo Basin is estimated is estimated to be 0.17 % and annual net degradation, 0.09 % between 2000 and 2005 (de Wasseige et al. 2012; Ernst et al. 2013). The highest rate was observed in DRC (0.32 % $year^{-1}$) followed by Cameroon (0.17 % $year^{-1}$), and the Republic of Congo (0.16 % $year^{-1}$). Annual mean net deforestation rate in the Congo Basin during the same 5-year period was estimated to be 0.17 % $year^{-1}$ (Ernst et al. 2013).

In Asia, 65 % of forest loss was a result of land use change to agriculture (FAO 2009), with 23 % of this loss resulting directly from intensification of slash-and-burn agriculture, and 13 % from direct land use conversions to small-size farms (FAO 2009). In Latin America and the Caribbean, forest areas have been reduced from 923 807 000 ha in 1990 to 859 925 000 ha in 2005. The majority (47 %) of these losses was due to conversion of forestland into large farms. Achard et al. (2002) reported a mean (± Standard Error, SE) annual forest loss between 1990 and 1997 of 2.5 ± 1.4 million ha (0.38 % $year^{-1}$) in Latin America, 2.5 ± 0.8 million ha in Southeast Asia (0.91 % $year^{-1}$), and 0.85 ± 0.30 million ha (0.43 % $year^{-1}$) in Africa. Most of this forest loss was due to land use conversion to agriculture. Hansen et al. (2008) reported a total loss of forest area in the tropics of 27.2 ± 2.2 million ha (1.39 ± 0.08 %) from 2000–2005. Brazil showed the highest rate of forest loss, 47.8 %, followed by Indonesia at 12.8 %. Africa accounted for 5.4 % of forest loss during the same period. Large-scale application of slash-and-burn agriculture by farmers was one of the principal causes of deforestation in the tropics.

In Brazil, natural forest areas have been cleared and converted to pastures for cattle grazing. In the mid-1990's, the increase in cattle to 10 million head, which is double the numbers that were reported in 1950, forced an increase in pasture area from 3.5 to 9.5 million ha (Kaimowitz 1995). Grazing occupied about 50 million hectares in the Amazon Basin (Chomitz and Thomas 2001). Unfortunately, pastures that are established after deforestation have a limited life span because soil nutrients leach out rapidly. Over 50 % of these pastures have had to be subsequently abandoned due to soil degradation (Steinfeld et al. 1997).

Logging has remained a serious cause of deforestation in the tropics. Africa supplied 19 % (658 million m^3) of world log production in 2006. Most of this production has come from natural forests, and the International Tropical Timber Organization (ITTO 2006) estimated that only 6 % of natural production forests standing on permanent public areas were managed sustainably. Moreover, 600 000 km^2 (representing 30 %) of forests were under logging concessions for industrial exploitation in Central Africa in 2003 (Laporte et al. 2007). Logging not only reduces forest area, but also selectively removes species of high commercial value. Species targeted by logging included *Aucoumea klaineana* Pierre (Burseraceae), *Entandrophragma spp.* (*E. angolense*, *E. candollei* et *E. cylindricum*, Meliaceae), *Lovoa trichilioides* (Meliaceae), *Lophira alata* (Ochnaceae), *Erythrophleum ivorense* (Leguminosae), *Millettia laurentii* De Wild. (Fabaceae), *Guibourtia tessmannii* Benn. (Fabaceae), *Pericopsis elata* (Harms) van Meuwen (Fabaceae), *Milicia excelsa* (Welw.) C.C Berg, (Moraceae), *Guarea cedrata* (A. Chev.) Pellegrin (Meliaceae), *Guarea laurentii* (Meliaceae), *Guarea thompsonii* Sprague & Hutch. (Meliaceae), *Gossweilerodendron balsamiferum* (Verms.) Harms (Fabaceae), *Pterocarpus soyauxii* Taub. (Fabaceae; in Africa), *Dalbergia spp.* (Fabaceae), *Tectona grandis* L.f. (Lamiaceae),

Pterocarpus spp. (Fabaceae; in Southeast Asia), *Symphonia globulifera* L.f. (Clusiaceae), *Swietenia macrophylla* King (Meliaceae), *Cedrela odorata* L. (Meliaceae), *Swietenia humilis* Zuccarini (Meliaceae), *Manilkara bidentata* (A.DC.) A.Chev. (Sapotaceae), *Dipteryx odorata* (Aubl.) Willd. (Fabaceae), and *Apuleia leiocarpa* (Vogel) J, F. Macbr., Fabaceae; in Latin America; Global Forest Watch 2000a, b; Hall et al. 2003). Selective logging caused an annual forest loss of 1,200 km^2 in areas under conservation, and loss of between 12 075 to 19 823 km^2 of forests in the Amazon in Brazil in 1992 (Asner et al. 2005). Another effect of logging is the reduction of fauna species diversity (Willott et al. 2000).

2.1.1.2 Grazing

Grazing affects savanna ecosystems, by altering floristic composition and physiognomy, and incurring ecosystem degradation (Skarpe 1991). Indeed, degradation of rangeland occurs under intensive livestock grazing. Dyksterhuis (1949) reported a replacement of relative palatable perennial grasses by less palatable or annual ones in savannas that were exposed to intensive grazing.

Browsing occurs in tropical savannas. Indeed, large herbivores including African elephant (*Loxodonta africana* Cuvier, Elephantidae), African buffalo (*Syncerus caffer* Sparrman, Bovidae), blue wildebeest (*Connochaetes taurinus* Burchell, Bovidae), cattle (*Bos taurus* L., Bovidae), giraffe (*Giraffa camelopardalis* L., Giraffidae), impala (*Aepyceros melampus* Lichtenstein, Bovidae), kudu (*Tragelaphus strepsiceros* Pallas, Bovidae), waterbuck (*Kobus ellipsiprymnus* Ogilby, Bovidae), and zebra (*Equus burchelli* Boddaert, Equidae), feed on leaves and twig of trees and shrubs in tropical savannas. Intensive browsing results in defoliation that may induce plant defenses through the secretion of substances reducing twig and leaf palatability; it may also result in further browsing, thereby reducing the woody layer of the savanna.

2.1.1.3 Effects of Unsustainable Use of Ecosystem Resources on Soil, Groundwater and Fauna

Slash-and-burn agriculture, the most common agricultural practice in the tropics, has negative effects on soil fertility and microfauna. A 30% reduction in carbon, nitrogen and phosphorus contents of a soil that had been cultivated for 6 years using this practice was reported by Tiessen et al. (1992). The authors found that 8–10 years of fallow were needed to restore fertility levels to those similar to the original site conditions prior to cultivation. Substantial annual losses of soil fertility in the sub-Saharan region have been reported. In the two-year period between 1982 and 1984, there was a loss per hectare per year of 22 kg of nitrogen, 2.5 kg phosphorus and 15 kg of potassium (Stoorvogel et al. 1993). Demographic growth exerts a pressure on land, and the length of the fallow period decreases, resulting in an increase of deforestation for agriculture. However, long-term fallows may be able to maintain fertility levels in soils under slash-and-burn agriculture for more than 200 years (Lawrence and Schlesinger 2001). Slash-and-burn agriculture also increases runoff and loss of nutrients from watersheds (Gafur et al. 2003). Chemical inputs like fertilizers and pesticides are harmful to microfauna and pollute watercourses and groundwater. Heavy machinery used for logging compacts soil, and fragments forests by opening skid trails for log extraction. Diesel and fuel oils used in vehicles and agricultural machines may contain heavy metals that pollute forest ecosystems.

Plant root systems redistribute water from the lower wetter layers to the upper drier layers in tropical savanna, a phenomenon that is referred to as 'hydraulic lift' (Jackson et al. 2000). Intensive grazing destroys the grass layer of savannas, which reduces water uptake by grasses, and may result in the modification of hydraulic lift. Hydraulic lift increases evapotranspiration in tropical savannas (Jackson et al. 2000; Ryel et al. 2002); therefore, any damage to grassland in tropical savannas will influence the local microclimate, and affects fauna and soil properties.

2.1.1.4 Effects of Unsustainable Use of Resources on Plants and Biodiversity

Intensive grazing for extended periods of time (hereafter, overgrazing) is widespread throughout the tropics and has deleterious effects on the ecosystem, such as biodiversity reduction. The restriction of cattle's grazing in pasture areas most often results in overgrazing. Ranchers are then forced to create new grazing areas to increase animal productivity. When no new pasture can be found, growth in animal production is made by increasing the size of the herd on the same grazing area, increasing the pressure on the land (Steinfeld et al. 1997). For that reason, overgrazing increases the risk of biodiversity loss and soil degradation. As demand grows for cow meat, economic pressure pushes production beyond the limits of the rangeland, thereby exceeding the carrying capacity of pasture areas. Overgrazing causes soil compaction, soil erosion and depletion of soil fertility.

Unsustainable use of resources also includes deforestation, which is one of the primary reasons for the loss of biodiversity in the tropics (Brooks et al. 2002; Pandit et al. 2007). Deforestation causes habitat loss for a number of animal species. Brooks et al. (2002) suggested that, owing to habitat loss, endemic plant species and diversity hotspots would be destroyed, resulting in the extinction of many species on the IUCN Red List. It is estimated that only 10% of the Indian Himalayas will still have dense forests by the year 2100 due to deforestation, and that 366 endemic vascular plants as well as 35 endemic vertebrate taxa will have their habitat destroyed (Pandit et al. 2007).

2.1.2 *History of Agroforestry*

The history of agroforestry up until 1993 has been well-documented in Nair (1993). In this section, we briefly review agroforestry history, drawing heavily upon Nair (1993) and focusing on new developments in this discipline since his review.

Agroforestry has arisen from a need to conserve tropical forest conservation and to implement sustainable use practices. Indeed, agroforestry is '*a new word for an old practice*'. People have always tried to benefit from forest ecosystems and trees. In the Christian Bible, an example of a homegarden (i.e., association of multipurpose trees and shrubs, annual or perennial plants and/or livestock within the household compound; Fernandes and Nair 1986) is given (Adam and Eve lived in the Garden of Eden). Homegardens have long been widespread in Africa, South and Southeast Asia, and Latin America. For instance, native peoples in Central America have long managed "conucos," which are gardens where agriculture has been practiced in a traditional manner (Esquivel and Hammer 1988) for centuries (Reynoso 1881, cited in Esquivel and Hammer 1988; Ortiz 1985; Valdés 1986). In Europe during the Middle Ages, farmers practiced slash-and-burn agriculture and integrated trees into their farms. Another example of an agroforestry system is the practice of 'retaining' trees, the products of which have food, medicinal or commercial value when clearing land for farming. Such trees are left in the fields that are used for food or cash crops, and their products are harvested yearly. This practice is widespread throughout the Congo basin. After clearing a patch of forest for crops, farmers would burn the cleared vegetation, and then plant trees in association with other species on the same land (King 1987). In Latin America, associating trees and crops was an old practice (Wilken 1977). For instance, natives in Central and South America practice the chacra system, which consists of small-scale shifting cultivation that has evolved into a shaded agroforestry system (Denevan 1971; Porro et al. 2012). In this system, food crops such as cassava (*Manihot esculenta* Crantz, Euphorbiaceae) and banana (*Musa* spp. L., Musaceae) are cultivated under cacao (*Theobroma cacao* L., Sterculiaceae) and other shade trees, and forest tree species, which provide timber and bark. This swidden fallow management also included plots of peanut or groundnut (*Arachis hypogaea* L., Fabaceae), and pineapple (*Ananas comosus* (L.) Merr., Bromeliaceae), and perennial fruit trees such as peach-palm (*Bactris gasipaes* Kunth, Aeraceae), star apple (*Chrysophyllum cainito* L., Sapotaceae), avocado (*Persea americana* Mill, Lauraceae), guava (*Psidium guajava* L., Myrtaceae), and uvilla or 'little grape' (Denevan et al. 1984).

Swidden cultivation, which is also known as slash-and-burn agriculture, has been widely practiced in tropical deciduous forests of Africa and Southeast Asia for centuries. In Southeast Asia, the *Taungya* management system is used as an alternative to slash-and-burn practices, through food crop integration into planted tree fields before canopy closure. *Taungya* replaced slash-and-burn agriculture in Myanmar (Burma) and India in the 1800s. Following numerous trials due to encroachment of forest reserves by farmers practicing slash-and-burn agriculture, the regeneration of teak (*Tectona grandis* L.f., Verbenaceae) was encouraged through the promotion of agriculture in forests (Blanford 1958; Nair 1993). Farmers had the right to cultivate food crops during the establishment and growth of trees, and could avoid prosecution for forest destruction. The *Taungya* system was created, which consisted of planting trees and food crops in the same area sequentially. Several years later, *Taungya* was introduced to other parts of Asia, and to Africa and Latin America.

The introduction of trees and/or livestock into agricultural plots appeared as an approach to preserving tropical forests. Worldwide awareness of the value of tropical forests, and the need for conservation through agroforestry practices increased the importance for the research, political, and financial support in the 20th century and beyond. In the late 1970s, the acceptance of agroforestry as a sustainable and promising land use system on both farms and in forests (Nair 1993), was facilitated by:

- A reappraisal of World Bank procedures;
- A re-examination of forest policies by the FAO;
- A growing interest in alley cropping and agro-pastoral systems;
- A deteriorating food supply in several developing countries;
- An increase in the spread of deforestation and degradation of forest ecosystems in the tropics;
- The energy crisis of the 1970is that led to increasing in commodity prices and a lack of fertilizers;
- The establishment by the International Development Research Centre (IDRC) of Canada of a project that was aimed at identifying research priorities for tropical forestry.

The World Bank recommended provision of financial aid to assist farmers in increasing food production (King 1979). International research centers in agriculture were established as a consortium in the 1960's to deal with problems of deforestation and ecological degradation, with the aim of improving the productivity of major crops or livestock in the tropics. As of today, the Consultative Group on International Agricultural Research (CGIAR) is a consortium of 15 international research centers. They are home to more than 8,000 scientists, researchers, technicians, and staff working to create a better future for the world's poor. These centers are: (1) Africa Rice Center, headquartered in Cotonou, Benin, (2) Bioversity International, headquartered in Rome, Italy, (3) Center for International Forestry Research (CIFOR), headquartered in Bogor, Indonesia, (4) International Center for Agricultural Research in the Dry Areas (ICARDA), headquartered in Beirut, Lebanon, (5) International Center for Tropical Agriculture (CIAT), headquartered in Cali, Colombia, (6) International Crops Research Institute for the Semi-Arid Tropics (ICRISAT) headquartered in Andhra Pradesh, India, (7) International Food Policy Research Institute (IFPRI) headquartered in Washington, USA, (8) International Institute of Tropical Agriculture (IITA), headquartered in Ibadan, Nigeria, 9) International Livestock Research Institute (ILRI), headquartered in Nairobi, Kenya, (10) International Maize and Wheat Improvement Center (CIMMYT), headquartered in Mexico, D.F., Mexico, (11) International Potato Center (CIP), headquartered in Lima, Peru, (12) International Rice Research Institute (IRRI), headquartered in Los Baños, the Philippines (13) International Water Management Institute (IWMI), headquartered in Colombo, Sri Lanka, (14) World Agroforestry Centre (ICRAF), headquartered in Nairobi, Kenya and (15) WorldFish, headquartered in Penang, Malaysia. Of the 15 research centers, which each has its own charter, board of trustees, director general, and staff, ICRAF is the only one with a mission to generate science-based knowledge regarding the diverse roles that trees play in agricultural landscapes, and to use its research to advance policies and practices that benefit the poor and the environment.

It is worth noting that the Green Revolution, which occurred during the same period, had focused on the increased use of fertilizers and other chemical inputs to raise productivity to the detriment of poorer farmers who could not afford these inputs. As alternatives to this expensive farming system, alley cropping and integrated farming systems gained popularity. The growing interest in these alternative systems was further strengthened by several studies that demonstrated the benefits of intercropping non-legumes or legumes with annual crops (Papendick et al. 1976; Kang et al. 1981; Nair 1983). These findings led scientists to further investigate the feasibility of land-use systems that allowed trees to remain in the field, while examining the role of trees and grasses in the maintenance of soil productivity and soil erosion control, and livestock management practices on farms.

The International Development Research Centre (IDRC) of Canada was faced with the problem of the growing rate of deforestation in the tropics and its negative consequences, such as reduced soil fertility and increased soil degradation. Further, the FAO (1982) demonstrated that slash-and-burn agriculture accounted for 70 % of deforestation. A mandate was given to John Bene, an IDRC official, to identify gaps in research and forestry education in the world, to formulate forestry research programs that would obtain results with considerable economic and social impact on developing countries, and to prepare an action plan for securing the support of donors (Nair 1993). Bene's team concluded that priority should be given to production systems that integrate forestry, agriculture and animals to optimize land use management in the tropics (Bene et al. 1977). The IDRC report strongly recommended the establishment of an international organization that would support, plan, and coordinate research involving land management systems in agriculture and forestry on a global scale. The International Council for Research in Agroforestry (ICRAF) was subsequently created in 1977, and was expanded in 1991 to become the International Centre for Research in Agroforestry. In 2002, a further expansion led to the establishment of the World Agroforestry Centre.

In the beginning, ICRAF focused its activities on creating an inventory of current agroforestry systems, collecting information, introducing new approaches and systems of agroforestry, fine-tuning existing agroforestry practices, and disseminating information on erosion control and soil fertility conservation and replenishment. Most research activities were focused on alley cropping, fallow systems with nitrogen-fixing species such as *Leucaena leucocephala* (Lam.) de Wit., *Calliandra calothyrsus* Meisn., and *Inga edulis* Mart., short fallows with pigeon pea (*Cajanus cajan* (L.) Millsp.), intercropping, promotion of multi-purpose species such as *C. calothyrsus* (which is used as a fodder source and as a nitrogen-fixing species in agroforestry systems), and the development of agro-pastoral systems that are adapted to the tropics.

Despite achievements in the development and popularization of intercropping and alley cropping, forest areas in the tropics have continued to shrink each year. The rural poor, who rely on agriculture and the forest for food, medicine and income, have placed increased pressure on natural forests due to the decline in cocoa and coffee prices in the late 1980's. The annual rate of deforestation in humid tropical forests of Africa between 1990 and 1997 reached 0.43 % $year^{-1}$, with an annual deforested area of 0.85 ± 0.30 (SE) million hectares, whereas the annual regrowth rate was estimated to be only 0.07 % during the same period (Achard et al. 2002). A solution was urgently needed to address this problem. The World Commission on Forests and Sustainable Development (WCFSD) began to provide more extensive support to community-based agroforestry to reduce the exploitation of primary forest for subsistence products (WCFSD 1999). This poverty-reduction and forest-protection strategy could be achieved through the development and cultivation of marketable and under-utilized "new crops" from the forests (Leakey et al. 2005). Surveys were carried out to identify and rank priority species that farmers would like to plant on their farms (Franzel et al. 1996; Leakey and Newton 1994; Simons and Leakey 2004). A worldwide domestication program of high-value, multi-purpose trees and indigenous tree species was created, which has been part of the main research focus of agroforestry since the mid-1990's. Research priorities of ICRAF in 2013 included (http://www.worldagroforestrycentre.org/research/overview accessed July 30, 2013):

- Agroforestry systems
- Tree products and markets
- Tree diversity, domestication and delivery
- Land health
- Climate change
- Environmental services

The ICRAF operates in 6 regions: West and Central Africa, East Africa, South Africa, South Asia, Southeast Asia, and Latin America. ICRAF has a partnership on research methods in agroforestry and livestock with the ILRI (International Livestock Research Institute), which is based in Nairobi, Kenya.

The future of tropical agroforestry has to deal with many issues, including which land tenure. Agroforestry is a system of natural resource management, and the right to land is most often different from that which is inherent to natural resources ownership in tropical countries. Further, land tenure is complicated because of the overlap between customary, colonial and post-colonial rights. Land tenure issues in tropical agroforestry will be discussed in detail in Chap. 17.

Bibliography

Achard F, Eva HD, Stibig HJ, Mayaux P, Gallego J, Richards T, Malingreau J-P (2002) Determination of deforestation rates of the world's humid tropical forests. Science 297:999–1002

Asner GP, Knapp DE, Broadbent EN, Oliveira PJC, Keller M, Silva JN (2005) Selective logging in the Brazilian Amazon. Science 310(5747):480–482

Bene JG, Beall HW, Côté A (1977) Trees, food and people. IDRC, Ottawa

Blanford HR (1958) Highlights of one hundred years of forestry in Burma. Emp For Rev 37(1):33–42

Brooks TM, Mittermeier RA, Mittermeier CG, Da Fonseca GA, Rylands AB, Konstant WR, Elick P, Pilgrim J, Oldfield S, Magin G, Hilton-Taylor C (2002) Habitat loss and extinction in the hotspots of biodiversity. Conserv Biol 16:909–923

Chomitz K, Thomas T (2001) Geographic patterns of land use and land intensity in the Brazilian Amazon. Working paper, Development Research Group. World Bank, Washington DC

Denevan WM (1971) Campa subsistence in the Gran Pajonal, Eastern Peru. Geogr Rev 61(4):496–518

Denevan WM, Treacy JM, Alcorn JB, Padoch C, Denslow J, Paitan SF (1989) Indigenous agroforestry in the Peruvian Amazon: Bora Indian management of swidden fallows. Interciencia 9(6):346–357

Ernst C, Verhegghen A, Bodart C, Mayaux P, de Wasseige C, Bararwandika C, Begoto C, Esono Mba F, Ibara M, Kondjo Shoko A, Koy Kondjo H, Makak JS, Menomo Biang JD, Musampa C, Ncogo Motogo R, Neba Shu G, Nkoumakali B, Ouissika CB, Defourny P (2010) Congo basin forest change estimate for 1990, 2000 and 2005 by Landsat interpretation using an automated object-based processedchain. The International Archives of the Photogrammetry Remote Sensing and Spatial Information Science XXXVIII 4/C7

Ernst C, Mayaux P, Verhegghen A, Bodart C, Musampa C, Defourny P (2013) National forest cover change in Congo Basin: deforestation, reforestation, degradation and regeneration for the years 1990, 2000 and 2005. Global Change Biol 19(4):1173–1187. doi:10.1111/gcb.12092

Esquivel M, Hammer K (1988) The "conuco"-an important refuge of Cuban plant genetic resources. Kulturpflanze 36:451–463

FAO (1982) Tropical Forest Resources. FAO, Rome

FAO (2009) ftp://ftp.fao.org/docrep/fao/011/i0350e/i0350e01a.pdf. Accessed 17 Jan 2012

Gafur A, Jensen JR, Borggaard OK, Petersen L (2003) Runoff and losses of soil and nutrients from small watersheds under shifting cultivation (Jhum) in the Chittagong hill tracts of Bangladesh. J Hydrol 274(1–4):30–46

Global Forest Watch (2000a) An overview logging in Cameroon. http://www.globalforestwatch.org/common/cameroon/. Accessed 17 Jan 2012

Global Forest Watch (2000b) A first look at logging in Gabon. http://www.globalforestwatch.org/common/gabon/english/report.pdf. Accessed 17 Jan 2012

Hall JS, Harris DJ, Medjibe V, Ashton PMS (2003) The effects of selective logging on forest structure and tree species composition in a Central African forest: implications for management of conservation areas. For Ecol Manag 183:249–263

ITTO (2006) Pérenniser les fôrets tropicales. http://www.itto.int/fr/. Accessed 17 Jan 2012

Jackson RB, Sperry JS, Dawson TE (2000) Root water uptake and transport: using physiological processes in global predictions. Trends Plant Sci 5:482–488

Kaimowitz D (1995) Livestock and deforestation in Central America. United Nations University Press, Tokyo

Kang BT, Wilson GF, Sipkens L (1981) Alley cropping maize (*Zea mays* L.) and leucaena (*Leucaena leucocephala* Lam.) in southern Nigeria. Plant Soil 63:165–179

King KFS (1979) Agroforestry: proceedings of the fiftieth symposium on tropical agriculture. Royal Tropical Institute, Amsterdam

King KFS (1987) The history of agroforestry. In: Steppler HA, Nair PKR (eds) Agroforestry: a decade of development. ICRAF, Nairobi, pp 1–11

Laporte NT, Stabach JA, Grosch R, Lin TS, Goetz SJ (2007) Exploitation of industrial logging in Africa. Science 316(5830):1451

Lawrence D, Schlesinger WH (2001) Changes in soil phosphorus during 200 years of shifting cultivation in Indonesia. Ecology 82(10):2769–2780

Leakey RRB, Tchoundjeu Z, Schreckenberg K, Shackleton SC, Shackleton C (2005) Agroforestry Tree Products (AFTPs): targeting poverty reduction and enhanced livelihoods. Int J Agric Sustain 16:5–16

Nair PKR (1979) Intensive multiple cropping with coconuts in India. Verlag Paul Parey, Berlin

Nair PKR (1983) Agroforestry with coconuts and other tropical plantation crops. In: Huxley PA (ed) Plant research and agroforestry. ICRAF, Nairobi, pp 79–102

Ortiz F (1985) Nuevo catauro de cubanismos. Ed. Ciencias Sociales, La Habana. pp 526
Pandit MK, Sodhi NS, Koh LP, Bhaskar A, Brook BW (2007) Unreported yet massive deforestation driving loss of endemic biodiversity in Indian Himalaya. Biodivers Conserv 16:153–163
Porro R, Miller RP, Tito MR, Donovan JA, Vivan JL, Trancoso RL, Van Kanten RF, Grijalva JE, Ramizet BL, Goncalves AL (2012) Agroforestry in the Amazon: a pathway for balancing conservation and development. In: Nair PKR, Garrity D (eds) Agroforestry—The future of global land use. Advances in Agroforestry 9. doi:10.1007/978–94-007–4676-3_20
Papendick RI, Sanchez PA, Triplett GB (1976) Multiple cropping. Special publication No 27. American Society of Agronomy, Madison
Puustjärvi E, Markku S (2002) Development of common framework for forest-related definitions. http://www.fao.org/docrep/005/Y4171E/Y4171E06.htm#TopOfPage. Accessed 14 Apr 2013
Ryel RJ, Caldwell MM, Yoder DK, Or D, Leffler AJ (2002) Hydraulic redistribution in a stand of *Artemisia tridentata*: evaluation of benefits to transpiration assessed with a simulation model. Oecologia 130:173–184
Spears J (1987) Agroforestry: a development-bank perspective. In: Steppler HA, Nair PKR (eds) Agroforestry: a decade of development. ICRAF, Nairobi, pp 53–66
Steinfeld H, de Haan C, Blackburn H (1997) Livestock-Environment interactions: issues and options. European Commission Directorate-General for Development, Brussels
Stoorvogel JJ, Smaling EMA, Janssen BH (1993) Calculating soil nutrient balances in Africa at different scales. I. Supra-national scale. Fertil Res 35:227–235
Tiessen H, Salcedo IH, Sampaio EVSB (1992) Nutrient and soil organic matter dynamics under shidting cultivation in semi-arid northeastern Brazil. Agric Ecosyst Environ 38(3):139–151
Valdés S (1986) La evolucion de los indoamericanismos en el espanol hablado en Cuba. Ed. Sociales, La Habana, pp 80
Wasseige C, Devers D, de Marcken P, Eba'a Atyi R, Nasi R, Mayaux P (2009) Les Forêts du Bassin du Congo—Etat des Forêts 2008. 426 pages, ISBN 978–92-79–132 11–7. doi:10.2788/32456, Office des publications de l'Union Européenne
WCFSD (1999) Our forests—our future. Summary Report: World Commission on Forests and Sustainable Development, WCFSD, Winnipeg, MB, Canada, p 38
Wilken GC (1977) Integrating forest and small-scale farm system in Middle America. Agrofor Syst 3:291–302
Willott SJ, Lim DC, Compton SG, Lutton SL (2000) Effects of selective logging on butterflies of a Bornean rainforest. Conserv Biol 14(4):1055–1065

Chapter 3
Definitions and Classification of Agroforestry Systems

Abstract Several definitions have been proposed to agroforestry, of which the most commonly used are those of Lundgreen and Raintree (Agricultural research for development: potentials and challenges in Asia, 1982, pp 37–49) and Leakey (Agroforest Today 8:1, 1996). Agroforestry is any land-use system, practice or technology, where woody perennials are integrated with agricultural crops and/or animals in the same land management unit, in some form of spatial arrangement or temporal sequence. Agroforestry is also a dynamic and ecologically -based natural resource management system. Agroforestry refers to the deliberate introduction or retention of trees on farms to increase, diversify, and sustain production for increased social, economic, and environmental benefits. Agroforestry system classification can be based on vegetation structure, function of woody perennials in the system, levels of management input, and environmental conditions and ecological suitability of the system. Agroforestry practices rather than systems are also used as the unit of an ecologically -based classification that is rooted in the role of trees in agricultural landscape.

3.1 Introduction

Agroforestry, together with forest management, is one of several viable alternatives to unsustainable management of natural resources that has been proposed for tropical forest ecosystems. Owing to its complexity and local specificity, several definitions have been suggested for agroforestry since its inception as a full agricultural science discipline (Lundgreen and Raintree 1982). These definitions take into account the different components of agroforestry, either through their spatial (simultaneous) or temporal (sequential) arrangements.

3.1.1 Definition of Agroforestry

Simply put, agroforestry is the introduction, or deliberate retention, of trees on farms through either spatial or temporal arrangements. It is commonly said that agroforestry is a 'new name for an old practice.' The word 'agroforestry' has its roots in the concepts of 'agrisilviculture' and 'agrosilviculture' (King 1968). According to Lundgreen

A. Atangana et al., *Tropical Agroforestry*, DOI 10.1007/978-94-007-7723-1_3,

and Raintree (1982), agroforestry is 'a collective name for land-use systems, practices or technologies, where the woody perennials (shrubs, trees, bamboo…) are deliberately integrated with agricultural crops and/or animals in the same land management unit, in some form of spatial arrangement or temporal sequence.' As stated in Nair (1993), this definition requires:

- The use of two or more species of animals or plants, and one of the latter is a woody perennial
- At least two production systems (e.g., tree-crop, crop-livestock or tree-livestock systems) allowing both ecological and economic interactions between different components to occur
- The cycle of an agroforestry system should last at least one year
- The agroforestry system to be more complex, both ecologically (structurally and functionally) and economically, than traditional monocropping systems.

Leakey (1996) defined agroforestry as 'a dynamic, ecologically based, natural resource management system that, through the integration of trees in farm and rangeland, diversifies and sustains production for increased social, economic and environmental benefits'. Wyant (1996) stated that, for an agroforestry system to become successful, it must comply with an inherent "ecosystem integrity": the state of system development in which the habitat structure, natural function and species composition of the system are interacting in ways that ensure its sustainability in the face of changing environmental conditions as well as both internal and external stresses. Khasa (2001) has refined Leakey's definition of the integrated concepts that are associated with agroforestry, by defining agroforestry as a collective term for dynamic natural resource management systems, where woody perennials are integrated spatially and/or temporally with valuable herbaceous or woody crops (food, industrial, horticultural, forage, botanical, cover, decorative, handicraft) and/or livestock, terrestrial and aquatic organisms, in order to diversify and sustain production to increase the wealth and well-being for land-users at all levels, depending on the ecological, socio-economic, political and cultural circumstances. The biophysical and socio-economic concepts that are included in this definition apply to different levels in the hierarchy of land-use management, including the micro (household management), meso (village, watershed, or local community) or macro (region or ecozone) levels. Agroforestry systems aim at improving livelihoods and ecosystems, and are characterized by:

- Productivity: sustained and increased crop production
- Sustainability: conservation of ecological functions of environmental components such as soil fertility and biodiversity
- Adoptability of the systems by users (landowners): at this level, the involvement of stakeholders throughout the process of development (diagnosis, solution development, implementation) is essential; agroforestry systems' research must necessarily be ***participatory***
- Simplicity and robustness: facilitate adoption of new agroforestry technologies by poor farmers

Agroforestry is different from *social forestry*, which is the practice of using trees and/or tree planting to pursue social objectives, through delivery of the benefits to the local people. The purpose of agroforestry is the ownership and control of forest resources and their management by the local people. Agroforestry is also different from forest management, which is the management of a forest ecosystem or wooded land to maintain biodiversity and the ecological niches of the various components of this system. Agroforestry is a dynamic system in which some plant species with high nutritional, medicinal, economic and ecological values are used by local people for their well-being, through integrating trees and/or animals on the farm. Community forestry, which is an aspect of social forestry, refers to tree planting activities undertaken by a community on communal lands. Community forestry is gaining popularity in the Congo Basin as more and more communities are expressing the need to strengthen their forests through enrichment planting of high-value, multi-purpose species. Agroforestry, which puts an emphasis on the interactive association between woody perennials and crops or animals for diversification and sustainability of production and profits, should also not be confused with farm forestry, which refers to tree planting on farmers' fields, mostly for the purpose of establishing woodlots.

3.1.2 Some Basic Concepts in Agroforestry

Agroforestry is based on the concept of associating trees with crops on the same piece of land simultaneously or sequentially for increased, diversified and sustained benefits, and for environment preservation. It is a natural resource management option that can be used to solve a specific problem in a manner that benefits farmers, consumers, and environmentalists. Key points to consider when developing an agroforestry system are:

- Use of trees in the system: the woody perennial to be introduced in farmlands should be adapted to the locality, and provide various benefits, such as food, income, medicine, and shelter to the local population for many years.
- The species should be known to positively interact ecologically and economically with staple-food crops of the locality. However, some agroforestry systems do not have a food crop component, like mixed orchards, and cocoa agroforests.
- All stakeholders (farmers, researchers, extension services, consumers, traders, and policy-makers) should be involved during the development and fine-tuning of the agroforestry system, from species identification to dissemination of the technology that increases the adoptability and spread of the system. The process should be *participatory*.

Important factors to be taken into consideration when designing an agroforestry system in a specific region are:

- Environment: ecology of the locality, such as soil type, climate, and relief
- Human environment: eating habits, cultivation practices, and demography

- Economy: market accessibility, economic value of products, competing products, periodicity, and opportunity costs of the system
- The difficulty of implementing the system.

3.1.3 Classification of Agroforestry Systems

The classification of agroforestry systems can be performed using several criteria, most of which have been outlined in Nair (1993), including:

- The structural composition and arrangement of the different components in the system
- The temporal sequence of introducting of different components into the system
- The function of woody perennials in the system, either as windbreaks, a source of shade, or for soil conservation
- The level of input in terms of the management of the system, or the scale of management for commercial purposes
- The environmental conditions and ecological suitability of the system, based on the assumption that certain types of systems are more appropriate in certain areas than others for various environmental reasons.

Agroforestry is a new name for an old practice, and agroforestry systems can therefore be classified as 'traditional' or 'science-developed' systems, depending on the locality. From 1982 to 1987, inventories of agroforestry systems and practices were conducted by ICRAF staff in developing countries (Nair 1987). Drawing upon the findings of these inventories, agroforestry systems have been developed, or adapted in a top-down approach, as poor farmers were generally not included in the design and planning process. Such an approach often creates difficulties in the stage of adaptation by farmers. For example, alley farming did not spread successfully in the humid tropic lowland rainforests of Africa. The difficulties that were encountered in farmers' adoption of new agroforestry practices in certain ecological regions led to the introduction of more participatory methods in agroforestry (bottom-up approach). Different approaches that are used to classify agroforestry systems are illustrated in Table 3.1.

Sinclair (1999) suggested a general classification of agroforestry practices that aims at identifying different types of agroforestry, and which groups together similar practices He suggested the use of the term 'practice' rather than 'system' (Sinclair 1999) as the unit of classification, where practices that are intended for similar ecologies and prospects for management can be grouped together. Sinclair's classification proceeds in two steps: (i) classification of major types of agroforestry practices according to the components that are involved and the predominant usage of land, and (ii) further classification of these components in terms of the arrangement, density and diversity of the tree component that is involved. This ecologically based -approach of classification is rooted in the role of trees in agricultural landscapes (Sinclair 1999).

Table 3.1 Major approaches to the classification of agroforestry systems (Adapted from Nair 1985)

Categorization of systems			Grouping of systems	
Structure		Function	Agroecological/ environmental adaptability	Socio-economic and management level
Nature of components	Arrangement of components	*Productive function*		
Agrisilviculture (crops and trees/ shrubs)	Spatial	Food (fruits)	*Systems in/for* Lowland humid tropics	*Based on level of technology input* Low input
Silvopastoral (animals and trees)	Mixed dense (Homegarden)	Fodder (*Calliandra* leaves)	Highland humid tropics (above 1,200 m above sea level)	Medium input
Agrosilvopastoral (crops, pasture/ animals and trees)	Mixed sparse (trees in pastures)	Fuelwood	Lowlands subhumid tropics (savanna)	High input
Other systems (multipurpose tree lots, semi-aquatic such as riparian systems, apiculture with trees/ shrubs such as *Calliandra calothyrsus*)	Strips (shelterbelts, riparian systems)	Biomass (for biofuel)	Highland subhumid tropics (Tropical Highlands; e.g., Kenya)	*Based on cost/ benefit relationships* Commercial
	Boundary (live fences)	Carbon		Intermediate
	Temporal -Concomitant -Overlapping -Coincident -Sequential -Interpolated	Timber and other wood products (for shelter…)		
		Non-Timber Forest Products (NTFPs)		Subsistence
		Protective function -Shelterbelt -Windbreak -Riparian system -Soil conservation -Moisture conservation -Soil improvement -Shade		

3.1.3.1 Classification Based on Vegetation Structure

Structure involves the nature and arrangement of elements that form components of the system. There are several main elements in a conventional agroforestry system: trees or shrubs, herbaceous crops or fodder plants, and/or animals. Generally, the herbaceous component is present in any agroforestry system, with the exception of beekeeping, aquaculture with trees, or associations of two woody perennials.

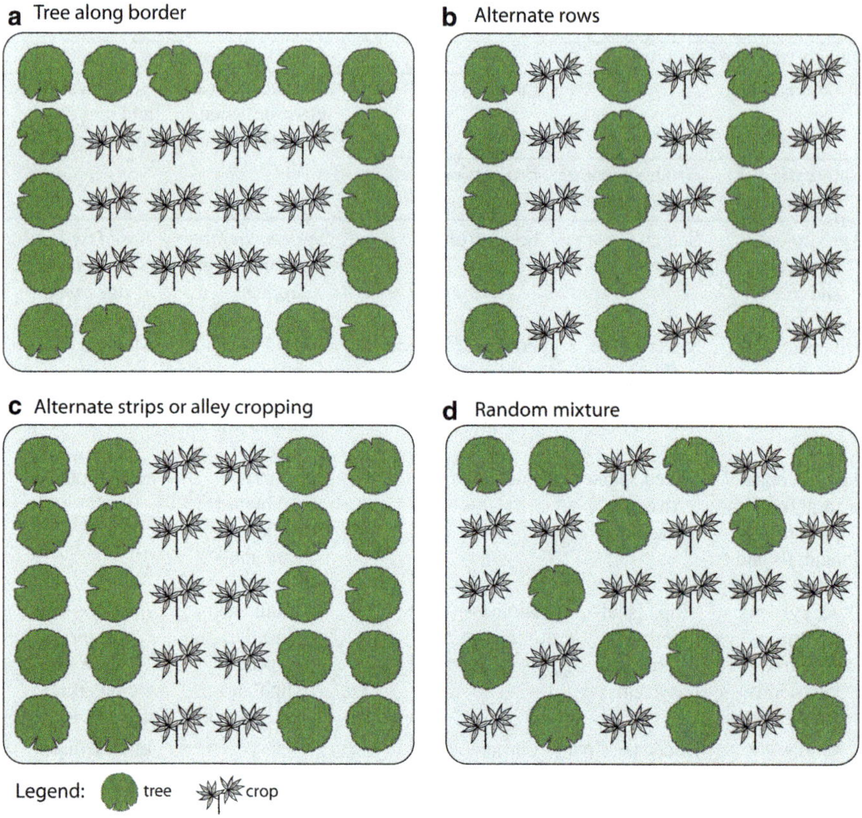

Fig. 3.1 Some spatial arrangements of crops in agroforestry (Adapted from Vergara 1981)

Nair (1985) illustrated the classification of agroforestry systems based on the nature of components (Fig. 3.3). Economic and ecological interactions between components of the system are a key element of agroforestry. Based on their structure, agroforestry systems can be classified into three groups: (1) silvopastoral systems that consist of pastures or animals and trees (for example, cattle and *Calliandra calothyrsus*); (2) agrosilvopastoral systems that consist of crops, pastures, or animals and trees; and (3) agrisilviculture, which includes crops and trees. To these groups is added aquasilviculture, which includes fish and trees. Components can be mixed spatially, either thickly or sparsely, or temporally deployed in the system. The different compositions of agroforestry systems are shown in Fig. 3.1. Spatial arrangements aim at optimizing land occupation. Arrangements can also consist of variations in the occupation of different strata. Sequential agroforestry systems range from 'conventional' slash-and-burn practices to *Cajanus* fallows (Fig. 3.2). Classification of agroforestry systems that is based on the nature of their components is included in a sequential agroforestry system (Fig. 3.3). What can also be included in this system is an agroforestry matrix for the humid and semiarid tropics (Fig. 3.4).

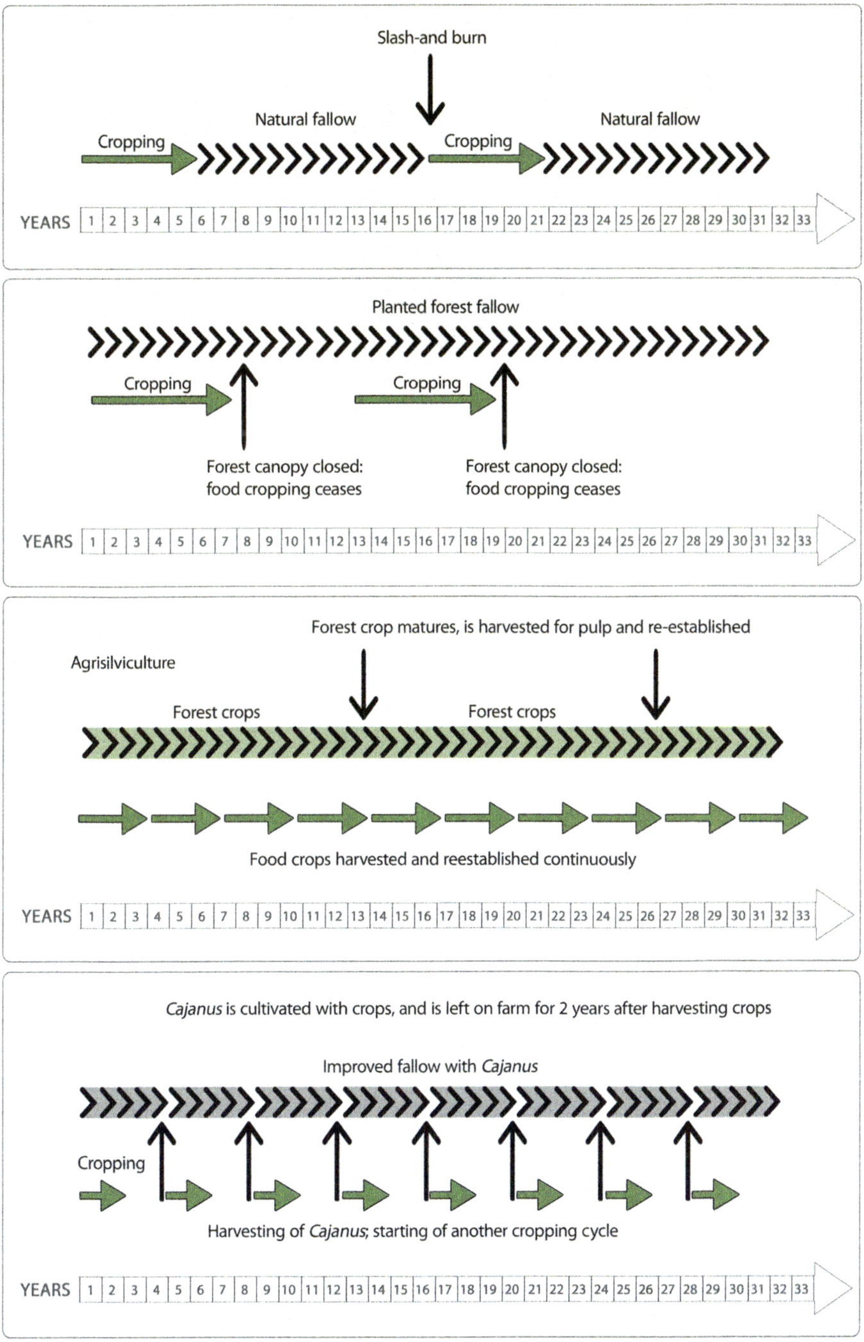

Fig. 3.2 Temporal arrangement of crops in agroforestry (Adapted from Vergara 1982)

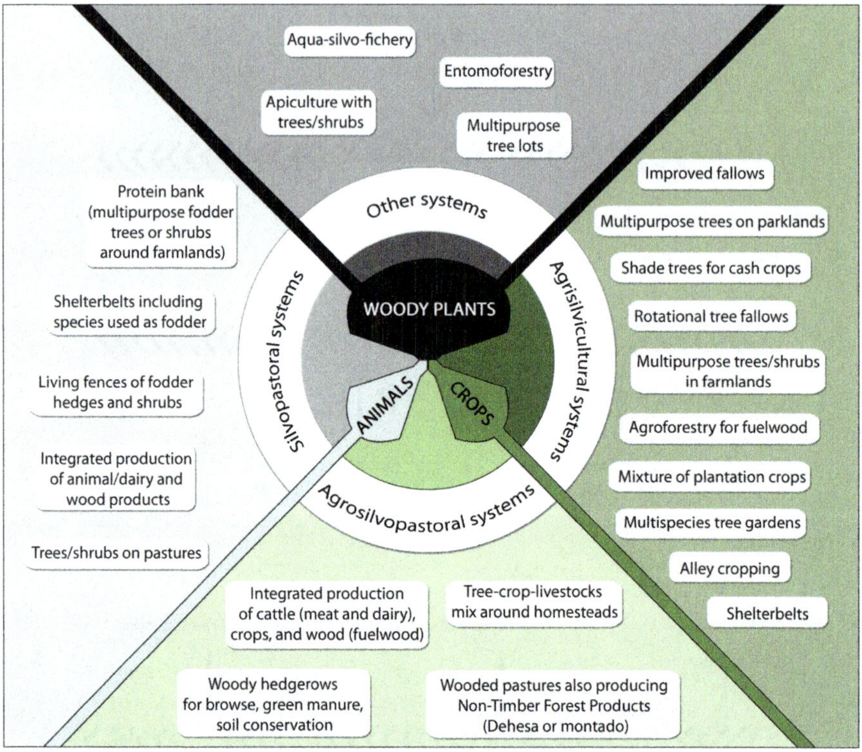

Fig. 3.3 Categorization of agroforestry systems based on the nature of components (Modified from Nair 1985)

3.1.3.2 Classification Based on Function

Functions that are used to classify agroforestry systems include productive functions, such as food production, and protective functions, such as shelterbelts (Nair 1985). Trees in agroforestry systems contribute significantly to the diversification and sustainability of production. For example, *Sesbania sesban* (L.) Merr. (Fabaceae), *Acacia nilotica* (L.) Willd. Ex Delile (Fabaceae) or *Leucaena leucocephala* (Lam.) de Wit (Fabaceae) are N_2-fixing species that enrich the soil with nitrogen in shrub/tree fallows, allowing replacement of nutrients that are exported by harvested crops. These species also contribute to the control of weeds (such as *Imperata cylindrica*). Woody perennials that are included in an agroforestry system can provide the following products:

- Timber for bioenergy, construction and other uses
- Food such as fruits, spices and nuts, and seeds used as soup-thickening agents
- Stimulants (*Garcinia kola* and *Cola acuminata* nuts)
- Aromas (*Inga edulis* fruits)
- Fats (*Allanblackia floribunda* and *Baillonella toxisperma* nuts)

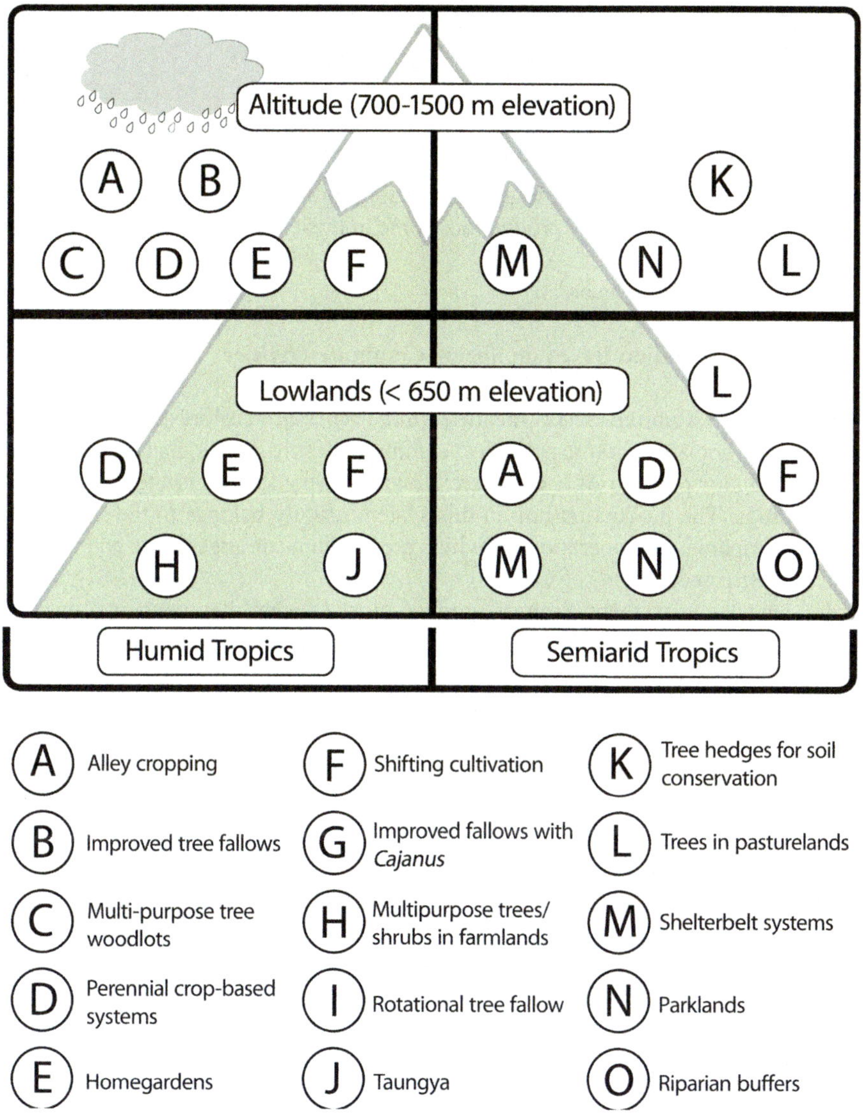

Fig. 3.4 An agroforestry matrix for the humid and semiarid tropics

- Medicinal products (*Annickia chlorantha* bark)
- Fodder (*Calliandra calothyrsus* leaves).

The sustainability of production of an agroforestry system can be improved through the following services that are provided by the trees:

- Weed control
- Nitrogen fixation

- Mulch
- Shade, particularly for some perennial crops such as cocoa
- Carbon sequestration
- Erosion control

Introduction of animals into agroforestry systems also provides benefits such as weed control through grazing, together with food products such as meat, milk and eggs, and dung for bioenergy production. Some of these benefits are unique to agroforestry systems.

3.1.3.3 Classification Based on Socio-economic Activities

Lundgreen and Raintree (1982) identified three types of agroforestry systems that are based on social economic activities: commercial, subsistence, and intermediate.

The primary purpose of a commercial agroforestry system is the sale of harvested crops. The production unit in this system usually belongs to the state or to private companies, the operations of which are medium- or large-scale, and labor is paid or contracted.

In *subsistence* systems, farmers produce most of what they consume, or consume most of their production. These farmers are generally poor, and those who cannot produce enough to meet the needs of their household are classed as subsistence farmers. In this system, the land is used primarily to meet the basic needs of the household. Subsistence systems include food crop farms (usually slash-and-burn practices), homegardens, and medium-sized cash crop plantations such as cocoa, coffee, oil palm, rubber, or tea among others. Labor is generally not hired but is done by members of the household. Any cultivation of cash crops such as cocoa, coffee, rubber or tea is performed in addition to subsistence farming activities.

An *intermediate* agroforestry system falls between subsistence and commercial systems both in terms of the intensity of production and its management. Cash crops are the main source of revenue, while food crops feed the household. Usually, farmers own the land that they farm or have long-term tenure of the land, and temporary labor is hired. This system is widespread in the tropics. In Cameroon, it was common to find farmers generating 3–5 million CFA (1 USD $\approx$ 500 CFA in May 2013) annually from cocoa or coffee beans in the 1980s. The major differences among intermediate, subsistence, and commercial agroforestry systems are the size of the area under cropping and the associated levels of economic prosperity.

This classification system can be useful for development efforts, but it has some drawbacks. The criteria for defining different classes are difficult to quantify, as standards for each category vary from one locality to another. What is considered an intermediate system in location A can be regarded as a subsistence system in location B. The barriers between different classes can also change over time.

3.1.3.4 Ecological Classification

Nair (1987) listed agroforestry systems that are practiced in the tropics, and characterized nearly all of these systems based on the ecological conditions of the different regions in which these systems were used. The systems are improved fallows in shifting cultivation, *Taungya* systems, tree gardens, alley cropping, multipurpose trees (MPTs) and shrubs on farmlands, crops in combination with plantation species, agroforestry fuelwood production, shelterbelts, windbreaks, soil conservation hedges, cut-and-carry fodder production, living fences for fodder production, trees and shrubs on pastures, woody hedges used for browsing, mulch and green manure, homegardens, aquaforestry (agro-silvo-fishery), various forms of shifting cultivation, and apiculture with trees and shrubs (Nair 1987).

The major types of agroforestry systems found in the tropics (Nair 1987) are shifting cultivation, homegardens comprised of intimate, multi-storey combinations of various trees, shrubs and crops around homesteads, *Taungya*, plantation-crop combinations, multilayer tree gardens, intercropping systems (in the humid lowlands of tropics), silvopastoral systems, windbreaks and shelterbelts, multipurpose trees on farmlands (in the semiarid lowlands of tropics), soil conservation hedges, silvopastoral combinations, and plantation-crop combinations (in the highlands of tropics). Some systems are well adapted to the highlands, while others perform well in the humid lowlands. Other agroforestry systems perform better in semi-arid zones. *Chagga* farms that are found on Mount Kilimandjaro in Tanzania and mountain plantations of western Nepal (Fonzen and Oberholzer 1984), integration of multipurpose trees in the mountains of Rwanda (Neumann 1983), and coffee-casuarina systems of Papua New Guinea (Bourke 1984) are examples of ecologically based agroforestry systems. Major agroforestry systems that are classified according to ecological zones are shown in Table 3.2. While agro-ecological characteristics can be used as a basis for designing agroforestry systems, agro-ecological zones alone cannot be used to classify systems, as most agroforestry systems are found in all ecological zones (Nair 1993). Similar ecological conditions can be found in different geographic areas. Consequently, an agroforestry system that is developed in Latin America can be adapted in the humid tropics of Africa. However, technology transfer should carefully consider socio-economic and cultural differences in the different geographic zones. Successful adoption of an agroforestry system requires that the system be designed in a participatory manner.

3.1.3.5 A Framework for Agroforestry Classification

Criteria that have been listed in Table 3.2 for the classification of agroforestry systems have limitations, and should be applied only to specific situations. No classification is universal, and each classification should be oriented towards a specific goal. Any classification should be done in two steps (Nair 1993), namely:

Table 3.2 Ecological spread of major agroforestry systems in the tropics and sub-tropics (Adapted from Nair 1987)

	Elevation (m asl)	Subhumid to perarid	Semihumid to semiarid	Prehumid to subhumid
Lowlands	<500	Arid-Semiarid		Subhumid-humid
		Dry months		
		1 2 3 4 5 6	7 8 9 10	11 12 13 14
		Annual rainfall (mm)		
		−500 to 500	500 to 1200	1200 and more
		Homegardens	Improved fallows in shifting cultivation	Homegardens
		Multipurpose woodlots for fuel	Trees on pasturelands	Improved fallows in shifting cultivation
		Multipurpose trees on croplands	MPT on woodlots for fuel	Trees in pisciculture
		Trees on pasturelands	Homegardens	Alley cropping
			Multilayer tree gardens	Multilayer tree gardens
			Plantation crop combinations	Improvement to Taungya
			Alley cropping	MPT woodlots for fuel
			Tree hedges for soil conservation	Plantation crop combinations
			Windbreaks/shelterbelts	MPT on croplands
			Improvement to Taungya	
Medium elevation	500 to 1200	MPT on woodlots for fuel	Plantation crop combinations	Alley cropping
		Windbreaks/shelterbelts	Trees in pasturelands	Improved fallows in shifting cultivation
		Trees on hedges for soil conservation	Improved fallows in shifting cultivation	Homegardens
		Multipurpose trees on croplands	Multilayer tree gardens	Multilayer tree gardens
		Trees on pasturelands	Improvement to Taungya	Plantation crop combinations
			MPT woodlots for fuel	MPT on croplands
			Homegardens	Improved fallows in shifting cultivation
			MPT on croplands	Improvement to Taungya

Table 3.2 (continued)

	Elevation (m asl)	Subhumid to perarid	Semihumid to semiarid	Prehumid to subhumid
			Plantation crop combinations	MPT woodlots for fuel
				Trees on pasturelands
Highlands	More than 1200	Trees on pasturelands	Improvement to Taungya	Homegardens
			MPT woodlots for fuel	Improvement to Taungya
			MPT on croplands	MPT on croplands
			Trees on pasturelands	Tree hedges for soil conservation
				Windbreaks/ shelterbelts

- Rank the system into the three major categories (agrosilvopastoral, silvipastoral and agrisilviculture) and an 'other category' for systems such as multipurpose tree lots
- Proceed with structural, functional, socio-economic or ecological classification. The criteria that are used include the arrangement of components, the function of components, agroecological zones, and socioeconomic aspects.

Bibliography

Bourke RM (1984) Food, coffee and Casuarina: an agroforestry system from the Papua New Guinea Islands. Agroforest Syst 2: 273–279

Fonzen P, Oberholzer E (1984) Use of multipurpose trees in the hill farming systems in western Nepal. Agroforest Syst 2: 187–197

Khasa DP (2001) Bringing agroforestry into the 21st century: an overview. In: Clason T (ed) Proceedings of the 6th North American Agroforestry Conference. CD-ROM. Hot Springs, Arkansas, USA. pp 19–27 (June 12–16, 1999)

King KFS (1968) Agri-silviculture: the Taungya system. Bulletin No 1, Department of forestry, University of Ibadan, Nigeria

Leakey R (1996) Definition of agroforestry revisited. Agrofor Today 8:1

Lundgreen BO, Raintree JB (1982) Sustained agroforestry. In: Nestel B (ed) Agricultural research for development: potentials and challenges in Asia. ISNAR, The Hague, pp 37–49

Nair PKR (1985) Classification of agroforestry systems. Agrofor Syst 3(2):97–128

Nair PKR (1987) Agroforestry systems inventory. Agrofor Syst 5:301–317

Nair PKR (1993) An introduction to agroforestry. Kluwer, The Hague

Neumann I (1983) Use of trees in smallholder agriculture in tropical highlands. In Lockeretz W (ed) Environmentally sound agriculture, Praeger, New York, pp 351–374

Sinclair FL (1999) A general classification of agroforestry practice. Agrofor Syst 46:161–180

Vergara NT (1981) Integral agroforestry: a potential strategy for stabilizing shifting cultivation and sustaining productivity of the natural environment. Working Paper Environment Policy Inst, East-West Center, Honolulu, Hawaii

Vergara NT (1982) New directions in agroforestry: the potential of tropical legume trees. Improving Agroforestry in the Asia-Pacific Tropics. pp 52

Wyant J (1996) Agroforestry-an ecological perspective. Agrofor Today 8:1

Chapter 4
Major Agroforestry Systems of the Humid Tropics

Abstract More than one hundred agroforestry systems (particular land-use systems involving integrating production of trees with crops and/or livestock, which are characterized by the environment, plant species and their arrangement, management, and socio-economic functions) have been recorded yet, with about 30 agroforestry practices (distinct arrangements of agroforestry components in space and time). Agroforestry systems and practices are often used simultaneously. Major agroforestry practices or technologies in the humid tropics include homegardens, perennial crop based systems, shifting cultivation, alley cropping, improved fallows and rotational tree fallows. Other agroforestry systems are valued in the humid tropics, including relay cropping, multilayer tree gardens, multipurpose trees on croplands and plantation-crop combinations. Since the mid-90s, the participatory domestication of high-value and multipurpose indigenous forest species using agroforestry techniques has been gaining momentum in the humid tropics.

4.1 Introduction

An agroforestry system is a particular land-use system involving integrated production of trees and crops and/or livestock, characterized by the environment, plant species and their arrangement, management, and socio-economic functions. In contrast, an agroforestry practice reflects a distinct arrangement of components in space and time. Similar practices are found in various systems under different situations. More than one hundred agroforestry systems have been identified in the tropics and temperate regions, together with about 30 agroforestry practices (Table 4.1). The distinction between agroforestry systems and practices is often unclear, and these terms are often used interchangeably, with both referring to forms of land use.

One particular agroforestry practice that has gained momentum in the tropics since the mid-1990's is the participatory domestication of high-value and multipurpose indigenous forest species. These species have provided local communities with food, income, medicine, and shelter (Leakey and Newton 2004; Leakey et al. 2005; Tchoundjeu et al. 2006). This practice is a form of agro-technology (a scientific term for an intervention that changes a practice or an existing system), which modifies the practice of introducing multipurpose trees and grasses on farms. The principles, rationale, and methods of agroforestry systems will be explained in Chap. 7. Several studies have been carried out since the 1980's to understand the

A. Atangana et al., *Tropical Agroforestry*, DOI 10.1007/978-94-007-7723-1_4,

Table 4.1 Major agroforestry practices or technologies and their main characteristics (Adapted from Nair 1993) (Source: Khasa (2001))

Agroforestry practice	Brief description or arrangement of components	Major groups of components (i.e., W: woody; H: herbaceous)	Agroecological adaptability
A. Agrisilvicultural systems (Agrisilviculture): woody perennials and agricultural crops			
1. Improved fallow	Tree or shrub species planted and left to grow during the fallow phase	W: fast-growing, preferably leguminous (e.g., *Leucaena leucocephala*) H: common agricultural crops	In shifting cultivation areas (Tropics)
2. *Taungya*	Combined stand of woody and agricultural species during early stages of establishment of plantations	W: usually plantation forestry species (*i.e., Tectona grandis*); H: common agricultural crops	In all ecological regions where *Taungya* is practiced; several improvements possible
3. Alley cropping (or hedgerow intercropping)	Woody species in hedges; agricultural species in alleys between hedges; microzonal or strip arrangement	W: fast-growing legumes, that coppice vigorously H: common agricultural crops	Subhumid to humid areas with high human population pressure and fragile soils
4. Relay cropping	Agricultural crops are interplanted with woody species which is harvested before the next rainy season and ramial chipped wood incorporated the soil	W: short rotation, preferable leguminous H: common agricultural crops	Tropical areas
5. Multilayer tree gardens	Multispecies, multilayer, dense plant associations with no organized planting arrangements	W: different woody components of various forms and growth habits; H: usually absent; shade-tolerant ones present	Areas with fertile soils, good availability of labor, and high human population pressure
6. Multipurpose trees on croplands	Trees scattered in cropland (e.g., maize in parkland) or according to some systematic patterns on bunds, terraces, or plot fields boundaries	W: multipurpose trees and other fruit trees Common agricultural crops	In all ecological regions, especially in subsistence; also commonly integrated with animals

Table 4.1 (continued)

Agroforestry practice	Brief description or arrangement of components	Major groups of components (i.e., W: woody; H: herbaceous)	Agroecological adaptability
7. Plantation crop combinations	1. Integrated multi-storey (mixed, dense) mixture of plantation crops 2. Mixtures of plantation crops in alternate or other regular arrangement	W: plantation crops such as cacao, coconut, rubber, coffee, palm, fruit trees, fuelwood or fodder species	In humid lowlands or tropical humid/subhumid highlands (depending on the plantation crop concerned); usually in small-holder subsistence system
8. Homegardens	Intimate, multi-storey combinations of various trees and crops around homesteads	W: Fruit trees and vegetables predominate H: shade tolerant agricultural species	In all ecological regions
9. Woody perennials in soil conservation and reclamation	Woody species on bunds, terraces, raisers, etc., with or without grass strips; woody species for soil reclamation	W: multipurpose and /or fruit perennials H: common agricultural species	In sloping areas, especially in highlands, reclamation of degraded, acid, alkali soils, and sand-dune stabilization
10. Sloping agriculture land technology	Planting of field and permanent crops in 3-5 m bands between double-contoured rows of woody plants to help control soil erosion and increase crop yields	W: woody species preferable leguminous H: agricultural crop of the locality	Southeast Asia, particularly in the hilly uplands
11. Hydro-agrisilviculture	Woody plant (e.g. *Sesbania rostrata*) is sequentially inter-cropped with wetland rice	W: flood-tolerant leguminous H: wetland rice	In Southeast Asia, also sometimes integrated with fishes (e.g. carp)
12. Irrigated agrisilviculture	Crop combination with fruit bearing woody perennials	W: fruit bearing woody plants (e.g. olive trees) H: agricultural crops of the locality (e.g. cereals)	Semi-arid, arid and Mediterranean regions (e.g., Sahel, Northern Africa, Middle East)
13. Shelterbelts and windbreaks, living hedges	Woody plants around farmlands/plots; multi-layers strips of trees and/or shrubs planted at several rows to alter wind flow	W: combination of tall-growing spreading types H: agricultural crops of the locality	In wind-prone areas

Table 4.1 (continued)

Agroforestry practice	Brief description or arrangement of components	Major groups of components (i.e., W: woody; H: herbaceous)	Agroecological adaptability
14. Fuelwood production	Interplanting firewood species on or around agricultural lands	W: firewood species H: agricultural crops of the locality	In all ecological regions, except in rainforest zones, where fuel wood is available (in the forest)
B. Silvopastoral systems (Silvopasture or silvopastoralism or silvipasture): woody perennials and pasture and or livestock			
15. Woody perennials on range land or pastures	Woody perennials scattered irregularly or arranged according to some systematic pattern	W: multipurpose trees, of fodder value Fo: present L: present	Extensive grazing in tropical and temperate areas
16. Protein banks	Production of protein-rich fodder/forage on farm/rangelands for cut-and-carry fodder/forage production	W: leguminous fodder woody perennial H: present Fo: present L: present	Usually in areas with high person: land ratio
17. Plantation crops with pastures and livestock	Livestock under woody perennials (e.g. coconuts, oil palms, pines and *Eucalyptus* spp.)	W: plantation crops Fo: present L: present	In areas with less pressure on plantation crop lands
18. Living fences of fodder woody perennials and hedges	Livestock browsing living hedges	W: fast-growing and coppicing fodder shrubs and trees L: present	In all ecological regions, very common in the tropics
C. Agrisilvopastoral or agrosilvopastoral systems			
19. Homegardens involving livestock	Intimate, multi-storey combinations of various woody species and crops, and livestock, around homesteads	W: fruit perennials predominate; also other woody species L: present	In all ecological regions with high density of human population
20. Multipurpose woody hedgerows	Woody hedges for browse, mulch, green manure, soil conservation	W: fast-growing and coppicing fodder shrubs and trees H: similar to alley cropping and soil conservation L: present	Humid to subhumid areas with hilly and sloping terrain
21. Dehesa or montado	Wooded pastures also producing Non-Timber Forest Products	W: multipurpose trees dominate	Spain and Portugal

Table 4.1 (continued)

Agroforestry practice	Brief description or arrangement of components	Major groups of components (i.e., W: woody; H: herbaceous)	Agroecological adaptability
Other Systems			
D. Forest farming systems			
22. Entomoforestry	Insects with woody species for honey production (apiforestry) or for pollination	W: melliferous woody perennials (other components may be present)	Depending on the feasibility of apiculture
23. Mycoforestry	Fungal spawn inoculated on trees of cultivated logs	W: living trees or logs Em: edible mushrooms	Temperate and Mediterranean regions and could be developed in tropical regions
24. Herboforestry	High-value specialty herbs cultivated under woody perennials	W: shade-woody species H: specialty shade-tolerant herbs	In all ecological regions
E. Aquaforestry			
25. Silvo-fish farming	Woody perennials liming fish ponds, tree leaves being used as forage for fish	W: woody perennials preferred by fish (other components may be present) Fi: fish and shrimps present H: grass planted on bunds	In all ecological regions especially in lowlands with a lack of animal proteins
26. Agrisilvopastoral fishery	Trees lining fish ponds, micro livestock raised around the ponds and temporal sequence for raising fish and agricultural crops in the ponds	W: woody perennials preferred by fish H: agricultural crops of the locality (e.g. vegetables and wetland rice) Fi: fish and shrimps present H: grass planted on bunds	In the humid Tropics, especially in Southeast Asia
27. Mangrove management	Plantation establishment and rehabilitation of degraded mangrove formations to mitigate erosion and reduce flooding, protect fish and shrimp ponds	W: mangrove woody perennials (e.g. *Rhizophora* spp., *Avicennia* spp.) Fi: fish and shrimp present	Mangrove ecosystems occurring in the intertidal zones along sheltered coasts and river banks in coastal areas in the tropics and sub-tropics

Table 4.1 (continued)

Agroforestry practice	Brief description or arrangement of components	Major groups of components (i.e., W: woody; H: herbaceous)	Agroecological adaptability
F. Hydroforestry			
28. Multispecies buffer strips riparian zone management	A buffer strip, which includes rows of trees and or shrubs, and a strip of native prairie grass to stabilize the stream, and serve as a sink for non-source pollutants from agricultural fields; woody perennials and grasses increased biodiversity for wildlife and provide biomass energy	W: short-rotation woody crops (e.g., *Populus* spp.) with shrub species H: native prairie grass	In all ecological regions in need to protect stream, riverbanks and lakes
29. Multispecies water catchment management	Planting of selected hydrophilic species for rehabilitation of water catchment	W: hydrophilic woody perennials (e.g., black spruce)	In swamped areas
30. Multipurpose woodlots	For various purposes (wood fodder, soil protection, soil reclamation)	W: multipurpose species; special location-specific species (other components may be present)	Various
31. Community forestry	Tree planting on common lands by local people	W: multipurpose tree species	In tropical regions

W *woody*; H *herbaceous*; Fo *fodder/forage for grazing*; Fi *fish*; L *livestock*; Em *edible mushrooms*

mechanisms underlying the functioning of existing agroforestry systems in order to fine-tune those systems. Systems that have been studied include alley cropping, improved fallows, *Taungya*, homegardens, windbreaks, and parklands together with cocoa, coffee and tea or rubber farms.

The spatial structure of farm compounds in forest areas in the humid tropics is as follows:

- Houses
- Homegardens
- Cash crops
- Food crops and fallows
- Forests

Fig. 4.1 An example of a homegarden in Cameroon (Source: Ann Degrande)

The most common agroforestry systems in the humid tropics include homegardens, perennial crop based systems, farm woodlots, alley cropping, improved fallows, and rotational tree fallows. Some agroforestry systems are specific to Amazonia (such as small-scale intensive farming systems, which is a form of homegarden) and to Southeast Asia (such as *Taungya*). Major agroforestry practices or technologies and their main characteristics are given in Table 4.1.

The aforementioned land-use systems that we have listed are described in detail in the sections that follow, starting with homegardens, which are an important component of homesteads in the tropics.

4.2 Homegardens

A homegarden can be defined as an intimate association of multipurpose trees and shrubs, annual or perennial plants, or livestock within the household compound, with the whole unit being managed by family labor (Fernandes and Nair 1986). Homegardens consist of an assemblage of trees, shrubs, and vines and herbaceous plants that are managed around the home compound (Fig. 4.1) by the household, and the products of which are used primarily for family consumption. This agroforestry system can also provide shade for livestock or serve ornamental purposes. Most homegardens are agrosilvopastoral systems. Kumar and Nair (2004) have suggested that homegardening is a generic concept (i.e., a group of terms), much like agroforestry itself. Homegardens are "structurally and functionally the closest mimics of natural forests yet attained" (Ewel 1999; Table 4.2).

Several terms have been used to describe agroforestry practices that are undertaken around homes, including mixed-garden horticulture (Terra 1954), homegarden (Ramsay and Wiersum 1974), mixed-garden or house garden (Stoler 1975), Javanese homegarden (Soemarwoto et al. 1976; Soemarwoto 1987), compound farm (Lagemann 1977), kitchen garden (Brierley 1985), household garden (Vasey

Table 4.2 A comparison of the ecological attributes of climax forests, homegardens and conventional agricultural systems (Kumar and Nair 2004)

Parameter	Natural climax vegetation (humid tropics)	Homegardens	Conventional agricultural systems
Biogeochemistry	Nutrient inputs equal outputs	Inputs and outputs balance each other	Outputs far exceed inputs
Biotic stress	Low	Low	High
Canopy architecture	Multistrata	Multistrata	One- or two-layered
Disturbance regimes	Rare except natural disturbances such as tree fall, wind throw etc.	Intermediate	High
Diversity	High	Intermediate	Low
Ecological succession	Normally uninterrupted, reaches a stable end-stage, e.g. climatic climax	Consciously manipulated	Arrested, succession does not proceed beyond the early stage
Entropy	Low	Low (?)	High
Floristic spectrum	Shade tolerant and intolerant	Shade tolerant and intolerant	Mostly shade intolerant
Input use	No external inputs	Low	High
Overall homeostasis	High (Odum 1969)	High	Low
Site quality	Progressive improvement (e.g., facilitation)	Progressive improvement	Steady decline
Standing biomass/net primary productivity (NPP)	Highest among the terrestrial ecosystems (mean NPP: 200 g m^{-2} $year^{-1}$)	Comparable to the climax formations but firm estimates are lacking[a]	Low (mean NPP: 650 g m^{-2} $year^{-1}$; Leigh 1975)
Sustainability	Sustainable	Sustainable	Unsustainable

[a]However, a lone report on this (Christanty et al. 1986), *cf* Torquebiau (1992), provides a value of 5.23 kJ (=1250 cal) m^{-2} yr^{-1} for the *pekarangan* gardens in Java. Clearly, this is lower than the annual energy fixation in both cultivated lands and tropical rainforest (i.e., 11.3 and 34.6 MJ m^{-2} yr^{-1}, respectively; Leigh 1975)

1985), and homestead agroforestry (Achuthan and Streedharan 1986; Leuschner and Khalique 1987).

Numerous types of homegardens have been described (Soemarwoto et al. 1976; Lagemann 1977; Bavappa and Jacob 1982; Wiersum 1982; Michon 1983; Fernandes and Nair 1986; Fernandes et al. 1984; Okafor and Fernandes 1986; Reynor and Fownes 1991; Tchatat et al. 1995), indicating that this system is widely distributed in the tropics and has been practiced for millennia. Because the primary function of a homegarden is subsistence, they most often contain vegetables, tuber crops, medicinal plants, multipurpose plants and indigenous fruit trees. Perennial crops such as cocoa, coffee or palms are frequently found in homegardens, but these gardens lack the operational size of cash crop farms. An inventory of the structures and functions of homegardens in the tropics was conducted by Fernandes and Nair (1986). The inventory subsequently was used to classify homegardens by region, depending on biophysical and socio-economic factors, and to describe the different compositions of homegardens (Table 4.3).

Table 4.3 Floristic elements of homegardens in different regions of the world (Kumar and Nair 2004)

Region/location and the floristic spectrum sampled	Number of species per garden	Total for geographical location	Source
South Asia			
Pitikele, Sri Lanka (edible species)	–	55	Caron (1995)
Kandy, Sri Lanka (woody species)	4–18	27	Jacob and Alles (1987)
Thiruvananthapuram, Kerala, India (all species)	–	107	John and Nair (1999)
Same as above	26-36	–	D. Jose (pers. comm., 1992)
Kerala, India (woody species)	3-25	127	Kumar et al. (1994)
Bangladesh (perennial species)	–	30	Leuschner and Khalique (1987)
Same as above	–	92	Millat-e-Mustapha et al. (1996)
Kerala, India (all species)	–	65	Achuthan and Sreedharan (1986)
Southeast Asia			
West Java (all species)	–	195	Abdoellah et al. (2001)
Northeastern Thailand (all species)	15–60	230	Black et al. (1996)
Chao Praya Basin, Thailand (all species)	26–53	–	Gajaseni and Gajaseni (1999)
West Java (all species)	–	602	Karyono (1990)
Central Sulawesi, Indonesia (all species)	28–37	149	Kehlenbeck and Maas (2004)
Cianjur, West Java (all species)	4–72	–	K Sakamoto (pers. comm., 2003)
Cilangkap, Java (all species)	42–58	–	Yamamoto et al. (1991)
South/CentralAmerica and the *Caribbean*			
Quintan Roo, Mexico (all species)	39	150	De Clerck and Negreros-Castillo (2000)
Cuba (all species)	–	80	Esquivel and Hammer (1992)
Central Amazon (woody species)	–	60	Guillaumet et al. (1990)
Belize (all species)	30	164	Levasseur and Olivier (2000)
Masaya, Nicaragua (all species)	–	324	Méndez et al. (2001)
Santa Rosa, Peruvian Amazon (all species)	18–74	168	Padoch and de Jong (1991)
Yucatan, Mexico (all species)	–	133–135	Rico-Gray et al. (1990)
As above	–	301	Rico-Gray et al. (1991)
Chiapas, Mexico (all species)	30	241	Vogl et al. (2002)

Table 4.3 (continued)

Region/location and the floristic spectrum sampled	Number of species per garden	Total for geographical location	Source
Cuba (all species)	18–24	101	Wezel and Bender (2003)
Other regions			
Catalonia, Spain (all species)	–	250	Agelet et al. (2000)
Southern Ethiopia (all species)	14.4	60	Asfaw and Woldu (1997)
Bungoma, Western Kenya (all species)	–	253	Bakes (2001)
Soqotra island, Yemen (all species)	3.9–8.4	–	Ceccolini (2002)
Democratic Republic of Congo (all species)	–	272	Mpoyi et al. (1994)
Bukoba, Tanzania (woody species)	–	53	Rugalema et al. (1994a)
Central, eastern, western and southern Ethiopia (all species)	–	162	Zemede and Ayele (1995)

[a] All except Catalonia are tropical

Homegardens contain a wide variety of species, which approximates the range of species encountered in natural forests (Gajaseni and Gajaseni 1999). In one example, 101 plant species were identified in 31 homegardens in Cuba, with each garden containing about 18 to 24 different species (Wezel and Bender 2003; Table 4.4). Similarly, the mean number of woody taxa that are found in homegardens in India can range from 11 to 39, with greater floristic diversity present in the smaller homesteads (Kumar et al. 1994). Homegarden diversity is strongly related to its age and other specific garden characteristics, household socio-economic features, and access to planting material (Coomes and Ban 2004). The average homegarden includes about four canopy strata (Tchatat et al. 1995; Gajaseni and Gajaseni 1999; Figs. 4.2 and 4.3), and their average area is frequently less than one hectare in size (Fernandes and Nair 1986).

Homegardens are carefully structured systems. For example, in Nigeria (West Africa), homegardens have a four-strata canopy that is dominated by fruit trees (Okafor and Fernandes 1987). Another example of a homegarden is the intensive small-scale system that is described in sect. 4.1.1. The structure and composition of a homegarden will depend upon its position in the overall farming system and on the livelihood strategies of its inhabitants. Rural transformation results in changes in livelihoods and farming systems, which has further impacts on homegarden function and composition (Wiersum 2006). Factors that affect the structure and composition of homegardens are listed in Table 4.5.

The choice of species to be included in a homegarden depends upon the products that these species provide (Gajaseni and Gajaseni 1999). The choice of species, together with their arrangement and management, can vary within a community or village (Méndez et al. 2001). Tchatat et al. (1995) have described the homegardens

Table 4.4 Factors impacting the structure and composition of homegardens with special reference to Indonesian homegardens (Wiersum 2006)

Factors	Conditions	Examples and remarks
Geographic location	Urban versus rural location	Urban homegardens often smaller and more aesthetically oriented
Environmental conditions	Climate conditions	Variation in annual crops cultivated only in favorable climatic seasons is mostly less pronounced than in permanent crops that have to be adapted to variable climatic conditions over much larger periods
	Soil conditions	With decreasing soil fertility crop diversity tends to decrease and the effect of competition by trees on understorey becomes more pronounced. Dense tree gardens occur mostly in volcanic soils, while on tertiary soils tree gardens are open
Role in farming systems	Degree of complementarity to open field cultivation systems	If homegardens are the only land asset more inclusion of staple food crops
	Established versus incipient farming system	Incipient gardens first dominated by annual crops, with time increased incorporation of tree crops
Socioeconomic conditions of the household	Wealth status	With increased wealth, increased importance of commercial and aesthetic crops
	Access to markets	Commercial crops stimulated by good market access
	Access to off-farm employment	In case of access to financially lucrative employment decreased, importance of commercial crops
	Gender-related issues	Gardens of female-headed households often more household oriented (for consumption)
Cultural factors	Food preferences	Cultural preferences in respect to consumption of vegetables and spices

of lowland rainforests of Cameroon as following a floristic and structural approach as well as a socio-economic approach. Homegardens in this area consist of a front yard for ornamental plants, and a larger backyard where food crops and fruit trees are grown. Three groups of species characterized the homegardens in this area, depending on the garden's life history and usage.

The first species group is primarily composed of maize (*Zea mays* L., Poaceae), which can be combined with other annual crops such as the common bean (*Phaseolus vulgaris* L., *Fabaceae*) and groundnut or peanut (*Arachis hypogaea* L., Fabaceae). The second group consists of multi-annual food crops such as plantain (*Musa* spp.), and cassava or manioc (*Manihot esculenta* Crantz, Euphorbiaceae). The third group contains mainly fruit species such as safou or African pear (*Dacryodes edulis* H.J. Lam; Burseraceae), mango (*Mangifera indica* L., Anacardiaceae), or citrus (Rutaceae) trees, and other trees with various uses. These homegardens are mainly intended to produce food for the household, whereas *Chagga* homegardens in Tanzania, for example, tend to be commercial and consist mainly

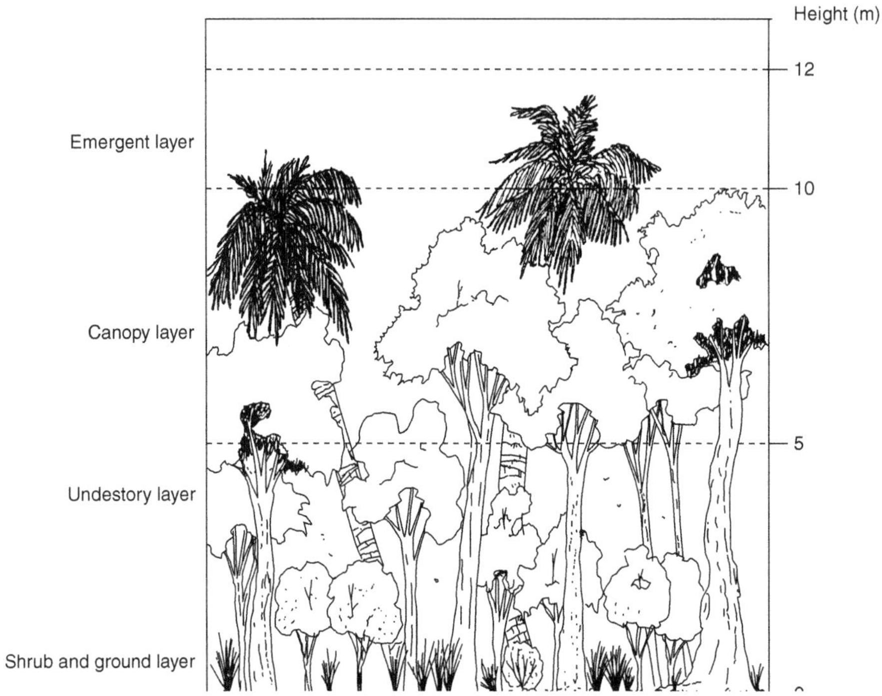

Fig. 4.2 Vertical physical structure of vegetation in a homegarden system of Sristachanalai, Thailand (Gajaseni and Gajaseni 1999)

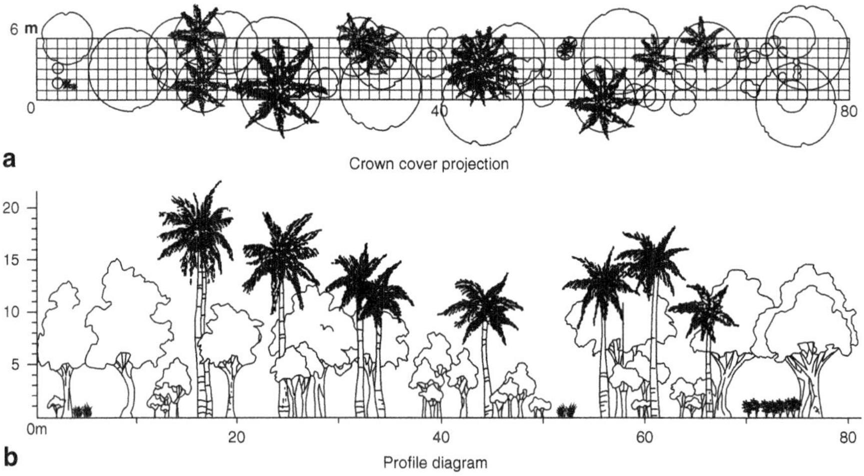

Fig. 4.3 Crown cover projection **(a)** and profile diagram **(b)** of homegarden systems in Sristachanalai, Thailand (Gajaseni and Gajaseni 1999)

Table 4.5 Annual litter and nutrient additions through multipurpose trees in homegardens in Kerala, India (Isaac and Achuthan 2006)

Species	Litterfall (kg m^{-2})	Nitrogen (g m^{-2})	Phosphorus (g m^{-2})	Potassium (g m^{-2})
Mangifera indica	0.87	8.8	0.42	2.8
Artocarpus heterophyllus	0.73	8.6		
Anacardium occidentale	0.72	8.8	0.17	1.1
Ailanthus triphysa	0.38	6.4	0.35	1.2
Artocarpus hirsutus	0.65	6.5	0.39	1.7
Swietenia macrophylla	0.64	6.8	0.42	2.6
Total	3.99	45.9	2.09	11.6

of arabica coffee (*Coffe arabica* L., Rubiaceae) and bananas (Fernandes et al. 1984). African homegardens that are intended for food production, as is the case in southeastern Nigeria, consist mainly of food crops such as yams (*Dioscorea* spp., Dioscoreaceae), manioc (*Manihot utilissima* Pohl, Euphorbiaceae), taro (*Colocasia esculenta* (L.) Schott, Araceae)), cocoyam or malanga (*Xanthosoma sagittifolium* (L.) Schott, Araceae), *Musa* spp., banana (*Musa* x *paradisiaca* L.), maize (*Zea mays* L.), okra (*Albemoschus esculentus* (L.) Moench = *Hibiscus esculentus* L., Malvaceae), squashes and pumpkin (*Cucurbita pepo* L., Cucurbitaceae), Thunberg's amaranth (*Amaranthus thunbergii* Moq., Amaranthaceae), and *Solanum* spp (Solanaceae). These homegardens also harbor trees and shrubs, small ruminants, poultry, and occasionally swine that are kept in pens. Manure from the animals is used as fertilizer for the plants. An analysis of nine fertility properties indicates that the soils in these gardens are healthier than those under fallow or surrounding secondary forests (Tchatat 1996; Tchatat et al. 2004). In central Sulawesi (Indonesia), 149 crop species were recorded in 30 homegardens that had been randomly selected from three villages, and the number of vegetation layers differed depending on the age and size of the homegarden (Kehlenbeck and Maass 2004). There is minimum export of soil nutrients in homegardens (Gajaseni and Gajaseni 1999), and production of nutrients from litter is very high (Fig. 4.4, Table 4.6), which indicates that litter from a homegarden has potential as an agricultural fertilizer.

Tropical homegardens have more favorable microenvironments than the surrounding areas, with lower soil and air temperatures and higher relative humidity. They also can be very productive, in India, for example, homegardens produced enough fuelwood to meet societal demands (Kumar et al. 1994).

4.2.1 Intensive Small-Scale Farming Systems

Altieri and Farrell (1984) reported on intensive small-scale farming systems that were located close to homegardens and food crop farms. Indeed, these farms are less than 1 ha in area and are located around homesteads. Common inter-cropping (maize and beans; garlic and onions, which are mixed with lettuce and cabbage; maize and potatoes) is practiced in these systems, which can include 5 to 10 tree

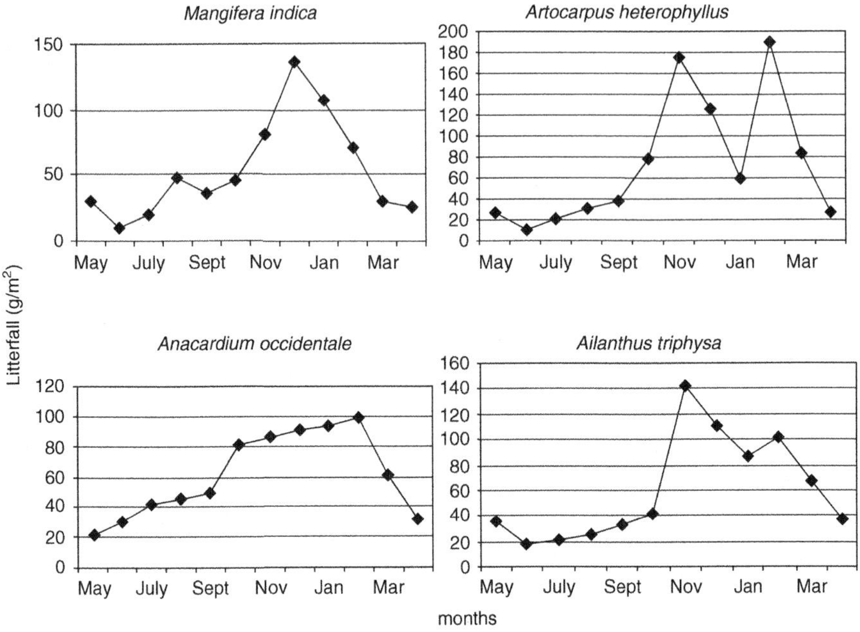

Fig. 4.4 Litterfall patterns of multipurpose trees in the homegardens of Kerala, India (Isaac and Achuthan 2006)

Table 4.6 Average species richness and diversity indices of trees per cocoa agroforest in three sub-regions of Southern Cameroon (Sonwa et al. 2007)

	Ebolowa (n=20 agroforests)	Mbalmayo (n=20 agroforests)	Yaoundé (n=60 agroforests)	Whole region	P-value
Species richness	21 (±1.8) a	26 (±2.3) a	15 (±1.2) b	21 (±1.1)	0.001
Shannon index	3.9 (±0.13) a	4.2 (±0.12) a	3.11 (±0.16) b	3.7(±0.96)	<0.001
Piélou equitability	0.73 (±0.02) a	0.72 (±0.02) a	0.60 (±0.03) b	0.70 (±0.02)	<0.001
Simpson index	0.10 (±0.01) a	0.08 (±0.01) a	0.19 (±0.23) b	0.12 (±0.01)	<0.001

n^*=20 agroforests; n^{**}=60 agroforests; P-values are the significance level of the Kruskal–Wallis test; n.s = not significant (p>0.05). Subregions not sharing a common letter in a row are significantly different at P=0.05

crops and 10 to 15 annual crops. This system also contains grape arbors to provide shade, and 3 to 5 animal species (chickens, ducks, rabbits and pigs). Such small-scale intensive farming systems can be found in the densely populated Kerala State of India (Achuthan and Sreedharan 1986). These intensive systems contain livestock, poultry, fisheries, and tree and plantation crops in mixtures on the same piece of land. Food crops are mostly arrow root (*Maranta arundinacea* L., Marantaceae), rice (*Oryza sativa* L., Poaceae), cassava, Chinese potato (*Coleus parviflorus* Benth., Lamiaceae), taro (*Colocasia* spp.), elephant yam (*Amorphophallus paeoniifolius* (Dennst.) Nicolson, Araceae) and sweet potato (*Ipomoea batatas* (L.)

Lam., Convolvulaceae) (Achuthan and Sreedharan 1986). Pulse crops are cultivated in these systems (for example: cowpea (*Vigna unguiculata* (L.) Walp., Fabaceae), pigeon pea (*Cajanus cajan* (L.) Millsp. Fabaceae), mung bean (*Vigna radiata* (L.) R. Wilczek, Fabaceae), together with breadfruit (*Artocarpus altilis* (Parkinson) Fosberg, Moraceae), annona (*Annona* spp., Annonaceae), banana, kokum (*Garcinia indica* Choisy, Clusiaceae), gooseberry (*Ribes uva-crispa* L., Grossulariaceae), guava (*Psidium guajava* L., Myrtaceae), and jackfruit (*Artocarpus heterophyllus* Lam., Moraceae)) (Achuthan and Sreedharan 1986). Nath et al. (2005) also reported small-scale homesteading in the densely populated Chittagong Hills Tracts of Bangladesh. Homesteading uses plantings of both horticultural (*A. heterophyllus*, *Citrus reticulata* Blanco, *Litchi chinensis* Sonn. (Sapindaceae), *Ananas comosus* (L.) Merr. (Bromeliaceae), banana, guava, and mango)) and timber species (*Gmelina arborea* Roxb. (Lamiaceae), *Tectona grandis* L.f. (Lamiaceae), *Albizia* spp. (Fabaceae), *Swietenia macrophylla* King (Meliaceae), and *Acacia* spp. (Fabaceae)).

4.3 Perennial Crop Based Agroforestry Systems

Agroforestry with a cultivated tree crop component is widespread in the tropics. In the humid tropic lowlands, it consists mainly of cocoa (*Theobroma cacao* L., Sterculiaceae) or robusta coffee (*Coffea canephora* Pierre ex A.Froehner, Rubiaceae), which is grown under the shade of several trees. In 2000, cocoa agroforests covered between 300,000 and 400,000 hectares in Cameroon (Kotto-Same et al. 2000). In the tropical highlands of Africa or South America, arabica coffee- or tea- (*Camellia sinensis* (L.) Kuntze, Theaceae) based systems are most common. Other examples of species that are used in perennial agroforestry systems in the tropics are: oil palm (*Elaeis guineensis* Jacq., Arecaceae), coconut (*Cocos nucifera* L., Arecaceae), cashew (*Anacardium occidentale* L., Anacardiaceae), rubber tree (*Hevea brasiliensis* Müll. Arg., Euphorbiaceae), black pepper (*Piper nigrum* L., Piperaceae), vanilla bean (*Vanilla planifolia* Jacks. Ex Andrews, Orchidaceae; in Madagascar), sisal (*Agave sisalana* Perrine, Asparagaceae), and carnauba palm (*Copernicia prunifera* (Mill.) H.Moore, Arecaceae; mostly in Latin America). They are mostly cash crops, and although some of these species originate in the tropics, their culture or their introduction coincided with the colonization of the region. Cacao, vanilla, and rubber originated in Latin America; while coffee likely originated in Ethiopia, the arabica species may be indigenous to Central Africa (Lashermes et al. 1999).

Over the last several years, research has been focused on finding methods to increase food crop production, but little has been done pertaining to the introduction of trees and animals into perennial crop farms where cocoa or coffee is grown. Farmers form the bulk of perennial commodity crop producers (cocoa in Ivory Coast, Ghana, Nigeria and Cameroon; coconut in Southeast Asia). Cocoa agroforests mix cacao with other crops and tree species; most of the latter are retained during land clearing. Forest trees that are retained when clearing land for farm establishment provide shade and other environmental services. In Ondo State, Nigeria, 487 non-cocoa trees belonging to 45 species and 24 families were recorded in 21 ha of cocoa agroforests, with 86.8 % of the trees producing edible fruits (Oke and Odebiyi 2007). Each cocoa farm has a multistrata structure (Fig. 3.7), and the flora is very diverse. In the humid forest zone of Cameroon, an inventory of 60 cocoa agroforest stands revealed the presence of 206 tree species, with an average of 21 species per agroforest (Sonwa et al. 2007), thereby demonstrating the high diversity found in the forests (Table 3.8). These figures are consistent with those of Bisseleua et al. (2008), who identified a total of 102 non-cocoa trees and 260 herbaceous species in five traditional cocoa agroforests in Cameroon. Food-producing tree species tend to be more frequently planted than other tree species in cocoa farms, and two-thirds of these food trees are native forest species (Sonwa et al. 2007).

Some fast-growing food crops, such as plantain, are used for shade during cacao establishment. Other food crops like maize, sweet potato, malanga, and cucumber (*Cucumis sativus* L.) are often associated with cacao in the early years of its growth, to provide for the household's food needs. This involves optimizing of airspace and soil management for the benefit of the cocoa plants. Plantains provide shade for the seedlings, and as the cocoa plants grow, the soil volume free for cocoa plant root expansion will increase as the food crops are reaped. Species commonly planted to provide shade for cocoa or coffee are typically local fruit trees (e.g., *Garcinia kola* Heckel (Clusiaceae), *Irvingia gabonensis* (Aubry-Lecomte ex O'Rorke) Baill. (Irvingiaceae), *Cola acuminata* Schott & Endl. (Sterculiaceae), *Ricinodendron heudelotii* (Baill.) Heckel, Euphorbiaceae), and *Dacryodes edulis*), timber species (e.g., *Baillonella toxisperma* Pierre (Sapotaceae), *Milicia excelsa* (Welw.) C.C. Berg, *Terminalia superba* Engl. & Diels (Combretaceae), *Cedrela* spp. (Meliaceae), or multipurpose trees (e.g., *Allanblackia floribunda* Oliv., Clusiaceae). Cocoa is usually seeded at a density of 1,100 trees per ha, with canopy trees planted at 30 individuals per ha. Shade is beneficial for cocoa, as biomass production is higher in the shade. Isaac et al. (2007a) found that cocoa produced 41.1 Mg ha^{-1} of standing biomass under *Milicia excelsa*, compared to 22.8 Mg ha^{-1} when the plants were not shaded. The study also indicated that soil exchangeable K increased under *Newbouldia*, and N and P uptake increased under shade. Cocoa agroforests also have beneficial effects on the soil and litter faunal communities. Richness of the fauna is greater in the litter than in the soil (da Silva Moço et al. 2009). In Ghana, Isaac et al. (2007b) reported suppression of K uptake in cocoa foliage by inter-cropping under *Terminalia superba* and *Newbouldia laevis* (P. Beauv., Bignoniaceae). The same study revealed that intercropping has no effect on cocoa biomass production in comparison to monoculture cocoa, whereas artificial shading stimulated foliage and root production.

In the tropical highlands, arabica coffee and tea are the most common perennial crops, regardless of the size of the industrial farm, while coconut plantations are most commonly encountered in the Philippines, Indonesia, Sri Lanka, Malaysia, and the Pacific Islands. Woody perennials are grown among the coffee or tea plants for firewood or honey production (e.g., *Calliandra calothyrsus* Benth.). Arabica coffee farms typically result in multi-strata agroforests. Examples of crop or livestock integration into coffee farms have been reported in Ethiopia, Colombia, and

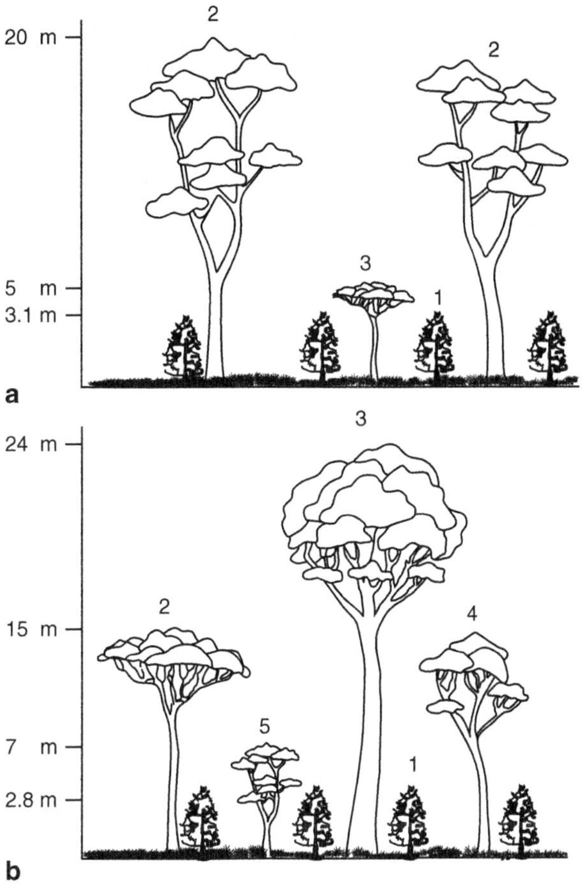

Fig. 4.5 The vertical structure of shade of two coffee systems in Chiapas, Mexico. **a** Most common species in an Inga system (IS): 1. Coffea arabica, 2. Inga latibracteata, 3. Heliocarpus appendiculatus. **b** Most common species in a rustic shade (RS) system: 1. Coffea arabica, 2. Inga latibracteata, 3. Heliocarpus appendiculatus, 4. Beliotia mexicana, 5. Croton draco. (Romero-Alvarado et al. 2002)

Kenya. However, in high altitudes, sensory attributes of Arabica coffee are negatively influenced by shade (Bosselmann et al. 2009). Coffee plants in mixed agroforests also have less branch growth and leaf production, and present earlier fruiting than coffee plants in monoculture systems (Campanha et al. 2004). In Costa Rica, Muschler (2001) found that, at low altitudes, an increase in shade results in an increase in fruit mass and bean size of *C. arabica.* He further postulated that at low altitudes, shade promotes slower and more balanced filling and uniform ripening of berries. Thus, the benefits of shade on coffee quality depend on altitude. Different shade structures can be encountered in coffee systems (Fig. 4.5) and, consequently, shade attributes and soil characteristics can vary according to the shade system (Table 4.7). Shade trees, especially indigenous trees with high leaf tannin concentrations, can also improve soil fertility in coffee systems (Teklay and Malmer 2004). Some timber species, such as silky oak (*Grevillea robusta* A.Cunn. ex R.Br., Proteaceae), can be inter-cropped with coffee without deleterious effects on coffee

Table 4.7 Mean values for shade attributes, soil variables, coffee-shrub characteristics and yields, in an Inga-coffee system (IS) and a Rustic-shade coffee system (RS) in Chiapas, Mexico (Romero-Alvarado et al. 2002)

Measured variables	IS	RS	Significance
Shade attributes			
Shade tree/ha	282±159	457±257	*
Shade-species richness in 100 m^2	1.6±0.7	3.0±1.4	***
Shade strata number	2.9±0.9	3.0±1.0	**
Shade cover (%)	76.9±5.7	71.2±8.8	*
Direct light (mol m^{-2} day^{-1})	18.2±4.4	21.9±6.3	*
Diffused light (mol m^{-2} day^{-1})	3.2±0.6	3.8±0.8	*
Shade-tree height (m)	11.3±3.9	11.0±4.7	N.S
Shade-tree diameter (cm)	27.0±14.1	23.6±11.4	N.S
Shade-tree basal total area (m^2)	0.7±0.8	1.0±1.3	N.S
Number of fruit trees ha^{-1}	100±74	200±10	N.S
Soil variables			
Litter density (cm)	3.2±1.7	2.2±0.9	*
pH	5.2±0.4	5.3±0.4	N.S
Organic matter (%)	6.7±0.9	6.0±1.5	N.S
Total nitrogen (%)	0.3±5.4	0.2±7.2	N.S
Extractable phosphorus (ppm)	4.8±4.2	2.4±2.0	N.S
Extractable potassium (meq 100 g^{-1})	8.7±2.9	12.0±4.8	N.S
Extractable calcium (meq 100 g^{-1})	9.0±8.7	9.9±9.5	N.S
Extractable magnesium (meq 100 g^{-1})	1.2±1.2	1.9±1.7	N.S
Exchangeable acidity (meq 100 g^{-1})	1.1±1.0	0.7±0.7	N.S
Aluminum (meq 100 g^{-1})	0.6±0.7	0.4±0.6	N.S
Coffee-shrub features			
Diameter at the central point of the stem (cm)	2.9±0.5	2.4±0.4	***
Central axis height (m)	3.1±0.4	2.8±0.4	*
Yields			
Clean coffee grains (kg ha^{-1})	1,900±1000	2,000±700	N. S

values in each column are means and their associated standard deviation * $P<0.05$; ** $P<0.001$; *** $P<0.001$; NS: $P>0.05$

production (at densities of 26, 34 and 48 stems ha^{-1}; Baggio et al. 1997). Shade trees that are grown in coffee farms are mostly used for firewood and as timber for local construction, as indicated by an inventory carried out in the Baoulé region of Côte d'Ivoire (Table 4.8).

Perennial crop farming is often associated with pastoral activities. In such cases, animal manure is used as fertilizer. For instance, in Malaysia, poultry farming is often practiced in coffee farms (Ismail 1986).

The choice of species to be introduced into a perennial crop farm depends on the canopy diameter of the trees and the rooting volume of the perennial crop during its growth phase, light levels (if the associated species is shade-tolerant), the possible interaction between the perennial species and the associated species, the life history of the perennial species, and the value (food, commercial, medicinal) of the associated species. Other factors such as parasites that are hosted by the associated crop are also considered.

Table 4.8 Tree species, according to their overall frequency in coffee and cocoa plantations (Modified from Herzog 1994)

Scientific name	Uses				Frequency (% of plantations)	
	W	F	M	C	Coffee	Cocoa
Planted trees						
Persea americana		*			42	67
Citrus reticulata		*			17	78
Mangifera indica		*			58	50
Citrus sinensis					8	67
Cola nitida	*	*	*		42	33
Cocos nucifera		*			25	0
Wild trees retained for their products						
Elaeis guineensis		*			83	100
Ricinodendron heudelotii	*	*	*		50	28
Alstonia congensis			*		8	6
Funtumia africana	*			*	0	6
Microdesmis puberula			*		8	0
Wild trees retained for shading						
Antiaris welwitschii var. africanum	*			*	67	28
Spondias mombin		*	*		67	11
Albizia adianthifolia				*	42	0
Cola cordifolia	*	*		*	8	17
Triplochiton scleroxylon	*	*	*	*	25	6
Anthocleista dialonensis	*	*			17	6
Musanga cecropioides	*				25	0
Spathodea campanulata	*	*			17	6
Sterculia tragacantha		*			8	11
Ficus mucuso	*	*			8	6
Lannea acida		*			17	0
Dialium guineense	*		*	*	8	0
Diospyros mespiliformis	*		*	*	8	0
Morus mesozygia	*	*		*	8	0
Terminalia superba	*	*		*	8	0
Wild trees too big or not worthwhile to be cut						
Ceiba pentandra			*		83	56
Milica excelsa	*				58	28
Bombax buonopozense		*	*		42	22
Dracaena manii		*			42	6
Ficus exaspera	*	*			17	6
Celtis mildbraedii	*			*	0	11
Cola gigantea				*	17	0
Cordia senegalensis	*			*	17	0
Blighia sapida	*	*		*	8	0
Bridelia ferruginea					8	0
Deinbollia pinnata			*		8	0
Holarrhena floribunda	*			*	0	6
Hunteria eburnea	*			*	8	0
Newbouldia laevis	*	*		*	8	0
Pterygota macrocarpa					0	6

W Fuelwood, *M* Medicine, *F* Food, *C* Construction, * use
(% of plantations): percentage of plantations on which a species was found

Fruit tree-based agroforestry systems are being more commonly adopted within the Congo Basin. In southern Cameroon, the majority of farmers grow safou or African pear (*Dacryodes edulis*) on their lands (Ayuk et al. 1999; Schreckenberg et al. 2002). Safou is a high value, indigenous fruit tree that produces a widely traded fruit. In the Makenene and Kekem areas of Cameroon, most farmers maintain orchards of safou. These orchards harbor other valuable timber and edible fruit trees, but safou is considered one of the main cash and staple food crops in these regions. Other valuable indigenous fruit trees, including *Irvingia* species, are common in orchards of lowland humid tropics of Africa. Agroforests that are based on bush mango (*Irvingia wombolu* Vermoesen = *Irvingia gabonensis* var. *excelsa*; also known as dika or ogbono) are found in the Ejagham region of Cameroon, as their seeds (dika nuts) are widely traded and consumed in the region. Fruit tree-growing strategies in the humid forest zone of Cameroon are strongly influenced by accessibility of markets to farmers (Degrande et al. 2006).

Trees with important food, medicinal, fodder, or timber values are retained when clearing land for food crop establishment in the tropics. This selective retention of high-value indigenous fruit trees is common in humid forest zones and tropical savannas. Therefore, croplands in the tropics most often harbor scattered tree species with food and commercial value. In the humid forest zone of Africa, indigenous tree species that are most frequently encountered in agricultural landscapes include *Irvingia wombolu, Dacryodes edulis, Baillonella toxisperma, Garcinia kola, Ricinodendron heudelotii* and *Cola* spp (mainly *C. acuminata, C. anomala* and *C. nitida*). In Southern African savannas, the most common indigenous fruit trees in farmlands include *Adansonia digitata* L. (Malvaceae), *Azanza garckeana* (F. Hoffm.) Exell et Hillc. (Malvaceae), *Ficus* spp. L. (Moraceae), *Diospyros mespiliformis* Hochst. Ex A. DC. (Ebeneceae), *Strychnos cocculoides* L. (Strychnaceae), *Strychnos madagascariensis* L. (Strychnaceae), *Strychnos spinosa* L. (Strychnaceae) and *Sclerocarya birrea* (A. Rich.) Hochst. (Anacardiaceae), whereas *Combretum imberbe* Wawra (Combretaceae), *Pericopsis angolensis* (Baker) Meeuwen (Fabaceae) and *Swartzia madagascariensis* Desv. (Fabaceae) are valued for timber (Campbell et al. 1991). Gradually, these tree-crop ecosystems are being transformed into tree-based agroforestry systems when food crops are harvested. Indeed, when the plot is left to fallow, the previous tree crop system becomes a tree fallow system, followed by a secondary forest system that is managed by farmers for the collection of non-timber or timber forest products. Such secondary forests with important fruit, nut, medicinal and timber trees are considered tree-based agroforestry systems.

Another system involving the association of annual crops and forest species during the early years of establishment of the plantation forestry is the *Taungya* system, the development of which was described by Nair (1993). The practice of '*Taungya*' (from the Burmese *taung* meaning hill and *ya* meaning cultivation) originated in Myanmar (Burma) and dates back to the early 20th century (Blanford 1958; Nair 1993). As Nair (1993) noted: "Originally this term designated shifting cultivation, and was subsequently used to describe afforestation methods. The land belongs to the state, which allows farmers to cultivate their species of interest (annual crops) in the plots, while dealing with forest tree seedlings such as *T. grandis*".

The agreement between government and the farmers would last two to three years, during which time the tree species would expand the canopy, soil fertility decreased, some surface soil was lost through erosion, and weeds infested the area, making the land less productive. Although wood is the ultimate product of a *Taungya* system, this system is an example of sequential culturing of woody plants and annual crops. The cultivation of annual crops in this system is dependent upon the availability of space and light, based on the spatial arrangement of trees. This system, which is native to southeast Asia, is basically an alternative to shifting cultivation, and is widespread throughout the tropics of Asia, Africa, and America, where it is known under different names, such as: *Tumpangsari* (Bahasa Indonesia) in Indonesia, *Kaingining* in the Philippines, *Ladang* (Bahasa Malaysia) in Malaysia, *Chena* in Sri-Lanka (Sinhalese), *Khumri, Jhooming* (or shifting cultivation Jhoom as practiced in northeastern India and Bangladesh), *Ponam*, *Taila* and *Tuckle* in India, *Shamba* in Kenya (meaning small farm in Swahili), *Parcelero* in Puerto-Rico, and *Consorciacao* (meaning intercropping in Portuguese) in Brazil (King 1968). *Taungya* systems can be classified as "partial" (participants' interests in crop establishment are primarily economic) or "full" (a more traditional system based on the lifestyle of the farmers). Forest plantations in the Congo Basin owe their origins to *Taungya*. The most common crops in a *Taungya* system are rice (*Oryza sativa*) in Asia, yams (*Dioscorea* spp.) and bananas in Africa, and maize (*Zea mays*) in the Americas. The major drawback to the system is the erosion of soil during early growth of forest species, which is during the years in which food crops are grown. However, there is potential for various combinations of species to sequester carbon through tree growth, such as coffee associated with food crops (Soto-Pinto et al. 2010).

4.3.1 Jungle Rubbers (Rubber Agroforests)

Noble and Dirzo (1997) defined jungle rubbers as "an enhancement of traditional slash-and-burn practices in which rubber trees, fruit and occasionally timber species are planted during the garden phase. Natural regeneration occurs, leading to an 'enriched' secondary forest." These perennial crop-based agroforestry systems (agroforests) are found in Southeast Asia, and produce fruits and rubber. However, rubber-based agroforestry systems are also encountered in Latin America and in the Congo Basin.

Private Dutch colonial estates introduced rubber trees (originally from Brazil, smuggled to Kew Gardens in the 18th century, distributed to India and other British colonies in the 19th century, and finally to Buitenzorg Botanical Gardens on Java in the early 1880's) into agricultural landscapes in Southeast Asia in the early 1900's, and farmers enriched their fallows with rubber trees. This cropping system became complex when farmers innovated with improved rubber farming practices and germplasm. Jungle rubbers are essentially rubber-based secondary forests, as their structure consists of a more or less closed canopy that is 20 m to 25 m in height and which is dominated by rubber trees and a dense undergrowth layer (Gouyon et al. 1993). These systems are very complex and high in terms of their biodiversity. Of the 268 plant species (other than rubber trees) that were recorded in a jungle rubber plot in Indonesia, 91 were tree species belonging to 22 families, including Anacardiaceae, Apocynaceae, Bombacaceae, Dilleniaceae, Euphorbiaceae, Fagaceae, Flacourtiaceae, Guttiferae, Lauraceae, Melastomaceae, Mimosaceae, Moraceae, Myrtaceae, Orchidaceae, Palmae, Papilionaceae, Proteaceae, Rubiaceae, Sapindaceae, Styracaceae and Theaceae (Gouyon et al. 1993). Jungle rubbers require low inputs, as tree species protect rubber from grasses.

Jungle rubbers, which are also known as rubber agroforests, are not specific to Southeast Asia. These agroecosystems were also reported in Amazonia (Schroth et al. 2003). Indeed, rubber plantations date back to the early 1900's in the Brazilian Amazon (Table 4.9); these rubber plantations, which resemble secondary forests, were planted on sandy riverbanks, or on humus or clay-rich soils. Cultivation in Amazonia is problematic though because of leaf blight (Dean 1987; Lieberei 2007).

4.4 Farm Woodlots

Woodlots are stands of trees that provide environmental services, including soil rehabilitation or fertilization, and wood for households, thereby replacing wood collected from off-farm stands or forests. Rotational woodlots are sequential agroforestry systems, as they involve three phases (Kwesiga et al. 2003):

- An establishment phase (trees are inter-cropped with annual food crops, i.e., maize, sorghum, millet, rice in tropical savannas or groundnut and cassava in humid forest zones);
- A tree fallow phase (no cropping);
- A cropping phase after harvest of trees.

The woody species used in rotational woodlots in the tropics of Africa include *Acacia auriculiformis*, *A. crassicarpa*, *A. julifera*, *A. leptocarpa*, *A. mangium*, *A. polyacantha*, *A. nilotica*, *Gliricidia sepium*, *Leucaena leucocephala*, *Senna siamea* and *Sesbania sesban* (Kwesiga et al. 2003; Nyadzi et al. 2003; Kimaro et al. 2007; Akinnifesi et al. 2008). Other examples of this successful sequential agroforesy system include the agroforestry Mampu village or the IBI clean development mechanism carbon sink project on the Bateke plateau in the Democratic Republic of the Congo (http://cdm.unfccc.int/filestorage/L/U/G/LUGSF92NSFDYLAC5AWRWQ-2IO1RJTW5/PDD.pdf?t=NTh8bWxuNHl0fDAI_g_yoXMpOCpFsGDFyf07, http://www.mampu.org/historique_en.html, Bolaluembe Boliale 2009; Peltier et al. 2010). The selection of woody species to be used in rotational woodlots should be made on the basis of many criteria, including wood production and crop yield (Nyadzi et al. 2003). For instance, *L. leucocephala* produced more wood than the other studied tree species in a rotational woodlot experiment in Northwestern

Table 4.9 Planting periods of rubber (*Hevea brasiliensis*) groves on the sandy river bank of the more fertile slope and plateau soils of the Eastern side of the Tapajós river, Brazilian Amazon (Schroth et al. 2003)

Village	Sandy river bank		Clayey soils on slope and plateau	
	Until 1950	After 1950	Until 1950	After 1950
Cajutuba	*1948*–1960	1970–1987, 1974–1977, 1978–1999	Early 1900, *1918–1950*	1968
Aramanai				1957, 1993
Santa Cruz		1955–1980, 1970–1975, 1972–1979, 1974–1977	*~1900, 1930,<1950*	1955–1985, <1959,1960–1970
São Domingos		1966–1967, 1967–1971		
Maguari		1956, 1960–2002, 1973, 1975–1980	*1920, <1930, 1945, <1950, 1950*–1955	1969
Acaratinga		1965–1984, 1972–1988	*<<1950*	
Jaguarari	*1944*			
Pedreira	*<1950*	1956–1961, 1972–1977, 1982		1963–1978
Piquiatuba	*<1930, 1948* (2), *1950*	1953, 1965, 1976, 1990–1992	*1939*	1956
Marituba	*<1950*	1967–1985, 1970, 1976, 1987	*<<1950*	
Bragança		1958, 1967, 1979		
Marai	*1930, 1946*			1960–1965, 1977–1981
Nazaré		1970, 1970–1974, 1987		
Taurari	*1920, <1950*	1958, 1962, 1991		
Igarapé de Matancin	*<1940, 1945*	1980, 1982 (2), 1990		
Prainha	*1936–1949*	1970–1991		

Each date represents one rubber grove. The villages are arranged from North to South; villages in the edaphic savanna without access to plateau soils were not included. For better visibility, plantings from 1950 or earlier were printed in bold; (2) indicates two planting dates; (~) approximately; (<) before; (<<) much before

Tanzania (Table 4.10); *L. leucocephala* also increased the subsequent maize yield over a three-year period in the same experiment (Fig. 4.6).

4.5 Annual or Biennial Food Crop Farms: Slash-and-Burn Agriculture

Shifting cultivation is widespread throughout the tropics (Grandstaff 1980; Padoch and De Jong 1987). The practice is described by various terms, depending on the locality (Nair 1993 page 56). Shifting cultivation consists of growing two or three seasons of food crops on a plot of land, and then leaving the plot in fallow for

Table 4.10 Growth and biomass of three tree species planted as woodlots at the age of seven years at Shinyanga, Tanzania (Nyadzi et al. 2003)

Woodlot species	3 years		7 years		
	Survival (%)	Biomass (Mg ha^{-1})	Height (m)	Wood (Mg ha^{-1})	
				Un-pruned	Pruned
Acacia nilotica	78	2.5	3.9	8.4	6.0
Acacia polyancantha	76	7.8	5.8	70.9	49.7
Leucaena leucocephala	80	15.4	7.7	88.9	34.6
SED	7.2	1.34	0.64	23.5	–

SED standard error of differences between means

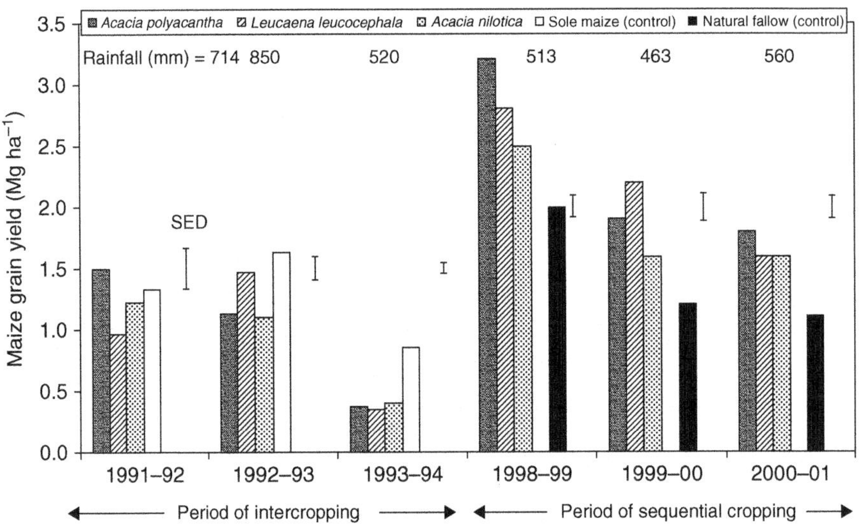

Fig. 4.6 Grain yield of maize intercropped between trees during the first three years of woodlots and after clearing woodlots at Shinyanga, Tanzania. Vertical bars are the standard errors of differences (SED) between treatment means in the respective years (Nyadzi et al. 2003)

several (7-10) years to restore land fertility, while other plots are being cultivated (crop rotation). Vegetation is removed by clearing and burning before food crop plantation. Crutzen and Andreae (1990) estimated that this method was being practiced by nearly 300-500 million people on 300 to 500 million ha of land in the tropics. Several crops are grown using this method in the tropics, the most important being cassava (*Manihot esculenta*), yams (*Dioscorea* spp.), rice (*Oryza sativa* and *O. glaberrima*), maize (*Zea mays*), groundnuts (*Arachis hypogaea*), plantains and bananas (*Musa* spp.), and cucumbers (*Cucumis sativus* L., Cucurbitaceae), all of which are primarily intended for consumption by farmers, and secondarily for sale. To this list, we could add tomatoes and other fruits (such as pineapple). Rice cultivation is predominant in Madagascar, as well as in the tropics of Asia and the Pacific Islands. Rice is also cultivated in West and Central Africa, whereas cultivation of cassava is widespread in Latin America and in the Congo Basin. Sometimes crop species are associated on farms. There is a spatio-temporal association of cassava, groundnut, and maize, which are grown at the same time on the same land in the rainforest zone of Africa. The groundnut fixes nitrogen in the soil, and has a short life cycle, allowing maize and groundnuts to be harvested after 3-4 months, while cassava is allowed to grow, and the leaves and tubers are harvested one year later. Very little shade is left intact, and previous vegetation is completely removed by the shifting cultivation. This technique is widespread in the tropics. The land previously occupied by dense vegetation, and containing large trees is cleared to make room for crops, and the wood ash from burning the cleared vegetation is used as fertilizer. However, this technique is destructive. Soil microfauna are destroyed, as is the humus. In addition, wood ash only acts as fertilizer for the first crop cycle, and a fallow period of at least 10 years is required to rebuild the original vegetation and restore soil fertility. This explains roaming techniques, whereby farmers cultivate on the same plot for 2-3 years, then move to an area that has been previously set-aside in primary or secondary forest.

Slash-and-burn agriculture is also a sequential rotation of trees and crops on the same plot. However, slash-and-burn techniques differ from *Taungya* because food crops are the primary objective of the system, and trees are not planted. A large number of cases (136) of slash-and burn were reported as part of a project on "alternatives to slash-and-burn", (Fujisaka et al. 1996). These cases were classified according to country and fallow length (Tables 4.11 and 4.12). Slash-and-burn agriculture is known under different names depending on the country: *Chitimene* in Zambia, *Tavy* in Madagascar, *Masole* in Central Africa, *Milpa* in Mexico and Central America, *Conuco* in Venezuela, Roca in Brasil, *Ladang* in Indonesia, etc.

Fallow, a period during which a plot is left to stand between two phases of culture, is often associated with food crops. Fallow periods may extend for up to 10 years or more. Depending on the duration of the fallow period, it can be considered either as a fallow or a secondary forest if the duration is long enough. Fallow allows the rebuilding of physio-chemical soil properties related to fertility for a new crop-growing cycle. In the rainforest region of Madagascar, fallow periods decreased over 30 years (between 1969 and 1999) from between 8 and 15 years to 3 and 5 years. Fallow vegetation changed within 5-7 fallow/cropping cycles after deforestation from trees and shrubs, to herbaceous fallows with *Imperata cylindrica* and *Aristida* spp. L. (Poaceae) grasslands (Styger et al. 2007). Frequent use of fire when converting fallow lands to crop production also encourages the replacement of native species with exotic ones and favors grasses over woody species.

4.5.1 Alley Cropping/Intercropping Systems

Alley cropping was developed as an alternative to bush fallowing in the tropics (Kang et al. 1981). It was popularized during the 1980's and 1990's by ICRAF. Alley cropping is an agricultural system in which agricultural crops are grown in

Table 4.11 Coding cases of slash-and-burn agriculture for Table 4.12 (Fujisaka et al. 1996)

Variable	Coding	Descriptor
Initial cover	1.	Primary forest
	2.	Secondary/degraded forest, bush fallow, agroforest
	3.	Grassland/savanna
User	1.	Indigenous
	2.	Government sponsored colonist
	3.	"Spontaneous" settlers and ranchers
"Final" cover	1.	Fallow, secondary regrowth
	2.	Pasture
	3.	Perennial crops, agroforest
	4.	Plantation crops, taungya
Fallow length	0.	Not a cyclical pattern
	1.	Short (1-2 years)
	2.	Medium (3–8 years)
	3.	Long (more than 8 years)
High value crops	+	Vegetables, high value annual crops
	++	Coffee, fruit, other perennials, betel nut[a]
	+++	Drugs: opium, coca

[a] It is worth noting that this study was published when tree domestication activities were just started by ICRAF. For this reason, important indigenous fruit trees that were enrolled in the tree domestication program are not mentioned here

alleyways formed by hedgerows of various leguminous plants including trees and shrubs, and grasses. These plants are usually trees and shrubs, and can include legumes (*Acacia* spp., *Leucaena leucocephala*, *Gliricidia sepium* (Jacq.) Kunth ex Walp. (Fabaceae), *Sesbania sesban*, *Sesbania grandiflora* (L.) Poiret (Fabaceae), and *Calliandra calothyrsus* Benth. (Fabaceae)) and actinorhizal plants (*Alnus nepalensis* D.Don (Betulaceae), *Alnus acuminata* Kunth (Betulaceae), *Casuarina equisetifolia* L. (Casuarinaceae), Asteraceae family (e.g., *Tithonia diversifolia* (Hemsl.) A.Gray). The plants are pruned during a crop's growth to prevent shade, and to reduce competition for light, nutrients, and soil moisture between crops and legumes. A distinction is sometimes made between alley farming, which involves livestock production, and alley cropping, which does not involve animals. In the intercropping system, coppicing whereby trees are cut down, allowing the stumps to regenerate for a number of years, is not allowed. Commercial trees are managed through pruning and thinning in order to obtain a good quantity and quality of wood at the rotation age (e.g. winter wheat-*Paulownia* in China, beans-*Eucalyptus grandis* in Brazil).

The principle of alley cropping encourages nitrogen fixation by the legumes or actinorhizal plants to maintain soil fertility during cultivation or to replace nutrients exported by crops, while avoiding or mitigating competition between the legumes/actinorhizal plants and crops. Pruning twigs and leaves can enrich the soil when added as mulch, or are used as a source of fuelwood for the household. The pioneers of alley cropping in sub-Saharan Africa are B.T. Kang and D.U.U. Okali (International Institute of Tropical Agriculture, Ibadan, Nigeria), Bahiru Duguma (ICRAF, Cameroon), and Freddy Kwesiga and Bashir Jama (ICRAF East and Southern

Table 4.12 Classification of 107 cases of slash-and-burn agriculture from secondary data according to initial vegetative cover, user, "final" vegetative cover, and fallow length (Fujisaka et al. 1996). For coding, see Table 4.11

Ref*.	Initial cover	User	Final cover	Fallow length	Country
a. Primary forest, indigenous users, secondary regrowth					
4	1	1	1	2	Sri Lanka
58	1	1	1	?	Venezuela
b. Primary forest, settlers, natural regeneration, long fallow					
59	1	3	1	3	Indonesia
c. Primary and secondary forests, indigenous users, natural regeneration					
65	1-2	1	1	2	Madagascar
71	1-2	1	1	2	Philippines
44	1-2	1	1	2++	Malaysia
10	1-2	1	1	?	Venezuela
60	1-2	1	1	?	Thailand
31	1-?	1	1	?	Indonesia
33	1-2	1	1	?	Indonesia
81	1-2	1	1	?	Philippines
101	1-2	1	1	?	Philippines
77	1-2	1	1	3	Ivory Coast
85	1-?	1	1	?	Zambia
80	1-?	?	1	3	Sierra Leone
47	1-2	?	1	3++	Papua New Guinea (PNG)
d. Secondary forest, indigenous users, natural regeneration					
54	2	1	1	1+++	Thailand
7	2	1	1	2	Laos
67	2	1	1	2	Philippines
87	2	1	1	2	India
76	2	1	1	2	Philippines
46	2	1	1	2	Colombia
95	2	1	1	2	Colombia
17	2	1	1	2	Colombia, Ecuador
24	2	1	1	2	Panama, Colombia
62	2	1	1	2	India
36	2	1	1	2	Guyana
83	2	1	1	2	India
43	2	1	1	2	Laos
43	2	1	1	2	Laos
48	2	1	1	2	Colombia
89	2	1	1-3	2	India
88	2	1	1-3	2	India
90	2	1	1-3	2	India
32	2	1	1	3	Indonesia
41	2	1	1	3++	Mexico
56	2	1	1	3	Thailand
9	2	1	1	3	Zambia
20	2	1	1	3	Zambia
102	2	1	1	3	Democratic Republic of Congo (DRC)
51	2	1	1	3+	Mexico
25	2	1	1	3	Malaysia
26	2	1	1	3	Sarawak
70	2	1	1	3	Brazil

Table 4.12 (continued)

Ref*.	Initial cover	User	Final cover	Fallow length	Country
44	2	1	1	3	Thailand
76	2	1	1	3	Ivory Coast
27	2	1	1	3	Venezuela
44	2	1	1-3	3	Colombia
84	2	1	1	?	Zambia
5	2	1	1	?	Mexico
72	2	1	1	?	India
57	2	1	1	?	Belize
21	2	1	1	?	Zambia
103	2	?	1	2-3	DRC
97	2	?	1-2	3	Tanzania
47	2	?	1	3++	PNG
47	2	?	1	3++	PNG
47	2	?	1	3	PNG
47	2	?	1	3	PNG
47	2	?	1	3	PNG
47	2	?	1	3++	PNG
80	2	?	1	?	Congo
80	2	?	1-2	2	Madagascar
e. Secondary forest, colonists, natural regeneration					
18	2	3	1	2	Laos
29	2	3	1	2	Laos
66	2	3	1	2	Peru
f. Primary and secondary forest, indigenous and settlers, conversion to agroforest					
100	1	1	3	?[a]	Colombia
99	1	1	3	?	Colombia
23	1	1	3	?	Venezuela
35	1	1	3	?	Colombia
69	1	1	3	?	Brazil
53	1	1	3	?	Java
80	1	1	3	?	Guinea
34	1-2	1	3	3	Colombia
93	1-2	1-3	3	?	Peru
28	1-2	3	3	3	Peru
66	1-2	3	3	3	Peru
40	1-2	3	3	0++	Philippines
22	2	1	3	2	Java
96	2	1	3	3	Colombia
75	2	1	3	3	Philippines
52	2	1	3	3	Brazil
24	2	1	3	3	Colombia
49	2	1	3	3	Vietnam
37	2	1	3	?	Colombia
61	2	1	3	0++	Indonesia
55	2	1	2-3	?	Laos
1	2-?	1	2-3	?	Sudan
3	2	?	3	?	Nigeria
45	2	?	3	2	Java
47	2	?	3	3	PNG
6	2	?	3	3	PNG
16	2	3	3	0	Ivory Coast
49	?	2[a]	3	?	Vietnam

Table 4.12 (continued)

Ref*.	Initial cover	User	Final cover	Fallow length	Country
g. Secondary forest, colonists, conversion to plantation crops or taungya					
63	2	2	4	0	Sri Lanka
12	2	2	4	0	Thailand
80	2	2	4	?	Sierra Leone
80	2	2	4	?	Guinea
53	?	2	4	?	Java
84	?	2	4	?	Kenya
31	2	1[b]	4	0++	Indonesia
h. Primary and secondary forest (mostly), settlers, conversion to pasture					
94	1	3	2	0	Nicaragua
91	1	3	2	0	Brazil
2	1	3	2-3	0+++	Colombia
68	1	3	2	?+++	Colombia
38	1	?	2	0	Brazil
44	1-2	3	2	0+++	Thailand
75	1-2	3	2	?	Philippines
15	1-2	3	?	0++	Brazil
11	1-2	1	2	0	Venezuela
8	2	3	2	0++	Brazil
i. Grassland, indigenous and settlers, natural regeneration, pastures					
32	3	1	2	2	Indonesia
39	3	1	2	2	PNG
55	3	1	2	?	Laos
13	3	1	2	2	PNG
39	3-?	1	?	2	PNG
70	3	1	3	0	Brazil
30	3	1	?	?	Borneo
74	3	?	1	?	Zambia
47	3	?	1	3++	PNG
47	3	?	1	3	PNG
47	3	?	1	3	PNG
j. Others					
Other indigenous users					
73	?	1	1	2	Sri Lanka
68	?	1	1	?	Colombia
70	?	1	3	0	Brazil
42	?	1	3	0++	Indonesia
44	?	1	1-3	?++	Philippines
50	?	1	?	?	Zambia
Other primary forest users					
92	1	?	1	?	Venezuela
78	1	?	1	?	Kenya
80	1	1	?	?	Cameroon
82	1	2	?	?++	Vietnam
79	1	3	?	?	Guatemala
98	1	3	?	?	Brazil
80	1	?	?	?	Ghana
Other					
14	?	?	4	0	Cuba

[a] These agroforests have continued to be exploited for the perennial crops and could be re-used for annual cropping, although years of such "fallows" were generally not specified;
[b] "Rare" case of adoption of rubber trees by indigenous group;
* Publication number reporting a case

Fig. 4.7 Establishment of alley cropping. **a** Tree establishment: *Calliandra calothyrsus* or calliandra is planted alongside maize (*Zea mays*); **b** After harvest of maize, calliandra is left to grow for about 2 years; **c** During the fallow period, beehives are placed in the calliandra tree plot; **d** After the fallow phase, calliandra is cut back; **e** The best branches of calliandra are removed and used for staking of yams, tomatoes, etc; **f.** After the cutting back the calliandra trees, crops are planted in the alleys (Source: ICRAF-Cameroon)

Africa). The design and implementation of an alley cropping system in a given region should be determined by the following conditions:

- Identify the legume/actinorhizal plant suitable for the ecological area of interest. *Cassia siamea* Lamk (Syn. *Senna siamea* Irwin et Barneby, Fabaceae) is more appropriate than *Leucaena leucocephala* in a semi-arid climate (Jama-Adan et al. 1993). Similar studies found that *Calliandra calothyrsus* was a more appropriate choice than *Leucaena leucocephala* in the humid lowland forest zone of Africa, where the latter species became an invasive weed (Kanmegne and Degrande 2002).
- Determine the appropriate number of hedgerows per hectare and quantify the "loss of farm space" due to these hedges. Different arrangements are possible; the most common being 4×0.25 m, which gives 10,000 trees ha^{-1}, which seems quite optimal. Double rows are also possible. In that case they are spaced more widely apart. It is possible to establish in one ha 20 rows of *L. leucocephala,* 100 m in length, separated by 5 m (Duguma et al. 1988).
- Determine the optimal fallow-cultivation period, and the time required before the first pruning on leguminous species in the system.
- Determine the intensity and frequency of pruning (Duguma et al. 1988).
- Consider the slope in the region, as hedges are effective against erosion.
- Consider the labor intensity provided by households for the establishment and maintenance of the system.

Establishment of alley cropping is illustrated in Fig. 4.7.

In general, legumes are initially pruned one year after the system has been implemented and crops are grown in the corridors. Leguminous trees fix nitrogen to the soil, and studies have shown that it increases soil concentrations of P, K, Ca, and Mg (Nair 1993, page 126). Alley cropping also helped to increase plant production by over 8 tons of dry matter per hectare per year (Nair 1993, page 125). The dry matter can be used as mulch to increase soil organic matter content and improve soil chemical properties (Nair 1993, page 127). Species such as *C. siamea* and *Inga edulis* can be used to control erosion and to increase soil organic matter, due to the slow rate of decomposition of their leaves. The more rapid the decomposition of the dry matter that has been used as mulch, the faster the nutrients are released into the soil. The loss of soil nutrients in alley cropping is less marked than in soils that have been planted with conventional crops (Nair 1993, page 128). Some legumes, such as *Flemingia macrophylla* (Willd.) Merr. (Fabaceae), have a positive effect on soil temperature and moisture conservation (Nair 1993, p. 129). Various studies have observed increasing crop production through the practice of alley cropping (Nair 1993, p. 130; Kang et al. 1989, 1990; Kang and Duguma 1985). Therefore, alley cropping appears to be an alternative solution to four problems of natural resource management in the tropics: land tenure; reduced soil fertility due to short fallows; soil erosion; and lack of fuel wood. In addition, alley cropping provides nutrients to the soil.

In Cameroon, *Calliandra calothyrsus* is the shrub most commonly used in alley cropping (refer to the work of Bahiru Duguma and colleagues in 1988-1998). This species was selected after screening 10 species suitable for agroforestry common in the humid lowland rainforest of Cameroon (Duguma and Tonye 1994). *Calliandra* was planted in 4×0.25 m spacing, and then pruned down to 0.50 m height one year after implementation, with the prunings used as mulch. Early results were disappointing. Problems included poor growth of shrubs, low biomass production due to pruning being done too early, low impact on weed control, and high demand for labor (Degrande et al. 2007). Changes to the original method led to improvements; such changes include (1) delaying the first pruning until after 2 years, (2) alternating the cropping period with a year of fallow, and (3) pruning *Calliandra* down to only 0.05 m height. These modifications led to the evolution from the original alley cropping to a rotational tree fallow system (Kanmegne and Degrande 2002). Tests conducted on-station in Yaoundé showed that this system maintains high maize production (Table 4.12). However, performance in a farm setting was not as good as what had been observed on-station (Degrande et al. 2007), resulting in a low rate of adoption of the system by farmers (Degrande and Duguma 2000; Degrande et al. 2007). About 52 % of households involved in testing alley cropping and rotational tree fallows indicated that more than half of their soils were fertile, and respondents did not perceive the decline in fertility as a problem. About 73 % of households reported that they had enough land to meet their households' needs (Degrande and Duguma 2000). In reality, numerous problems arose with the adoption of alley cropping and rotational tree fallow in the area, especially in terms of labor demand.

The system is labor-intensive, with high labor requirements, especially for tasks such as filling polyethylene bags for seedlings, watering plants, especially during the dry season, and pruning. In addition, erosion is not a problem in humid lowland

Table 4.13 Fallow cropping cycle and maize grain yields (Mg ha^{-1}) on station, Yaoundé (Degrande et al. 2007)

Treatment	1990		1991		1992		1993		1994		1995		1996	
	1	2	1	2	1	2	1	2	1	2	1	2	1	2
T1a	1.52	NF	2.98	NF	3.54	NF	2.54	NF	2.17	NF	2.33	NF	2.69	NF
T1b	1.52	NF	2.98	NF	3.54	NF	2.54	NF	NF	NF	NF	NF	3.58	NF
T2	2.13	TF	3.70	TF	4.79	TF	5.09	TF	4.55	TF	3.33	TF	3.68	TF
T3	TF	TF	TF	TF	6.28	TF	6.09	TF	TF	TF	TF	TF	6.51	TF
T4	2.72	TF	4.48	TF	TF	TF	TF	TF	5.27	TF	4.82	TF	TF	TF
SED	0.38	–	0.28	–	0.14	–	0.44	–	0.14	–	0.36	–	0.35	–

NF natural fallow, *TF* tree fallow of *Leucaena leucocephala* and *Gliricidia sepium* mixture, *T1* control treatment of continuous maize cropping with 1 season of maize and 1 season natural fallow each year; in 1994 plots were split to allow the comparison with a 2-year natural fallow (T1b) in addition to continuous cropping (T1a), *T2* continuous maize cropping with 1 season of maize grown between the rows of trees (regularly pruned back as hedgerows) and 1 season of tree fallow during which the hedges were allowed to grow unchecked, *T3* 2 years of tree fallow followed by 2 years of cropping, as in treatment 2; T4 same as treatment 3, but starting with the cropping cycle

zones of Cameroon, where there are very low slopes and many trees. Livestock production is also not popular in the region due to the presence of the tsetse fly (*Glossina palpalis* Wiedemann, Glossinidae), which limits the usefulness of *Calliandra* leaves as fodder, one of the major benefits associated with this plant. Constraints on adoption of alley cropping and tree fallows are shown in Table 4.13. Additionally, there are constraints imposed by land rights. Poor farmers, who do not have long-term tenure, are unlikely to invest in improved fallows, because benefits are obtained only after a longer period of time. Wood production for fuel is also not a constraint in the humid tropic lowlands, due to the presence of a large amount of dead wood in farmers' fields. To overcome these difficulties, an alternative land-use system was suggested to farmers: improved fallows with legumes shrubs, such as pigeon pea (*Cajanus cajan*).

In Southern Africa, herbs or leguminous trees like *Sesbania sesban*, and *Tephrosia vogelii* have been used to restore soil fertility (Table 4.14). A fallow of 2 to 3 years with *S. sesban* planted at a spacing of 0.5 × 0.5 m proved effective in the maintenance of soil fertility in Southern Africa (Kwesiga and Coe 1994). This agroforestry system accumulates more nitrogen than do herbaceous fallows. It has proven to be successful in Zambia, and the number of farmers that were reported as using this practice increased from 200 in 1994 to 3,000 in 1997 (Kwesiga et al. 1999). *Sesbania sesban* not only restores soil fertility, but also provides fodder for cattle and wood for fuel. In Malawi, the adoption of this practice is low (Mafongoya et al. 2007), perhaps due to problems related to land tenure, labor shortages, and a long waiting time for benefits (2–3 years). The *Sesbania* fallow should be performed in rich soil (Goma 2003). Other species that provide organic inputs to the soil, such as *Gliricidia sepium*, *Leucaena leucocephala*, and *Calliandra calothyrsus*, are better suited for alley cropping in the region (Fig. 4.10).

Alley cropping with *Leucaena leucocephala* and *Calliandra calothyrsus* at densities of 6,680 trees per ha has been implemented successfully by farmers in East Africa (Shepherd et al. 1997). *Leucaena*, in association with maize (six rows

Table 4.14 Land-use constraints, farm conditions, and potential agroforestry solutions in West province, Cameroon (Degrande et al. 2007)

Constraint	Site	Solution
Poor soil fertility	Flat land: <0.5 ha	a. Improved fallow of *Sesbania*, *Cajanus* and *Tephrosia*
	Flat land: >0.5 ha	b.Improved fallow of *Calliandra*
Soil erosion	Slope	c. Rotational hedgerow
Poor soil fertility + erosion	Slope	d. Combination of (a) and (c) on same plot
Destruction of crops by wind	Flat land	e. Windbreak of *Calliandra* planted at 8-10 × 0.25 m
Destruction of crops by wind + soil erosion	Slope	f. Rotational hedgerow + windbreak of *Calliandra* planted 4 m × 0.25 m (alternate rows are managed as hedges and the other as windbreaks or for pole production)
Bee migration + need for income diversification	Flat land or slope	g. *Calliandra* plantation, serving as a constant source of nectar
Dry season fodder shortage	Pasture land	h. Enrichment planting
		i. Fodder bank or feed garden

of maize spaced at 75 × 25 cm between two rows of *Leucaena* spaced at 4 × 0.5m) gave better results than *Calliandra* planted at the same density as *Leucaena* (Mugendi et al. 1999a, b) on the humid highland slopes of Mount Kenya. *Leucaena* has also been used for alley cropping in the wet uplands of Western Kenya (Imo and Timmer 2000). This technique seems appropriate for the African highlands (Kang 1993), where problems include erosion and lack of suitable firewood, and the agricultural system is characterized by the cultivation of cereals (mainly maize) in combination with cattle grazing.

Mulching twigs from pruning provides excellent fertilizer for tropical soils (Mureithi et al. 1994). However, mulch from *Leucaena* pruning did not increase production of crops in alley cropping in the semi-arid tropics of India, where the primary benefit of alley cropping was fodder production during the dry season (Singh et al. 1988). This practice increases competition for soil water between *Leucaena* and crops, which negatively affects crop production (Singh et al. 1989). Soil moisture availability is a very important factor to consider in the implementation of alley cropping with shrubs in semi-arid tropics. The number of prunings conducted on the legume should be limited to three times, implying that nitrogen harvest would be sufficient to achieve substantial benefits for the crop (Nair 1993). The relationship between rainfall and alley cropping is illustrated by Nair (1993, p. 136). Alley cropping studies with *Faidherbia albida,* in association with maize and common bean, were also tested in coastal Kenya, with inconsistent results (Jama and Getahun 1991). Alley cropping would be appropriate in the highlands of the humid tropics, where agriculture and husbandry are practiced in rural areas, but inappropriate for the lowlands.

Other soil-improving agroforestry technologies that are well-suited to situations of poverty and other demographic pressures on the land include, simultaneous inter-cropping or coppiced fallows (trees are planted in between maize, and once properly established, the trees are cut back and the biomass is incorporated into

Table 4.15 N_2 fixed on smallholder farms in Southern Africa by various legumes (Mafongoya et al. 2007)

Legume	N_2 fixed (kg ha^{-1})	Source
Bambara nut (*Vigna subterranea* (L.) Verdc.)	52	Rowe and Giller (2003)
Cowpea (*V. unguiculata*)	47	Rowe and Giller (2003)
Groundnut (*Cajanus cajan*)	33	Rowe and Giller (2003)
Pigeon pea	39	Rowe and Giller (2003)
Pigeon pea	3-82	Mapfumo et al. (2000)
Pigeon pea	97	Chikowo et al. (2004)
Cowpea	28	Chikowo et al. (2004)
Acacia angustissima	122	Chikowo et al. (2004)
Sesbania sesban	84	Chikowo et al. (2004)
Gliricidia sepium	212	Mafongoya PL (Unpublished)
Acacia angustissima	210	Mafongoya PL (Unpublished)
Leucaena colinsii	300	Mafongoya PL (Unpublished)
Tephrosia candida	280	Mafongoya PL (Unpublished)
Tephrosia vogelii	157	Mafongoya PL (Unpublished)

the soil), annual relay (fallow) cropping of trees (i.e., fast-growing trees or shrubs are planted after the food crop is established), and biomass transfer (cut-and-carry system requiring separate areas where shrubs and trees are planted) (de Wolf 2010).

4.6 Improved Fallows and Rotational Tree Fallows

Improved fallows, which are sometimes called a sequential agroforestry system (Rae et al. 1998), consist of planting selected species or retaining them from natural regeneration. Short-duration fallows are characterized by fast-growing leguminous trees or shrubs for replenishment of soil fertility to support food crop production. Medium- to long-duration fallows harbor diverse species that have been established for amelioration of degraded and abandoned lands as well as for the use of tree products (Rao et al. 1998). Improved fallows tend to attain the objectives of natural fallows in a shorter time through the choice of tree species, spacing, density, pruning, and establishment. Fallow systems overcome constraints on crop production through maintenance of soil fertility during the cropping period by recycling and conserving nutrients, restoring the soil's physical properties, and controlling soil-borne pests and weeds (Buresh and Cooper 1999). Fallow processes for overcoming constraints to crop production that are used in the tropics were ranked by Buresh and Cooper (1999; Table 4.15). These processes can be achieved through the choice of an appropriate fallow system. Rotational tree fallow and short-rotation fallow are the most popular improved fallows in the tropics. Improved fallows should be distinguished from enriched fallows, which consist of planting certain tree species at low density into natural fallows in an effort to produce high-value products such as fruits, medicines, or high-grade timber that provide economic benefits to households during the fallow period (Brookfield and Padoch 1994). A summary of case studies of improved fallows in the tropics is illustrated in Table 4.16.

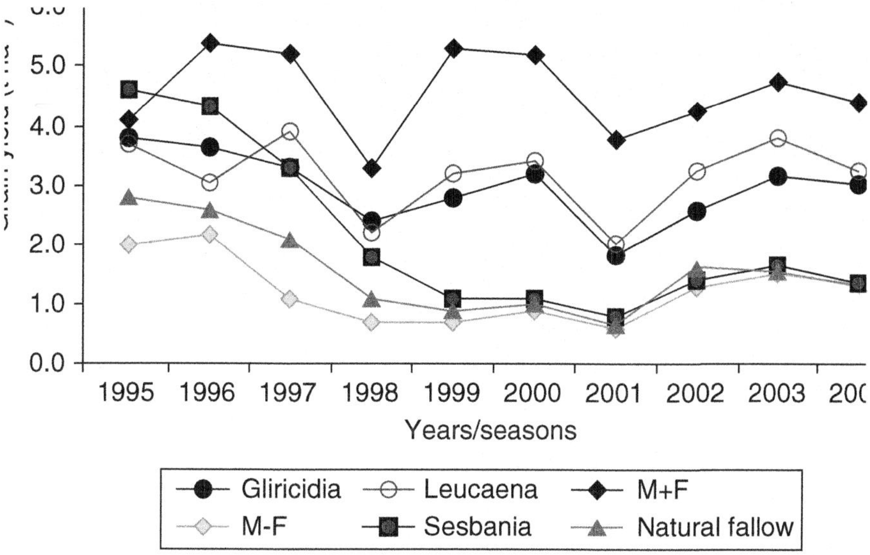

Fig. 4.8 Grain yield (Mg ha^{-1}) obtained from various fallow species for ten seasons at Msekera Research Station, Zambia. (Mafongoya et al. 2007)

It is important to note that, in certain cases, alley cropping is classified within improved fallows. This is the case when trees are allowed to grow, and a fallow period occurs between crops. Alley cropping is then referred to as rotational hedgerow inter-cropping or tree-based improved fallow if the trees are not in hedges but are planted in spacing of 1 × 1 m or 2 × 2 m. In the humid lowland forest zone of Cameroon, evaluation of alley cropping revealed several difficulties encountered by farmers. The system evolved into a rotational tree fallow, following the introduction of a fallow phase of at least one year (Kanmegne and Degrande 2002).

4.6.1 Improved Fallows with Herbaceous Legumes: the Case of Cajanus Cajan

Improved fallows agroforestry using *Cajanus cajan* shares similarities with the widespread groundnut (*Arachis hypogaea*, or peanut) farms in the humid tropic lowlands of Africa. A groundnut (or peanut) farm is a food crop farm where peanuts are grown in combination with cassava, maize, and a few other food crops. The primary crop consists of groundnuts, which are harvested after 3-4 months. The groundnut farm is a phase in the shifting cultivation system. The system cycle is (i) fallow or forest, (ii) farm establishment after clearing, logging, and burning, (iii) groundnut farm, in combination with cassava and maize, (iv) harvest of groundnut and maize, (v) cassava farm maintenance until the next year, (vi) harvest of cassava tubers and leaves, (vii) fallow. Sometimes, a second crop cycle is planted before the plot is left to fallow. The peanut, which is a legume, enriches the nitrogen in the soil.

Table 4.16 A ranking of fallow processes, listed in decreasing order of importance, for overcoming constraints to crop production in the tropics (Buresh and Cooper 1999)

High base status, N deficient soil		High base status, N and P deficient soil		Low base status, high Al soil	
Constraint/process	Potential of fallow	Constraint/process	Potential of fallow	Constraint/process	Potential of fallow
1) N supply		1) N supply		1) N supply	
N_2 fixation	High	N_2 fixation	High	N_2 fixation	High
Retrieval from subsoil	High	Retrieval from subsoil	High	Retrieval from subsoil	Low, ?
Capture of subsurface lateral flow	?	Capture of subsurface lateral flow	?	Capture of subsurface lateral flow	Low, ?
2) Weeds		2) P supply		2) P supply	
Fool propagules	?, L1	Chemical transformations	Low	Chemical transformations	Low
Reduce seed pools, germination	High	Reduce P complexation	Low	Reduce P complexation	High
Reduce weed vigor	High	Special acquisition mechanisms	?	Special acquisition mechanisms	?
3) Soil structure	Low, ?	3) Weeds		3) Cation supply	
4) Soil pests	?	Fool propagules	?	Retrieval from subsoil	Low, ?
		Reduce seed pools, germination	High	Al-organic acid interactions	Intermediate, ?
		Reduce weed vigor	High	CEC of soil organic matter	High, ?
		4) Soil structure	High	4) Weeds	
		5) Soil pests	?	Fool propagules	L1, ?
				Reduce seed pools, germination	High
				Reduce weed vigor	High
				5) Soil structure	High
				Soil pests	?

L1 may be locally important, *?* importance unknown

Improved fallow with *Cajanus cajan* (pigeon peas) has pigeon peas planted at 1×0.40 m spacing, and then maize is planted between rows of pigeon peas at the same spacing. After harvesting the maize, the pigeon peas are left on the plot for a second year. The pigeon peas are harvested the next year, the residues are burned or incorporated in the soil, and food crops (e.g., cassava, maize, peanut) grown. In the third year, the cycle restarts with the cultivation of pigeon peas being inter-cropped with maize. *Cajanus* fallow is illustrated in Fig. 4.9. Farmers in Edo State, Nigeria, combine pigeon peas with *Dioscorea*, maize or cassava in their homegardens and farms, and the occurence of *Cajanus*/cassava combinations can go up to 35 % in

Fig. 4.9 *Cajanus cajan* fallow, one year after the pigeon peas were planted. (Source: Ann Degrande)

farms (Table 4.17 and 4.18). Pigeon pea is advantageous because it does not lower crop production. There is even an increase in crop production (80 % for maize and 97 % for peanut) after a *Cajanus* fallow. This increase has had a positive effect of the adoption of this technology (Degrande et al. 2007). Other reasons for adoption are soil fertility improvement and weed suppression (Degrande et al. 2007). Advantages listed by farmers include the reduction of the fallow period, the availability of pigeon pea beans for consumption, the ease of clearing of a *Cajanus* fallow, especially for the women, the ease of planting peanuts on a plot where *Cajanus* had previously been cultivated, and the direct seeding of *Cajanus,* that requires less physical effort than alley cropping establishment (Degrande et al. 2007). In addition, the increased crop production from the practice occurs quickly, and its profitability has been demonstrated (Degrande 2001).

In Nigeria, *Cajanus* fallows increased maize production by 200 % and that of groundnut by 350 % over 6 years. A *Cajanus* fallow, pruned at 60 cm, was also found to be suitable for livestock production in savanna zones (Agyare et al. 2002). In the same region, *Cajanus* fallows were found to increase maize grain yield between 0.43 and 2.39 Mg per ha in the first year after fallow, but yield decreases in the second year by 17.6–50 % (Abunyewa and Karbo 2005). The same study

Table 4.17 A summary of case studies of short-duration, improved fallows in the tropics in 1999 (Buresh and Cooper 1999)

Type	Location	Species	Duration	Products and services	Stage in the development and adoption process
Tree fallow	Eastern Zambia	*Sesbania sesban*	2–3 years	Soil fertility, fuelwood, weed control	Advanced farmer testing
		Tephrosia vogelii	2–3 years	Soil fertility, insecticide	
Tree fallow	Zimbabwe	*Cajanus cajan*, *Sesbania sesban*	2–3 years	Soil fertility, fuelwood, weed control	Early farmer testing
		Acacia angustissima	2–3 years	Soil fertility, fodder	
Tree fallow	Southern Malawi	*Sesbania sesban*, *Cajanus cajan*	8 months	Soil fertility, insecticide	Early farmer testing
(Relay cropping)		*Tephrosia vogelii*	8 months	Soil fertility, insecticide	
Tree fallow	South Cameroon	*Calliandra calothyrsus*	2 years	Soil fertility, fodder, honey	Early farmer testing
Tree fallow	Philippines	*Leucaena leucocephala*	4 years	Soil fertility, erosion control, fodder, fuelwood	Farmer adopting
Tree fallow (woodlots)	Southern Ghana	*Senna siamea*, Leucaena *leucocephala*,	> 3years	Fuelwood, soil fertility	Early farmer testing
Tree fallow (woodlots)	Tanzania	*Acacia* spp.	> 3years	Fuelwood, soil fertility	Early farmer testing
Enrichment	Eastern Amazonia, Brazil	*Acacia angustissima*, *Inga edulis*, *Acacia mangium*, *Clitoria racemosa*, *Sclerolobium paniculum*	2 years	Soil fertility	Researcher design
Alley farming—contour hedgerow	Southwestern Nigeria	*Leucaena leucocephala*, *Gliricidia sepium*, *Senna spectabilis*	0.5–10 years	Soil fertility, fodder, fuelwood, soil conservation, honey	Early farmer testing
Alley farming—contour hedgerow	Northwestern Cameroon	*Leucaena leucocephala*, *Gliricidia sepium*, *Senna spectabilis*	0.5–10 years	Soil fertility, fodder, fuelwood, soil conservation, honey	Advanced farmer testing
Herbaceous cover crop	Eastern Zambia	*Mucuna pruriens*, *Crotalaria* spp.	2–8 months	Soil fertility, weed suppression, fodder	Early farmer testing

Table 4.17 (continued)

Type	Location	Species	Duration	Products and services	Stage in the development and adoption process
Herbaceous cover crop	Tanzania	*Crotalaria* spp.	2–8 months	Soil fertility, weed suppression	Farmer adopting
Herbaceous cover crop	Uganda	*Mucuna pruriens, Crotalaria ochroleuca, Dolichos lablab*	2–8 months	Soil fertility, weed suppression, food, fodder	Early farmer testing
Herbaceous cover crop	Kenya	*Mucuna pruriens, Crotalaria* spp.	2–8 months	Soil fertility, weed suppression	Early farmer testing
Herbaceous cover crop	Benin	*Mucuna pruriens*	2–8 months	Weed suppression, soil fertility	Farmer adopting
Herbaceous cover crop	Honduras	*Mucuna pruriens*	2–8 months	Soil fertility, weed suppression	Farmer adopting

Table 4.18 Frequency of crop combinations in different farming systems in Edo State, Nigeria (Ogbe and Bamidele 2007)

Normal	Homegarden (%)	Near farm (%)	Distant farm (%)
Cajanus/Manihot	6	23	35
Zea/Cajanus	10	3	37
Discorea/Cajanus	4	18	25

revealed that after two years of a fallow period, there was an increase in organic carbon in the soil, as well as an improvement of total nitrogen by 48.5 %, and CEC (Cation Exchange Capacity) by 17.8 % (Abunyewa and Karbo 2005). There are two major constraints upon the adoption of this technique: seed supply, and storage of *Cajanus* seeds (Degrande et al. 2007). Cajanus fallow, along with other rotational fallows, has also been found to increase soil infestation of snout beetle (weevil, Curculionidea) in maize farms in Eastern Zambia (Sileshi and Mafongoya 2003). Snout beetle is a major pest for maize production; therefore, some landowners are likely to be discouraged from adopting *Cajanus* fallows because of this negative factor.

Other agroforestry practices used in the tropics include shelterbelts or windbreaks (rows of trees planted around farms to protect crops, animals and soil from natural hazards), silvopasture (Alavalapati et al. 2005) and contour tree buffer strips. For more details, please refer to Nair (1993).

Bibliography

Abunyewa AA, Karbo KN (2005) Improved fallow with pigeon pea for soil fertility improvement and to increase maize production in a smallholder crop-livestock farming system in the subhumid zone of Ghana. Land Degrad Dev 16:447–454

Achuthan NM, Sreedharan C (1986) Agroforestry farming systems in the homesteads of Kerala, southern India. Agrofor Syst 4:339–363

Agyare WA, Kombiok JM, Karbo N, Larbi A (2002) Management of pigeon pea for crop-livestock production systems in the Guinea savanna zone of northern Ghana. Agrofor Syst 54:197–202

Alavalapati JRR, Mercer DE, Montambault JR (2005) Agroforestry systems and valuation methodologies. In: Alavalapati JRR, mercer DE (eds) Valuing agroforestry systems. Kluwer Academic Publishers, The Netherlands, pp 1–8

Akinnifesi FK, Sileshi G, Ajay OC, Chirwa PW, Kwesiga FR, Harawa R (2008) Contributions of agroforestry research and development to livelihood of smallholder farmers in Southern Africa: 2. Fruit, medicinal, fuelwood and fodder tree systems. Agricultural Journal 3(1):76–88

Altieri MA, Farrell J (1984) Traditional farming agroforestry systems of south-central Chile, with special emphasis on agroforestry. Agrofor Syst 2:3–18

Ayuk ET (1997) Adoption of agroforestry technology: the case of live edges in the Central Plateau of Burkina Faso. Agric Syst 54(2):189–206

Ayuk ET, Duguma B, Franzel S, Kengue J, Mollet M, Tiki-Manga T, Zekeng P (1999) Uses, management and economic potential of *Dacryodes edulis* (Burseraceae) in the humid lowlands of Cameroon. Econ Bot 53(3):292–301

Baggio AJ, Caramori PH, Filho AA, Montoya L (1997) Productivity of Southern Brazilian coffee plantations shaded by different stockings of *Grevillea robusta*. Agrofor Syst 37:111–120

Bavappa KVA, Jacob VJ (1982) High-intensity multispecies cropping: A new approach in small-scale farming in the tropics. World Crops (March/April) 47–50

Bolaluembe Boliale PC (2009) Problématique de la gestion durable des forêts artificielles en Afrique centrale: Cas de la forêt de Mampu à Kinshasa, en Réepublique Démocratique du Congo. Mémoire de Maîtrise, Sciences et gestion de l'environnement dans les pays en développement, Université de Liège et Faculté des sciences agronomiques de Gembloux, Belgique

Bosselmann AS, Dons K, Oberthur T, Smith Olsen C, Raebild A, Usma H (2009) The influence of shade trees on coffee quality in small holder coffee agroforestry systems in Southern Colombia. Agr Ecosyst Environ 129(1–3):253–260

Bourke RM (1984) Food, coffee and casuarina: and agroforestry system from the Papua New Guinea highlands. Agrofor Syst 2:273–276

Brierley JS (1985) The West Indian kitchen gardens: a historical perspective with current insights from Grenada. Food Nutr Bull (UNU) 7(3):52–60

Brookfield H, Padoch C (1994) Agrodiversity. Environment 36(5):7–11, 37–45

Buresh RJ, Cooper PJM (1999) The science and practice of short-term improved fallows: symposium synthesis and recommendations. Agrofor Syst 47:345–356

Campanha MM, Silva Santos RH, de Freitas GB, Martinez HEP, Garcia SLR, Finger FL (2004) Growth and yield of coffee plants in agroforestry and monoculture systems in Minas Gerais, Brazil. Agrofor Syst 63:75–82

Campbell BM, Clarke JM, Gumbo DJ (1991) Traditional agroforestry practices in Zimbabwe. Agrofor Syst 14:99–111

Coomes TO, Ban N (2004) Cultivated plant species diversity in home gardens of an Amazonian peasant village in northeastern Peru. Econ Bot 58(3):420–434

Crutzen PJ, Andreae MO (1990) Biomass burning in the tropics: Impact on atmospheric chemistry and biogeochemical cycles. Science 250:1669–1678

da Silva Moço MK, da Gama-Rodrigues EF, da Gama-Rodrigues AC, Machado RCR, Baligar VC (2009) Soil litter and fauna of cacao agroforestry systems in Bahia, Brazil. Agrofor Syst 76:127–138

Dean W (1987) Brazil and the struggle for Rubber. Cambridge University Press, Cambridge

Degrande A (2001) Farmer assessment and economic evaluation of shrub fallows in the humid lowlands of Cameroon. Agrofor Syst 53:11–19

Degrande A, Duguma B (2000) Adoption potential of rotational hedgerow intercropping in the humid lowlands of Cameroon. ODI Network paper No 103, DFID, UK

Degrande A, Schreckenberg K, Mbosso C, Anegbeh P, Okafor V, Kanmegne J (2006) Farmer's fruit tree growing strategies in the humid forest zone of Cameroon and Nigeria. Agrofor Syst 67:159–175

Degrande A, Asaah E, Tchoundjeu Z, Kanmegne J, Duguma B, Franzel S (2007) Opportunities for and constraints to adoption of improved fallows: ICRAF's experience in the humid tropics of Cameroon. In: Bationo A, Waswa B, Kihara J, Kimetu J (eds) Advances in integrated soil fertility management in sub-Saharan Africa: challenges and opportunities. Springer, Dordrecht, pp 901–910

de Wolf JJ (2010) Innovative farmers, non-adapting institutions: a case study of the Organization of agroforestry research in Malawi. In: German L et al. (eds) Beyond the biophysical: knowledge, culture, and power in agriculture and natural resource management. Springer, Germany, pp 217–239. doi:10.1007/978-90-481-8826-0 10

Duguma B, Kang BT, Okali DUU (1988) Effect of pruning intensities of three woody leguminous species grown in a alley cropping with maize and cowpea on an alfisol. Agrofor Syst 6:19–35

Duguma B, Tonye J (1994) Screening of multipurpose trees and shrub species for agroforestry in the humid lowlands of Cameroon. For Ecol Manag 64(2–3):135–143

Duguma B, Tonye J, Kanmegne J, Enoch T (1994) Growth of ten multipurpose tree species on acid soils in Sangmelima, Cameroon. Agrofor Syst 27:107–119

Ewel JJ (1999) Natural systems as models for the design of sustainable systems of land use. Agrofor Syst 45:1–21

Fernandes ECM, Nair PKR (1986) An evaluation of the structure and function of tropical homegardens. Agric Syst 21:279–310

Fernandes ECM, O'Kting'aiti A, Maghembe J (1984) The Chagga homegardens: a multistoried agroforestry cropping system on Mt Kilimandjaro, Northern Tanzania. Agrofor Syst 2:73–86

Fonzen PF, Oberholzer O (1984) Use of multipurpose trees in hill farming systems in western Nepal. Agrofor Syst 2:187–197

Fujisaka S, Hurtado L, Uribe R (1996) A working classification of slash-and-burn agricultural systems. Agrofor Syst 34:151–169

Gajaseni J, Gajaseni N (1999) Ecological rationalities of the traditional homegarden system in the Chao Praya Basin, Thailand. Agrofor Syst 46:3–23

Goma HC (2003) Potential for changing traditional soil fertility management systems in the wet miombo woodlands of Zambia: the chitemene and fundikila systems: In: Gichuru MP, Bationo A, Bekunda MA, Goma HC, Mafongoya PL, Mugendi DN, Murwira HK, Nandwa SM, Nyathi P, Swift MJ (eds) Soil fertility management in Africa: a regional perspective. pp 187–218

Gouyon A, de Foresta H, Levang P (1993) Does 'jungle rubber' deserve it name? An analysis of rubber agroforestry systems in Southeast Asia. Agrofor Syst 22:181–206

Grandstaff TB (1980) Shifting cultivation in northern Thailand: possibilities for development. Resource Systems Theory and Methodology Series No. 3. United Nations University, Tokyo

Herzog F (1994) Multipurpose shade trees in coffee and cocoa plantations in Cote d'Ivoire. Agrofor Syst 27:259–267

Imo M, Timmer VR (2000) Vector competition analysis of a *Leucaena*-maize alley cropping system in western Kenya. For Ecol Manag 126(2):255–268

Isaac Sheeba R, Achuthan NM (2006) Litter dynamics of six multipurpose trees in a homegarden in southern Kerala, India. Agroforest Syst 67:203–213

Isaac ME, Timmer VR, Quashie-Sam SJ (2007a) Shade tree effects in an 8-year-old cocoa agroforestry system: biomass and nutrient diagnosis of *Theobroma cacao* by vector analysis. Nutr Cycl Agroecosyst 78:155–165

Isaac ME, Ulzen-Appiah F, Timmer VR, Quashie-Sam SJ (2007b) Early growth and nutritional response to resource competition in cocoa-shade intercropped systems. Plant Soil 298:243–254

Ismail T (1986) Integration of animals in rubber plantations. Agrofor Syst 4:55–66

Jama-Adan (1993) Soil fertility and productivity aspects of alley cropping with Cassia siamea and Leucaena leucocephala under Semi-arid conditions in Machakos, Kenya. Ph.D. dissertation, University of Florida, Gainesville

Jama B, Getahun A (1991) Intercropping *Acacia albida* with maize (*Zea mays*) and green gram (*Phaseolus aureus*) at Mtwapa, coast province, Kenya. Agrofor Syst 14:193–205

Kang BT (1993) Alley cropping: past achievements and future directions. Agrofor Syst 23:141–155

Kang BT, Duguma B (1985) Nitrogen movement in alley cropping systems. In: Kang BT, van den Helde J (eds) Nitrogen in farming systems in the humid and subhumid tropics. Institute of soil fertility, Haren, pp 269–284

Kang BT, Kruijs ACBM van der, Attah-Krah AN (1989) Alley cropping for food crop production in the humid and subhumid tropics. International Workshop on Alley Farming in the Humid and Subhumid Tropics, 10–14 Mar. 1986, Ibadan, NG. IDRC, Ottawa, ON, Canada

Kang BT, Reynolds L, Atta-Krah AN (1990) Alley farming. Adv Agron 43:315–359

Kanmegne J, Degrande A (2002) From alley cropping to rotational fallow: farmer's involvement in the development of fallow management techniques in the humid forest zone of Cameroon. Agrofor Syst 54:115–120

Kehlenbeck K, Maass BL (2004) Crop diversity and classification of homegardens in central Sulawesi, Indonesia. Agrofor Syst 63:53–62

Kimaro AA, Timmer VR, Mugasha AG, Chamshama SAO, Kimaro DA (2007) Nutrient use efficiency and biomass production of tree species for rotational woodlot systems in semi-arid Morogoro, Tanzania. Agrofor Syst 71:175–184

King KFS (1968) Agri-silviculture: the Taungya system. Bulletin No 1, Department of forestry, University of Ibadan, Nigeria

Kotto-Same J, Moukam A, Njomgang R, Tiki-Manga T, Tonye J, Diaw C, Gockowski J, Hauser S, Weise S, Nwaga D, Zapfack L, Palm C, Woomer P, Gillison A, Bignell D, Tondoh J (2000) Alternatives to slash-and-burn: summary report and synthesis of phase II in Cameroon. ASB Coordination Office, ICRAF-Nairobi, Kenya, pp 72

Kumar MB, Nair PKR (2004) The enigma of tropical homegardens. Agrofor Syst 61:135–152

Kumar MB, Jacob George S, Chinnamani S (1994) Diversity, structure and standing stock of wood in the homegardens of Kerala in peninsular India. Agrofor Syst 25:243–262

Kwesiga F, Chisumpa S (1990) Ethnobotanical survey in Eastern province, Zambia. ICRAF, Nairobi

Kwesiga F, Coe R (1994) The effect of short rotation *Sesbania sesban* planted fallows on maize yield. For Ecol Manag 64:199–208

Kwesiga F, Franzel S, Place F, Ohiri D, Simanza CP (1999) *Sesbania sesban* improved fallows in eastern Zambia: Their inception, development and farmer enthusiasm. Agrofor Syst 47:49–66

Kwesiga F, Akinnifesi FK, Mafongoya PL, McDermott MH, Agumya A (2003) Agroforestry research and development in southern Africa during the 1990s: review and challenges ahead. Agrofor Syst 59:173–186

Lagemann J (1977) Traditional Farming Systems in Eastern Nigeria. Weltforum-Verlag, Munich

Lashermes P, Combes MC, Robert J, Trouslot P, Dhont A, Charrier AA (1999) Molecular characterization and origin of the *Coffea arabica* L. genome. Mol Gen Genet 261(2):259–266

Leakey RRB, Newton AC (1994) Tropical trees: the potential for domestication and the rebuilding of forest resources. HMSO, London, p 284

Leakey RRB, Tchoundjeu Z, Schreckenberg K, Shackleton S, Shackleton C (2005) Agroforestry Tree Products (AFTPs): Targeting Poverty Reduction and Enhanced Livelihoods. International Journal of Agricultural Sustainability 16:5–16

Lefroy FC, Dann PR, Wildin JH, Wesley-Smith RN, McGowan AA (1992) Trees and shrubs as sources of fodder in Australia. Agrofor Syst 20:117–139

Lieberei R (2007) South American Leaf Blight of the rubber tree (*Hevea* spp.): New steps in plant domestication using physiological features and molecular markers. Ann Bot 100:1125–1142

Leuschner WA, Khalique K (1987) Homestead agroforestry in Bangladesh. Agrofor Syst 5:139–151

Mafongoya PL, Bationo A, Kihara J, Waswa BS (2007) Appropriate technologies to replenish soil fertility in southern Africa. In: Bationo A, Waswa B, Kihara J, Kimetu J (eds) Advances in integrated soil fertility management in sub-Saharan Africa: challenges and opportunities. Springer, Dordrecht, pp 29–44

Méndez VE, Lok R, Somarriba E (2001) Interdisciplinary analysis of homegardens in Nicaragua: micro-zonation, plant use and socioeconomic importance. Agrofor Syst 51:85–96

Michon G (1983) Village-forest-gardens in West Java. In: Huxley PA (ed) Plant research and agroforestry. ICRAF, Nairobi, pp 13–24

Mugendi DN, Nair PKR, Mugwe JN, O'Neill MK, Woomer PL (1999a) Alley cropping of maize with Calliandra and Leucaena in the subhumid highlands of Kenya Part 1. Soil-fertility changes and maize yield. Agrofor Syst 46:39–50

Mugendi DN, Nair PKR, Mugwe JN, O'Neill MK, Woomer PL (1999b) Alley cropping of maize with calliandra and leucaena in the subhumid highlands of Kenya Part 2. Biomass decomposition, N mineralization and N uptake by maize. Agrofor Syst 46:51–64

Mureithi JG, Tayler RS, Thorpe W (1994) The effects of alley cropping with *Leucaena leucocephala* and of different management practices on the productivity of maize and soil chemical properties in lowland coastal Kenya. Agrofor Syst 27:31–51

Muschler RG (2001) Shade improves coffee quality in a sub-optimal coffee zone of Costa Rica. Agrofor Syst 85:131–139

Nair PKR (1987) Agroforestry systems inventory. Agrofor Syst 5:301–317

Nair PKR (1993) An introduction to agroforestry. Kluwer Academic Publisher, The Hague

Nair PKR, Fernandes ECM, Wambugu PN (1984) Multipurpose leguminous trees and shrubs for agroforestry. Agrofor Syst 2:145–163

Nair VD, Graetz DA (2004) Agroforestry as an approach to minimizing nutrient loss from heavily fertilized soils: The Florida experience. Agrofor Syst 61:269–279

Nath TK, Inoue M, Myant H (2005) Small-scale agroforestry for upland community development: a case study from Chittagong Hill Tracts, Bangladesh. J Forest Res 10(6):443–452

Neumann I (1983) Use of trees in smallholder agriculture in tropical highlands. In: Lockeretz W (ed) Environmentally sound agriculture. Prager, New York, pp 351–374

Nyadzi GI, Otsyina RM, Banzi FM, Bakengesa SS, Gama BM, Mbwambo L, Asenga D (2003) Rotational woodlot technology in northwestern Tanzania: Tree species and crop performance. Agrofor Syst 59:253–263

Ogbe DFM, Bamidele JF (2007) Potential of pigeon pea (*Cajanus cajan*) for planted fallow in Edo State, Nigeria. Asian J Plant Sci 6(3):490–495

Okafor JC, Fernandes ECM (1987) The compound farms of southeastern Nigeria: A predominant agroforestry homegarden system with crops and small livestock. Agrofor Syst 5:153–168

Oke DO, Odebiyi KA (2007) Traditional cocoa-based agroforestry and forest species conservation in Onda State, Nigeria. Agriculture, Ecosystems and. Environment 122:305–311

Padoch C, De Jong W (1987) Traditional agroforestry practices of native and Ribereno farmers in the lowland Peruvian Amazon. In: Gholz HL (ed) Agroforestry: realities, possibilities and potentials. Martinus Njhoff, Dordrecht, pp 179–194

Peltier R, Bisiaux F, Dubiez E, Marien J-N, Muliele J-C, Proces P et Vermeulen C. (2010) De la culture itinérante sur brûlis aux jachères enrichies productrices de charbon de bois, en Rep. Dem. Congo. ISDA 2010, Montpellier

Ramsay DM, Wiersum KF (1974) Problems of watershed management and development in the Upper Solo river basin. Conference on Ecologic guidelines for forest, land or Water Resources, Institute of Ecology, Bandung, Indonesia (Mimeo)

Rao MR, Nair PKR, Ong CK (1998) Biophysical interactions in tropical agroforestry systems. Agrofor Syst 38:3–50

Reynor WC, Fownes JH (1991) Indigenous agroforestry of Pohnpei. Parts 1 and 2. Agrofor Syst 16:139–157
Romero-Alvarado Y, Soto-Pinto L, Garcia-Barros L, Barrera-Gaytán JF (2002) Coffee yields and soil nutrients under the shades of *Inga* sp. vs multiple species in Chiapas, Mexico. Agrofor Syst 54:215–224
Roothaert RL, Franzel S (2001) Farmers' preferences and use of local fodder trees and shrubs in Kenya. Agrofor Syst 52:239–252
Schreckenberg K, Degrande A, Mbosso C, Boli Baboule Z, Boyd C, Enyong L, Kanmegne J, Ngong C (2002) The social and economic importance of Dacryodes edulis (G. Don) H.J. Lam in southern Cameroon. Forests Trees and Livelihoods 12:15–40
Schroth G, Coutinho P, Moraes VHF, Albernaz AL (2003) Rubber agroforests at the Tapajós river, Brazilian Amazon—environmentally benign land use systems in an old forest frontier region. Agr Ecosys Environ 97:151–165
Shepherd KD, Ndufa JK, Ohlsson E, Sjogren H, Swinkels R (1997) Adoption potential of hedgerow intercropping in maize-based cropping systems in the highlands of Western Kenya. 1. Background and agronomic evaluation. Experimental Agriculture 33:197–209
Sileshi G, Mafongoya PL (2003) Effects of rotational fallows on abundance of soil insects and weeds in maize crops in eastern Zambia. Appl Soil Ecol 23(3):211–222
Singh RP, Van del Beldt RJ, Hocking D, Korwar GR (1988) Alley cropping in the semi-arid regions of India. Proceedings of workshop on alley cropping, International Institute of Tropical Agriculture, Ibadan, Nigeria, March 10–14
Singh RP, Ong CK, Saharan N (1989) Above and below ground interactions in alley cropping in semi-arid India. Agrofor Syst 9:259–274
Soemarwoto O (1987) Homegardens: a traditional agroforestry system with a promising future. In: Steppler HA, Nair PKR (eds) Agroforestry: a decade of development. ICRAF, Nairobi, Kenya, pp 157–170
Soemarwoto O, Soemarwoto Z, Karyono, Soekartadiredja EM, Ramlan A (1976) The Javanese homegarden as an integrated agro-ecosystem. Science for a better environment. Science Council of Japan, Tokyo
Sonwa DJ, Nkongmeneck BA, Weise SF, Tchatat M, Adesina AA, Janssens MJJ (2007) Diversity of plants in cocoa agroforests in the humid forest zone of Southern Cameroon. Biodivers Conserv 16:2385–2400
Soto-Pinto L, Anzueto M., Mendoza J, Jimenez Ferrer G, de Jong B (2010) Carbon sequestration through agroforestry in indigenous communities of Chiapas, Mexico. Agrofor Syst 78:39–51
Spaan W, Bodnár F, Idoe O, de Graaff J (2004) Implementation of contour vegetation barriers under farmer conditions in Burkina Faso and Mali. Q J Int Agric 43(1):21–38
Stoler A (1975) Garden use and household consumption pattern in a Javanese Village. Ph.D. Dissertation, Columbia University, Department of Anthropology, New York, USA
Styger E, Rakotondramasy HM, Pfeffer MJ, Fernandes ECM, Bates DM (2007) Influence of slash-and-burn farming practices on fallow succession and land degradation in the rainforest region of Madagascar. Agr Ecosyst Environ 119(3–4):257–269
Tejwani KG (1994) Agroforestry in India. Oxford and IBH Publishing Co Pvt Ltd, New Delhi, India
Terra GTA (1954) Mixed-garden horticulture in Java. Malaysian J Trop Geogr 4:33–43
Tchatat M (1996) Les jardins de case agroforestiers des basses terres humides du Cameroun: Etude de cas des zones forestières des provinces du Centre et du Sud. These de Doctorat, Université de Paris 06, Paris, France, p 290
Tchatat M, Puig H, Tiki-Manga T (1995) Les jardins de case des provinces du Centre et du Sud du Cameroun: description et utilisation d'un système agroforestier traditionnel. J Agric Trad Bot Appl (JATBA) 37:165–182
Tchatat M, Ambassa-Kiki R, Mbila D, Fabre A (2004) Assessing soil chemical fertility in homegardens in forested areas of southern Cameroon. Afr Crop Sci J 12(2):163–170

Tchoundjeu Z, Asaah EK, Anegbeh PO, Degrande A, Mbile P, Facheux C, Tsobeng A, Atangana AR, Ngo Mpeck M-L, Simons AJ (2006) Putting participatory domestication into practice in West and Central Africa. Forests, Trees and Livelihoods 16:53–69

Teklay T, Malmer A (2004) Decomposition of leaves from two indigenous trees of contrasting qualities under shaded-coffee and agricultural land-uses during the dry season at Wondo Genet, Ethiopia. Soil Biol Biochem 36(5):777–786

Vasey DE (1985) Household gardens and their niche in Port Moresby, Papua New Guinea. Food Nutr Bull 7(3):37–43

Udawatta RP, Anderson SH (2008) CT-measured pore characteristics of surface soils influenced by agroforestry and grass buffers. Geoderma 145(1–3):381–389

Vergara NT (1981) Integral agroforestry: a potential strategy for stabilizing shifting cultivation and sustaining productivity of the natural environment. Working paper. Environment and Policy Institute. East-West Center, Honolulu, Hawaii, USA

Wezel A, Bender S (2003) Plant species diversity of homegardens of Cuba and its significance for household food supply. Agrofor Syst 57:39–49

Wiersum KF (1982) Tree gardening and taungya in Java: Examples of agroforestry techniques in the humid tropics. Agrofor Syst 1:53–70

Wiersum KF (2006) Diversity and change in homegarden cultivation in Indonesia. In: Kumar BM, Nair PKR (eds) Tropical homegardens: a time-tested example of sustainable agroforestry. Springer, The Netherlands, pp 13–24

Chapter 5
Major Agroforestry Systems of the Semiarid Tropics

Abstract Agroforestry practices of the semiarid tropics are most often designed to control erosion, and to make water and moisture available to plants throughout the cropping cycle. Agrosilvopastoral systems, i.e, the deliberate introduction of livestock into tree-crop systems, are more common in the semiarid tropics than the humid tropics. For that reason, the most common agroforestry practices of the semiarid tropics include homegardens, shifting cultivation, parklands, alley cropping, shelterbelts systems, woody perennials on rangelands and pastures, plantation crops with pastures and livestock, living fences of woody perennials and hedges that are useable as fodder, multipurpose woody hedgerows, parklands, tree woodlots, woody perennials for soil conservation or reclamation and sloping agriculture land technology, multi-purpose tree based systems and perennial crop based systems. Other agroforestry systems are valued in semiarid tropics, including protein banks and apiforestry. The most important food crops that are grown in tree-crop systems in the semiarid tropics are cereals (maize, rice and various species of millet) or potatoes.

5.1 Introduction

Agroforestry systems that are found in the semiarid tropics are generally designed to deal with agroclimatic conditions of tropical savannas. Water availability throughout the cropping cycle and erosion control are two important factors that threaten agricultural production in tropical savannas. For this reason, agroforestry systems in the semiarid tropics most often include agricultural practices that are aimed at regulating water and moisture availability in root crop systems over the cropping cycle and protecting the soils under food crops against wind erosion damages. Some agroforestry systems, such as windbreaks, live fences, scattered trees on farms and buffer strips, are used in the semiarid tropics to solve specific problems that are related to soil erosion. More details on the effects of agroforestry practices on erosion control are provided in Chap. 9.

In densely populated highland savannah areas, family compounds consist of homesteads that are delineated by hedges and a forested area. Each family compound includes houses, a homegarden and an agropastoral area. Cash crop fields are divided by hedges to indicate boundaries between landowners. Silvopastures

A. Atangana et al., *Tropical Agroforestry*, DOI 10.1007/978-94-007-7723-1_5,

are widespread in the semarid tropics owing to increased livestock production. The present chapter will focus on shifting cultivation as practiced in the semi-arid tropics. Agroforestry parklands, silvopastoral systems and shelterbelt systems, which are the most common agroforestry systems encountered in the semiarid tropics, with homegardens being described in details elsewhere (Chap. 4). The present chapter will also introduce live fence and buffer strip systems.

5.2 Annual or Biannual Food Crop Farms in the Semiarid Tropics: Shifting Cultivation

Shifting cultivation, as practiced in the semiarid tropics does not differ from that practiced in the humid tropics; only the food crops that are grown differ. The most important food crops in humid savannas are potatoes (*Solanum tuberosum* L., Solanaceae), common beans (*Phaseolus vulgaris* L., Fabaceae), maize (*Zea mays*) and tomatoes (*Lycopersicon esculentum*). Various cereals constitute the main annual crops that are grown in the semiarid tropics that surround the humid tropics. These cereals include rice (*Oryza sativa* L.), white fonio (*Digitaria exilis* (Kippist) Stapf, Poaceae), maize, several species of millet such as caracan millet (*Eleusine coracana* Gartn.), barnyard millet (*Echinochloa crus*-galli (L.) Beauv), pearl millet (*Pennisetum glaucum* (L.) R.Br.), proso millet (*Panicum miliaceum* L.), little millet (*Panicum sumatrense* Roth ex. Roem & Schult.) and koda millet (*Paspalum scrobiculatum* L.), *Setaria* grasses and sorghum (*Sorghum bicolor* (L.) Moench). Cotton (*Gossypium hirsutum* L., Malvaceae), rice and maize are the main cash crops in the semiarid tropics worldwide. Other important annual crops in the Sahelian zone of Africa include peanuts (*Arachis hypogaea*), onions (*Allium cepa* L., Amaryllidaceae) and garlic (*Allium sativum* L., Amaryllidaceae).

Shifting cultivation is the most common agricultural technique that is employed by farmers in the semiarid tropics of Africa. Shifting cultivation is also practiced in Southeast Asia and South America, and contributes substantially to biomass burning in the tropics (Hao and Liu 1994). For instance, farmers use savanna fires to clear land for agricultural purposes in the tropics. Some multipurpose tree species are usually spared when clearing land, which is usually done in the dry season, for crop establishment. These remnant trees may facilitate regeneration in fallow land (Carriere et al. 2002).

Soil quality of land in fallow is influenced by remnant tree species. Soils under remnant guinea plum (*Parinarium excelsum* Sabine, Rosaceae) and néré (*Parkia biglobosa* (Jacq.) R.Br. ex G.Don, Fabaceae) trees had higher concentrations of organic C, total N, extractable P, exchangeable K, Ca and Mg, total P and Ca, and CEC than did the open microsites (Sirois et al. 1998). The same study also revealed that extractable P, exchangeable K, Ca and Mg, total P and Ca, and CEC concentrations were greater under *Parinari excelsa* Sab. (Chrysobalanaceae) than under ehomi (*Erythrophleum guineense* G.Don., Fabaceae) in the Fouta-Djallon region of Guinea. Research has investigated several possible cultural associations based on agro-ecological zones in the tropics.

5.3 Multipurpose Trees on Farmlands: Agroforestry Parklands

Agroforestry parklands are a traditional farming system in the semi-arid and sub-humid zones of West Africa, and are characterized by the deliberate retention of wild multipurpose trees on cultivated or recently fallow land. Bonkoungou et al. (1994) defined agroforestry parklands as 'land use systems in which woody perennials are deliberately preserved in association with crops and/or animals in a spatially dispersed arrangement and where there is both ecological and economic differences between the trees and other components of the system'. Thus, parklands consist of intercropping crops under scattered trees, at a density of 20–50 trees per ha. Trees and animals are part of the parkland system, and they act as a source of food, fuel, fodder, medicinal products, materials for shelter and commodities with commercial value (Boffa 1999). Parklands provide numerous productive functions, including food production, increased soil fertility, wood production, wood fibers for handicrafts and clothing, and browse (Boffa 1999). Fat and oil production is also important in these systems, with shea butter extracted from karié or *Vitellaria paradoxa* C.F. Gaertn (Sapotaceae), oil produced from *Balanites aegyptiaca* (L.) Delile (Zygophyllaceae), and wine production from the sap of African Fan Palm (*Borassus aethiopum* Mart. Arecaceae). Sometimes, parklands are also used in livestock production, through a slow process of species selection, and management of tree density over decades. The practice is found in the Sahel and in southern Africa, including Malawi and Zimbabwe. Some species, such as *V. paradoxa* in farmlands, have become semi-domesticated, due to long-term anthropogenic selection in areas such as Northern Ghana (Lovett and Haq 2000). Yet, parklands differ from perennial crop-based systems like cocoa farms because the woody component of the system is non-cultivated trees.

The woody component of parkland varies according to region or country, as well as species association. Parklands constitute the predominant agroforestry system in semi-arid West Africa (Nair 1993). *Parkia biglobosa* (néré) and *Vitellaria paradoxa* (Fig. 5.1) are dominant tree species in the agroforestry parklands of sub-Saharan Africa (Oni 1997; Boffa 1999). Other species that are commonly found in parklands include *Prosopis africana* (Guill. & Perr.) Taub. (Fabaceae), *Adansonia digitata*, *Tamarindus indica* L. (Fabaceae), *Ziziphus mauritiana* Lam. (Rhamnaceae), *Faidherbia albida*, *Acacia senegal*, *Balanites aegyptiaca*, *Bombax costatum* Pellegr. & Vuill. (Malvaceae), *Borassus aethiopum*, *Elaeis guineensis*, *Lannea microcarpa* Engl. & K.Krause (Anacardiaceae), *Sclerocarya birrea*, *Vitex doniana*, *Hyphaene thebaica* (L.) Mart. (Arecaceae), *Diospyros mespiliformis* Hochst. Ex A. DC. (Ebenaceae), *Ceiba pentandra* (L.) Gaertn. (Malvaceae) and *Anogeissus leiocarpus* (DC.) Guill. & Perr. (Combretaceae; Boffa 1999). Also found in agroforestry parklands in West Africa are *Khaya senegalensis* (Desr.) A.Juss. (Meliaceae), *Acacia tortilis*, *Celtis integrifolia* Lam. (Ulmaceae), *Parinari macrophylla* Sabine (Chrysobalanaceae), and *Cordyla africana* Lour. (Fabaceae; Boffa 1999). Parklands can be characterized by dominant species, which vary according to climatic zone (Pullan

Fig. 5.1 *Vitellaria paradoxa* parkland (Source: Jules Bayala, ICRAF)

1974). For instance, in the Sahel, the dominant parkland species in parklands are *Acacia tortilis* subsp. *raddiana* (Savi) Brenan, *Balanites aegyptiaca, Acacia senegal, Hyphaene thebaica, Tamarindus indica, Borassus aethiopum* and *Pilostigma reticulatum* (DC.) Hochst., whereas the dominant species in Northern Sudan are *Faidherbia albida, Vitellaria paradoxa, Parkia biglobosa, Adansonia digitata, Borassus aethiopum, Cordyla pinnata* (Lepr. Ex A.Rich.) Milne-Redh., *Tamarindus indica* and *Sclerocarya birrea* (Boffa 1999). In Senegal, *Faidherbia* parklands are widespread. Most perennial species found in parklands provide food from fruits, seeds and leaves, together with firewood. These trees, such as *Adansonia digitata, Parkia biglobosa, T. indica, V. paradoxa, Z. mauritiana*, are part of tree domestication programs (Raebild et al. 2011). Tree density in parklands varies according to species. For néré, the density ranges between 2 and 3 per ha, while karité tends to grow in densities of five to ten trees per ha (Kater et al. 1992; Kessler 1992). *Faidherbia* and *Prosopis* tend to grow at even higher densities, i.e., 5–50 and 10–45 trees per ha, respectively (Depommier et al. 1992; Tejwani 1994). Optimum tree density for the best annual crop production varies with size and age, due to competition for soil resources (Singh et al. 2007).

Annual crops that are grown in agroforestry parklands include pearl millet, cotton (*G. hirsutum*; Fig. 5.2) and sorghum. In the Sahel parklands, the crops are usually associated with néré and karité (Boffa et al. 2000). Other crops encountered in the agroforestry parklands in Sahel include common beans (*Phaseolus vulgaris*), peanuts, cotton and maize. However, grain yield of sorghum under karité and néré trees is reduced by an average of 50–70 % in comparison with open fields (Kessler 1992). Management practices in parklands such as pruning trees could provide some benefits to soil fertility, and can help mitigate yield loss. Pruning trees in néré-karité agroforestry parklands has been shown to be beneficial for millet production. The highest millet grain yields and total dry matter were produced under fully pruned trees in a study undertaken in Burkina-Faso (Bayala et al. 2002). Pruning

Fig. 5.2 Cotton associated with *Faidherbia albida* parkland (Source: Jules Bayala, ICRAF)

trees also reduces root density in the crop's rooting zone, reducing the belowground competition in the néré-karité parklands of Burkina-Faso (Bayala et al. 2004).

Phosphorus and nitrogen availability is higher under canopy trees than in areas beyond canopies in a *F. albida – V. paradoxa* parkland in Burkina Faso (Gnankambary et al. 2008). *Faidherbia albida* trees, owing to their reverse phenology, minimize competition and enhance the fertility effect (reviewed by Sanchez 1995). Introducing trees into parklands reduces water loss and nutrient leaching in the system (Breman and Kessler 1997), hence allowing better nutrient utilization. Trees in parklands have a positive contribution to soil carbon content (Bayala et al. 2006), indicating that agroforestry parklands have potential as carbon sinks.

5.4 Silvopastures

Simply put, silvopastoral systems are the introduction of trees into livestock systems. The benefits may be phytoremediation (Michel et al. 2007), minimization of soil nutrient loss (Nair and Graetz 2004), and production of fodder, fruit, wood and shade (protecting livestock from heat). Silvopastures also, provide favorable wildlife habitats. For that reason, silvopastures usually have at least two layers: an upper canopy with trees providing shade, and a grass layer for grazing; when trees and shrubs are used for browsing, a third layer may be used between the trees and shrubs. In the tropics, silvopastures occur mostly in semiarid ecosystems. For instance, animal husbandry (cattle, sheep and goats) is widespread in the Sahel, Australian savannas and Llanos. The Brazilian Cerrado is one example of a tropical savanna in which cultivated pastures for beef and cattle production is intensive. In the Cerrado, eucalyptus-based silvopastoral systems are established with understory forage crops such as *Brachiaria*. Crops are sometimes introduced in pasturelands. In South America, maize, rice and soybeans are most often grown in silvopastoral systems in the first two years, followed by *Brachiaria*. In South America, maize is frequently included in silvopastures. The same scenario occurs in East African savannas. However in Central and South America, rainforests have been increasingly cleared for cattle ranching since the 1980s. For instance, Egler et al. (2013) reported that, during agricultural two agricultural censuses (1995/1996–2006), deforestation in the South border of Brazilian Amazon biome was closely related to increase in number of cattle and pasture area.

Selection of tree species is an important consideration when establishing silvopastures. Desirable characteristics of trees most often include production of trees and tree products such as barks, fruits and nuts, and leaves. Trees in the genera *Acacia* and *Eucalyptus* are common in tropical silvopastoral systems. *Pinus radiata* D.Don (Pinaceae), and exotic species introduced from California, is also common in pasturelands in Australia, whereas *Parkia biglobosa, Balanites aegyptica, Adansonia digitata, Bauhinia rufescens* Lam. (Fabaceae), *Faidherbia albida* and *Sclerocarya birrea* are found in silvopastoral systems in Africa.

Numerous herbaceous species are used as forage in the tropics, including *Brachiaria humidicola* (Rendle) Schweick., *B. brizantha* (Hochst. Ex A.Rich.) Stapf., *B. decumbens* Stapf., *B. mutica* (Forsk.) Stapf., *Cenchrus ciliaris* L., *Chloris gayana* Kunth, *Cynodon plectostachyus* (K.Schum.) Pilg., *Dichanthium aristatum* (Poir.) C.E. Hubbard, *Digitaria decumbens* Stent, *Panicum maximum* Jacq., *Paspalum dilatatum* Poir., *Pennisetum purpureum* Schumach. and *P. clandestinum* Hochst. Ex Chiov. These species most often constitute the grass layer of silvopastoral systems in the tropics.

5.4.1 Fodder Trees and Shrubs

Several leguminous shrub and tree species are used as sources of fodder for cattle and small ruminants in semiarid and subhumid tropical zones. Fodder species in the tropics belong to many families, including the Acanthaceae, Anacardiaceae, Annonaceae, Asclepiadaceae, Bignoniaceae, Bombaceae, Boraginaceae, Burseraceae, Capparidaceae, Caesalpiniaceae, Chenopodiaceae, Combretaceae, Convolvulaceae, Euphorbiaceae, Fabaceae, Labiatae, Mimosaceae, Rubiaceae and Tilliaceae (Nair et al. 1984; Lefroy et al. 1992; Topps 1992; Cajas-Giron and Sinclair 2001; Roothaert and Franzel 2001). The most frequently browsed species in the tropics are listed in Table 5.1.

5.5 Windbreaks

A shelterbelt or windbreak consists of one or more rows of trees and/or shrubs, which are planted across croplands or grasslands to redirect and reduce wind flow, thereby changing microclimate, and subsequently protecting homes, crops, grasses,

Table 5.1 Some important fodder trees and shrub species in the tropics (Adapted from Nair et al. 1984; Lefroy et al. 1992; Topps 1992; Roothaert and Franzel 2001; Cajas-Giron and Sinclair 2001; Devendra and Savilla 2002)

Species	Family	Common name	Origin
Faidherbia albida (Delile) A. Chev (=*Acacia albida*)	Fabaceae	Apple-ring acacia	Africa
Acacia angustissima (Mill.) Kuntze	Fabaceae	Prairie acacia	Central and North America
Acacia aneura F. Muell. ex Benth	Fabaceae	Mulga	Australia
Acacia ataxancatha DC	Fabaceae	Flame thorn	Africa
Vachelia farnesiana (L.) Willd. (=Acacia farnesiana)	Fabaceae	Needle bush	Central America
Acacia mellifera (Vahl) Benth	Fabaceae	Blackthorn	Africa
Acacia microbotrya Benth.	Fabaceae	Manna wattle	Australia
Acacia nilotica (L.) Willd. Ex Delile	Fabaceae	Gum arabic	Africa
Acacia saligna (Labil.) H.L. Wendl	Fabaceae	Golden wreath wattle	Australia
Acacia siberiana DC	Fabaceae	Paperback thorn	Africa
Acacia tortilis (Forks.) Hayne	Fabaceae	Umbrella thorn acacia	Africa
Acalypha fruticosa Forsk	Euphorbiaceae	Birch-leaved acalypha	Africa
Afzelia africana Sm	Fabaceae	Lemgue or Doussi	Africa
Albizia caribaea (Urb.) Britton & Rose (=Albizia niopoides (Spruce ex. Bewth) Burkart	Fabaceae	Albizia	Central and South America
Albizia chinensis (Osbeck) Merrill	Fabaceae	Silk tree, Chinese albizia	Southeast Asia
Albizia lebbeck (L.) Benth	Fabaceae	Indian siris	India, Southeast Asia, Australia,
Albizia saman F. Muell	Fabaceae	Saman	Central and South America
Anacardium excelsum L.	Anacardiaceae	Wild cashew	Central and South America
Aspilia mossambicensis (Oliv.) Wild	Compositae	Wild sunflower	Africa
Atriplex amnicola Paul G. Wilson	Amaranthaceae	River saltbush	Australia
Atriplex lentiformis (Torr.) S. Wats	Amaranthaceae	Big saltbush	Southwestern USA, Mexico
Atriples undulata (Moq.) D.Dietr	Amaranthaceae	Wavy-leaved saltbush	Souuth America
Balanites pedicellaris Mildbr. & Schlechter	Balanitaceae (Zygophyllaceae)	Small torchwood	Africa
Bridelia micrantha (Hochst.) Baill	Phyllanthaceae	Coastal golden-leaf	Africa
Bituminaria bituminosa C.H. Stirt	Fabaceae	Arabian pea	Mediterranean basin

Table 5.1 (continued)

Species	Family	Common name	Origin
Bulnesia arborea (Jacq.) Engl	Zygophyllaceae	Verawood	South America
Cajanus cajan (L.) Millsp.	Fabaceae	Pigeon pea	Africa, Asia
Caesalpinia coriaria (Jacq.) Willd	Fabaceae	Divi-divi	Central and South America
Calliandra calothyrsus Meisn	Fabaceae	Red calliandra	Central America
Cassia grandis L.f	Fabaceae	Pink shower tree	Central and South America
Ceiba pentandra (L.) Gaertn	Malvaceae	Kapok	Central and South America, Africa
Chamaecytisus palmensis (Christ.) Hutch	Fabaceae	Tagasaste, Tree lucerne	Africa (Canary Islands)
Commiphora zimmermanii Engl.	Burseraceae	Mururi	Africa
Cordia africana Lam.	Boraginaceae	Sudan teak	Africa
Cordia dentata Poir.	Boraginaceae	White manjack	Central and South America
Cordia alliodora (Ruiz & Pavon) Oken	Boraginaceae	Spanish elm	Central and South America
Crescentia cujete L.	Bignoniaceae	Calabash tree	Central and South America
Crotalaria goodiformis Vatke	Fabaceae		Africa
Cytisus mollis (Cav.) Pau (=*Chamaecytisus mollis* (Cav.) Greuter & Burdet)	Fabaceae		Mediterranean basin, Europe
Desmodium cinerium (Kunth) DC. rensonii	Fabaceae	Tick trefoil	Central America
Enteroobium cyclocarpum (Jacq.) Griseb	Fabaceae	Guanacaste, Elephant ear	Central and South America
Erythrina variegata L.; Erythrina spp	Fabaceae	Tiger's claw	Africa, Southeast Asia, Australia
Ficus spp	Moraceae		Africa
Flemingia macrophylla (Willd.) Merr	Fabaceae		Southeast Asia
Gliricidia sepium (Jacq.) Kunth ex Walp	Fabaceae		Central and South America, Southeast Asia
Grewia tembensis Fresen	Malvaceae		Africa
Guazuma ulmifolia Lam	Malvaceae	Bastard cedar, pigeon wood	Central and South America
Indigofera lupatana Baker f	Fabaceae	Muguti	Africa
Lantana camara L.	Verbenaceae	Spanish flag	Central and South America
Leucaena spp	Fabaceae		Africa, Southeast Asia, Australia
Maytenus putterlickioides (Loes) Exell and Mendonça	Celastraceae	Large-flowered spike thorn	Africa
Medicago arborea L	Fabaceae	Tree medick	Europe

Table 5.1 (continued)

Species	Family	Common name	Origin
Melia volkensii Guerke	Meliaceae	Melia	Africa
Milletia dura Dunn	Fabaceae	Milletia	Australia
Pachira quinata (Jacq.) W.S. Alverson	Malvaceae	Pochote	Central and South America
Platymiscium pinnatum (Jacq.) Dugand	Fabaceae	Quira, Amazon rosewood	Central and South America
Prosopis juliflora (Sw.) DC	Fabaceae	Mesquite	Africa, Latin America
Rhus natalensis Krauss	Anacardiaceae	Natal rhus	Africa
Senna spectabilis (DC.) Irwin & Barneby	Fabaceae	Cassia	Central and South America
Sesbania grandiflora (L.) Poiret	Fabaceae	Agati	Southeast asia, Australia
Sesbania sesban (L.) Merr.	Fabaceae	Egyptian riverhemp	Africa, Southeast Asia
Sterculia apelata (Jacq.) Karst.	Malvaceae	Panama tree	Central and South America
Tabebuia rosea DC; T. billbergii	Bignoniaceae	Pink trumpet	Central and South America
Tabebuia billbergii (Bureau & K. Schum.) Stabdl.	Bignoniaceae	Yellow poui	Central and South America
Tamarindus indica L.	Fabaceae	Tamarind	Africa
Genistamonspessulana (L.) O. Bol. & Vigo	Fabaceae	French broom	Mediterranean basin

livestock, and/or soils from wind erosion and providing wildlife with favorable habitats (Droze 1977; Brandle et al. 1988; Cleugh et al. 2002; Wright and Stuhr 2002). Windbreaks usually consist of multi-layer strips of trees and shrubs planted at least three rows deep. They are planted with the tree rows perpendicular to the prevailing wind direction, and are more effective when oriented at right angles to prevailing winds. Figures 5.3 and 5.4 illustrate windbreak layout, structure, and composition. Windbreaks are most often planted in semiarid areas in the tropics, where special importance is given to plant/water relationships as a consequence of limitations of available moisture (Smith and Jarvis 1998; Smith et al. 1998).

Windbreaks protect crops from wind erosion, and can result in 10–20 % increases in total crop yield (Wright and Stuhr 2002). Figure 5.5 illustrates the mechanisms by which windbreaks affect microclimate and soil productivity. Windbreaks are used to control blowing snow and to improve animal health by reducing stress on animals (Wright and Stuhr 2002), thereby increasing animal survival under winter conditions (Brandle et al. 1988, 2004). However, windbreaks can also create competition for water and nutrients between trees and crops (Smith et al. 1998).

Shelterbelts can be beneficial or harmful to crops. In terms of benefit, shelterbelts reduce wind velocity and air temperature in croplands (Ujah and Adeoye 1984), thereby reducing evapotranspiration and improving soil water availability ton crops in semiarid areas. In contrast, *Eucalyptus tereticornis* shelterbelts were

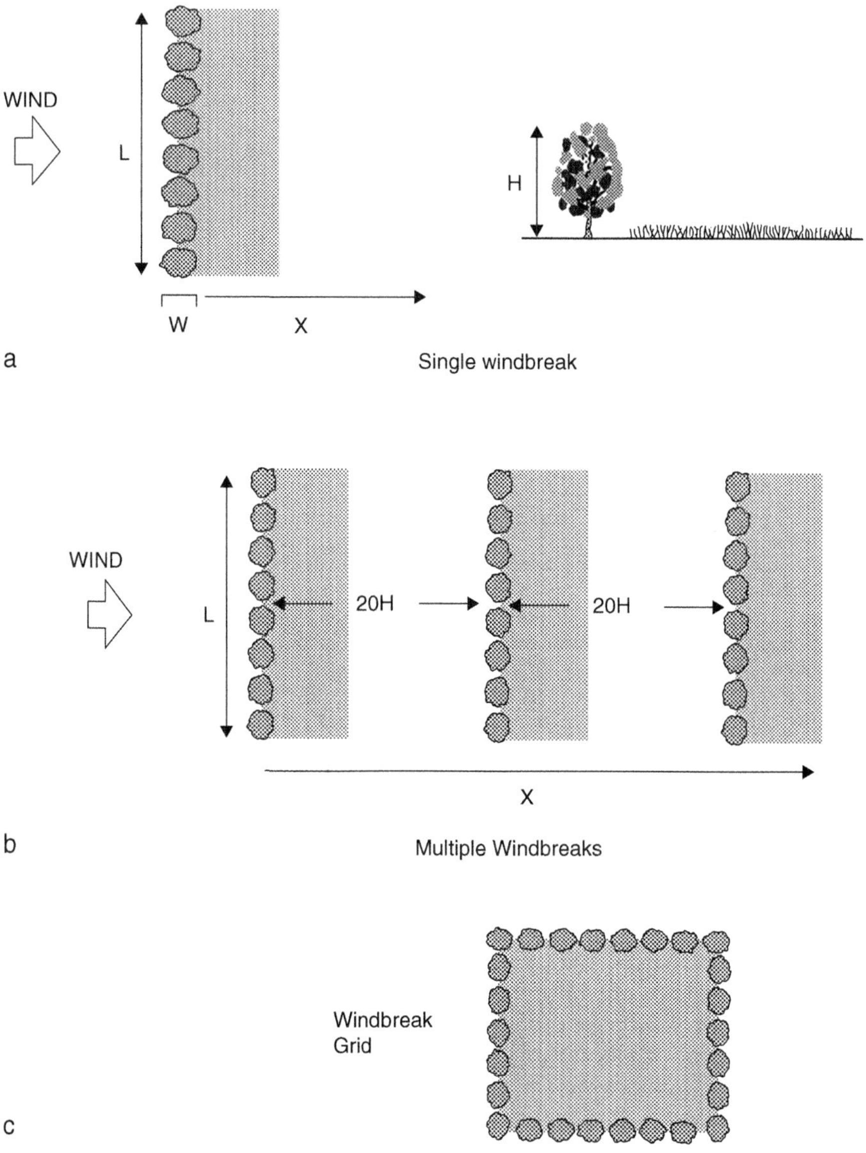

Fig. 5.3 Windbreak layout and dimensions (Cleugh 1998)

found to be harmful to crops, as eucalypts can release phytotoxins into soils (Singh and Kohli 1992). These phytotoxins have allopathic effects on crops, in that they reduced chickpea yield to the maximum extent, and impaired the performance of *Lens esculentum* (Singh and Kohli 1992). Therefore, prior to any establishment of

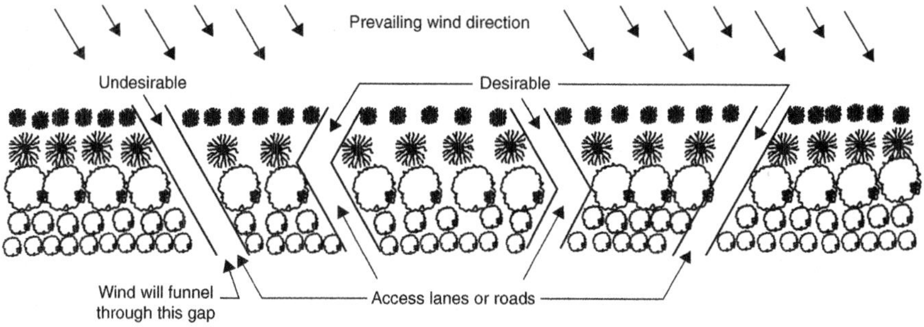

Fig. 5.4 Windbreak structure (Wright and Stuhr 2002)

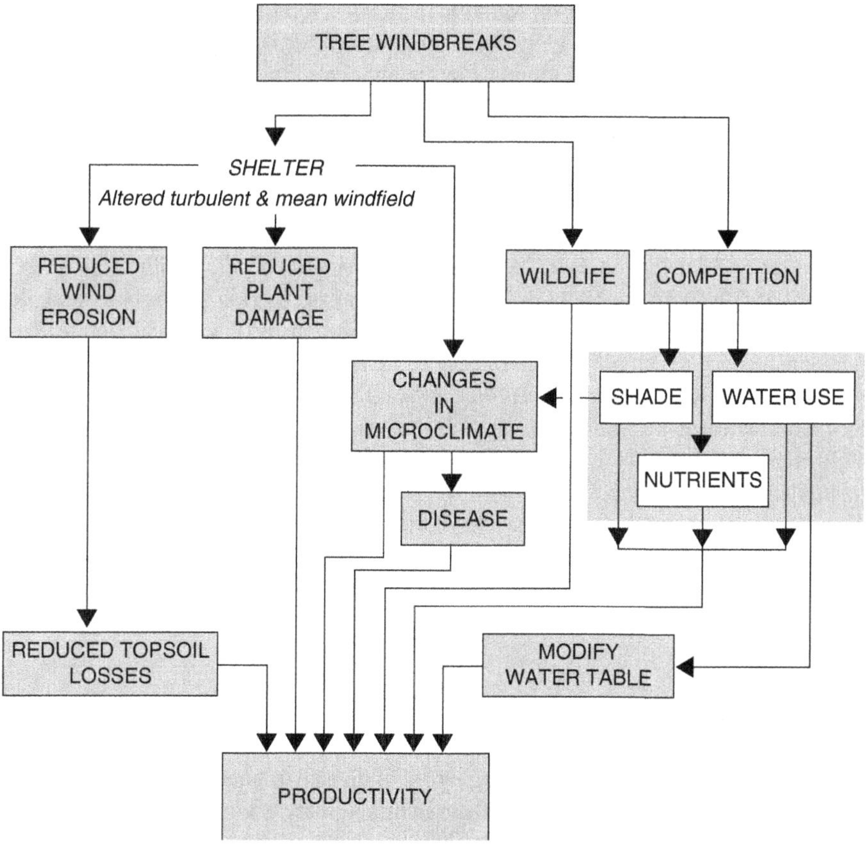

Fig. 5.5 Mechanisms by which windbreaks affect microclimate and plant productivity (Cleugh 1998)

shelterbelts, the compatibility of woody species and crops should be tested to avoid any negative interaction effects.

5.5.1 Trees Used as Windbreaks

Trees used in windbreaks can provide wood products such as firewood, lumber and posts, fruit and nuts, habitats for wildlife, fiber, honey, and fodder (e.g., calliandra leaves). Diversifying the species that are used in windbreaks can increase the range of agroforestry products. However, some constraints do exist with regard to the selection of species that would be included in windbreaks. Environmental risks such as insect attacks (especially termites), the presence of wild and domestic animals, poor soil conditions, and drought will reduce the range of choice as well as growth rate of crops. Water management is also important when establishing trees in a dry environment. Smith et al. (1998) found that *Azadirachta indica* (L.) Adelb. (Meliaceae) used less water than *Acacia nilotica* and *Acacia holosericea*, indicating that *A. indica* is more suitable for windbreaks in the Sahel region, when considering water competition between trees and crops was considered.

Woody species used in shelterbelts include species in the genera *Acacia*, *Albizia*, *Azadirachta*, *Callistemon*, *Casuarina*, *Cupressus*, *Dalbergia*, *Leucaena*, *Melaleuca*, *Pinus*, *Populus*, *Prosopis*, and *Tamarix* (Ujah and Adeoye 1984; Singh and Kohli 1992; Onyeotu et al. 1994; Chauchan 2003; Singh et al. 1998; Sun and Dickinson 1994; Sun and Dickinson 1997; Maiers and Harrington 1999; Smith et al. 1998; Singh et al. 1999). In Australia, woody species that are used in shelterbelts include *Callistemon salignus* Sm. Sweet (Myrtaceae), *C. viminallis* (Gaertn.) G.Don, *Eucalyptus microcorys* F.Muell., *E. intermedia* R.T.Baker, *E. torelliana* F. v. Mueller, *E. tessellaris* F.Muell., *Melaleuca armillaris* (Sol. Ex Gaertn.) Sm. and *M. linariifolia* Sm., and, whereas *E. tereticornis* Sm., *Dalbergia sissoo* Roxb., *Populus deltoides* W.Bartram ex Humphry Marshall, *Prosopis juliflora*, *Azadirachta indica*, *Acacia tortilis* and *Albizia lebbeck* (L.) Benth. are frequently used in windbreaks in India. *Eucalyptus camaldulensis* Dehnh. and *Acacia* spp. are common in shelterbelts in Africa, whereas Eucalypts, *Pinus brutia* Tenore and *Cupressus arizonica* Greene are frequently used in Latin America.

5.6 Live Fences

Live fences generally consist of a single row of densely planted trees or shrubs that are established to protect croplands from animals in areas where agrosilvopastoral systems are widespread. Harvey et al. (2005) reported narrow (3.76 m) and densely planted (323 trees km^{-1}) in Central America. Live fences are also used to delineate homesteads and to control erosion. Advantages of live fences as they are perceived by farmers in Sahel include erosion control, durability, protective efficiency, windbreak function, low management requirements, low costs and fodder supply (Ayuk 1997). Other benefits of live fences include provisions for firewood, timber and fruit (Harvey et al. 2005).

In fragmented landscapes, live fences increase total tree cover, divide pastures into smaller areas and provide direct physical connections to forest patches, thereby enhancing landscape connectivity (Chacón León and Harvey 2006). Live fences also provide habitats and resources for animals, although mean woody species richness in each live fence can be low (1.4–7.5 species per fence; Harvey et al. 2005). In contrast, numerous woody species were recorded in live fences in Central America (Baggio and Heuveldop 1984; Harvey et al. 2005), including *Acacia* spp., *Albizia guachapele* (Kunth) Dugand, *Annona* spp., *Azadirachta indica, Bursera simaruba* (L.) Sarg., *Calliandra calothyrsus, Cedrela odorata, Ceiba pentandra, Citrus aurantium* L., *Citrus lemon* (L.) Burm.f., *Citrus paradise* Macfad., *Citrus sinensis* (L.) Osbeck, *Cordia alliodora, Cordia dentate* Poir., *Erythrina costaricensis* Michell, *Eucalyptus saligna* Sm., *Eugenia salamensis* Doon. Smith, *Ficus werckleana* Rossberg, *Gliricidia sepium, Guazuma ulmifolia* Lam., *Inga spectabilis* (Vahl) Willd., *Inga vera* Kunth, *Pachira quinata* (Jacq.) W.S.Alverson, *Persea americana* Mill., *Spondias mombin* L., *Spondias purpurea* L., *Tabebuia rosea* A.P. de Candolle, *Tamarindus indica* in Central America (Baggio and Heuveldop 1984; Harvey et al. 2005) and *Acacia nilotica, A. seyal* Del., *Bauhinia rufescens, Euphorbia balsamifera* Aiton, *Prosopis juliflora, Ziziphus mauritiana* in Sahel (Ayuk 1997), respectively. The importance of *Acacia* species for utilization in live fences in Australia was highlighted by Thomson et al. (1994).

5.7 Buffer Strips

Runoff has negative effects on watersheds that are planted with row crops, as it increases erosion, soil loss and pollution of local surface waters. Agroforestry buffer strips, which intersperse crops with rows of trees and grass buffers, can reduce runoff and soil loss from watersheds through increased water infiltration and water storage (Anderson et al. 2009). Indeed, buffer strips improve soil water transport and retention (Udawatta and Anderson 2008). Buffer strips also trap sediments in agricultural watersheds (Yuan et al. 2009). Agroforestry buffer strips are suited in areas such as Southeast Asian uplands where contour farming is practiced (Garritty 1999). In semiarid areas of Africa, species that are targeted by farmers for inclusion in buffer strips provide additional benefits, such including food, fodder and medicine (Spaan et al. 2004). Indeed, the species that are preferred for buffer strips by farmers in Burkina Faso and Mali have included *Andropogon guayanus* Kunth (efficient against wind erosion), *Euphorbia balsamifera* (medicinal properties), *Jatropha curcas* L. (seeds used in traditional medicine and cattle feeding), *Piliostigma reticulatum* (the leaves have medicinal properties) and *Ziziphus mauritiana* (fruits; Spaan et al. 2004).

Bibliography

Fig. 5.3 Windbreak layout and dimensions (Cleugh 1998)Anderson SH, Udawatta RP, Seobi T, Garrett HE (2009) Soil water infiltration in agroforestry buffer strips. Agrofor Syst 75(75):5–16

Ayuk ET (1997) Adoption of agroforestry technology: The case of live edges in the Central Plateau of Burkina Faso. Agric Syst 54(2):189–206

Baggio A, Heuveldop J (1984) Initial performance of *Calliandra calothyrsus* Meissm, in live fence for the production of biomass. Agrofor Syst 2:19–29

Bavappa KVA, Jacob VJ (1982) High-intensity multispecies cropping: a new approach in small-scale farming in the tropics. World Crops (March/April) 47–50

Bayala J, Teklehaimanot Z, Ouedraogo SJ (2002) Millet production under pruned tree crown in a parkland system in Burkina-Faso. Agrofor Syst 54:203–214

Bayala J, Teklehaimanot Z, Ouedraogo SJ (2004) Fine root distribution of pruned trees and associated crops in a parkland system in Burkina Faso. Agrofor Syst 60:13–26

Bayala J, Balesdent J, Marol C, Teklehaimanot Z, Ouedraogo SJ (2006) Relative contribution of trees and crops to soil carbon content in a parkland system in Burkina Faso using variations in natural ^{13}C abundance. Nutr Cycl Agroecosys 76:193–201

Boffa J-M (1999) agrofory parklands in sub-Saharan Africa. FAO conservation guide 34. FAO, Rome

Boffa JM, Taonda SJB, Dickey JB, Knudson DM (2000) Field-scale influence of karite (*Vitellaria paradoxa*) on sorghum production in the Sudan zone of Burkina Faso. Agrofor Syst 49:153–175

Bonkoungou EG, Alexandre DY, Ayuk ET, Depommier D, Morant P, Ouadba JM (1994) Agroforestry parklands of the West African semi-arid lands, Conclusions and recommendations of an international symposium, ICRAF-SALWA, 25–27 Oct. 1993, Ouagadougou, Burkina-Faso pp 18

Brandle JR, Hintz DL, Sturrock JW (1988) Windbreak technology. Elsevier, Amsterdam pp 598

Brandle JR, Hodges L, Zhou XH (2004) Windbreaks in North America. New Vistas in Agroforestry, Advances in Agroforestry 1: 65–78

Breman H, Kessler JJ (1997) The potential benefits of agroforestry in the Sahel and other semi-arid regions. Develop Crop Sci 25:39–47

Brierley JS (1985) The West Indian kitchen gardens: a historical perspective with current insights from Grenada. Food and Nutrition Bulletin (UNU) 7(3):52–60

Cajas-Giron YS, Sinclair FL (2001) Characterization of multistrata silvopastoral systems on seasonally dry pastures in the Carribean Region of Columbia. Agroforest Syst 53: 215–225

Carriere SM, Letourmy P, Mckey DB (2002) Effects of remnant trees in fallows on diversity and structure of forest regrowth in a slash-and-burn agricultural system in southern Cameroon. J Tropical Ecol 18:375–396

Chacón LM, Harvey CA (2006) Live fences and landscape connectivity in a neotropical agricultural landscape. Agrofor Syst 68:15–26

Chauchan SS (2003) Desertification control and management of land degradation in the Thar Desert of India. The Environnementalist 23: 219–227

Cleugh HA (1998) Effects of windbreaks on airflow, microclimates and crop yields. Agroforest Syst 41:55–84

Cleugh HA, Prinsley R, Bird PR, Brooks SJ, Carberry PS, Crawford MC, Jackson TT, Meinke H, Mylius SJ, Nuberg IK, Sudmeyer RA, Wright AJ (2002) The Australian National windbreaks program: Overview and summary of results. Aus J Exper Agric 42:649–664

Crutzen PJ, Andreae MO (1990) Biomass burning in the tropics: Impact on atmospheric chemistry and biogeochemical cycles. Science 250:1669–1678

Droze WH (1977) Trees, prairie and people: A history of tree planting in the Plains States. USDA Forest service and Texas' Woman's University Press, Denton pp 313

Depommier D, Janodet E, Oliver R (1992) Faidherbia albida parks and their influence on soils and crops at Watinoma, Burkina Faso. In: Vandenbeldt RJ (ed) Faidherbia albida in the West African Semi-arid Tropics. ICRISAT, Patancheru-502 324, A.P. pp 111–116

Devendra C, Sevilla CC (2002) Availability and use of feed resources in crop-animal systems in Asia. Agric Syst 71(1–2):59–73

Egler M, Egler CAG, Franz B, de Araujo MSM, de Freitas MAV (2013) Indicators of deforestation in the Southern Brazilian Pre-Amazon. Regional Environ Change 13:263–271
Garrity DP (1999) Contour farming based on natural vegetative strips: expanding the scope for increased crop production on sloping lands in Asia. Environ. Develop Sustain 1:323–336
Gnankambary Z, Ilstedt U, Nyberg G, Hien V, Malmer A (2008) Nitrogen and phosphorus limitation of soil microbial respiration in two agroforestry parklands in the south-Sudanese zone of Burkina Faso: The effects of tree canopy and fertilization. Soil Biol Biochem 40:350–359
Grandstaff TB (1980) Shifting cultivation in northern Thailand: possibilities for development. Resource Systems Theory and Methodology Series No. 3. United Nations University, Tokyo
Hao WM, Liu MH (1994) Spatial and temporal distribution of tropical biomass burning. Global Biogeochem Cy 8(4):495–503
Harvey CA, Villanueva C, Villacís J, Chacón M, Muñoz D, López M, Ibrahim M, Gómez R, Taylor R, Martinez J, Navas A, Saenz J, Sánchez D, Medina A, Vilchez S, Hernández B, Perez A, Ruiz F, López F, Lang I, Sinclair FL (2005) Contribution of live fences to the ecological integrity of agricultural landscapes. Agr Ecosys Environ 111:200–230
Ismail T (1986) Integration of animals in rubber plantations. Agrofor Syst 4:55–66
Kater LJM, Kante S, Budelman A (1992) Karité (*Vitellaria paradoxa*) and néré (*Parkia biglobosa*) associated with crops in South Mali. Agrofor Syst 18:89–105
Kessler JJ (1992) The influence of karité (*Vitellaria paradoxa*) and néré (*Parkia biglobosa*) trees on sorghum production in Burkina Faso. Agrofor Syst 17:97–118
Lefroy FC, Dann PR, Wildin JH, Wesley-Smith RN, McGowan AA (1992) Trees and shrubs as sources of fodder in Australia. Agrofor Syst 20:117–139
Leuschner WA, Khalique K (1987) Homestead agroforestry in Bangladesh. Agrofor Syst 5:139–151
Lovett PN, Haq N (2000) Evidence for anthropic selection of the Sheanut tree (*Vitellaria paradoxa*). Agrofor Syst 48:273–288
Maiers RP, Harrington JT (1999) Windbreak tree establishment in semi-arid agricultural regions of New-Mexico. New Mexico Journal of Science 39: 214–228
Michel GA, Nair VD, Nair PKR (2007) Silvopasture for reducing phosphorus loss from tropical sandy soils. Plant Soil 297:267–276
Nair PKR (1987) Agroforestry systems inventory. Agrofor Syst 5:301–317
Nair PKR (1993) An introduction to agroforestry. Kluwer Academic Publishers, The Hague
Nair PKR, Fernandes ECM, Wambugu PN (1984) Multipurpose leguminous trees and shrubs for agroforestry. Agroforest Syst 2: 145–163
Nair VD, Graetz DA (2004) Agroforestry as an approach to minimizing nutrient loss from heavily fertilized soils: The Florida experience. Agrofor Syst 61:269–279
Neumann I (1983) Use of trees in smallholder agriculture in tropical highlands. In: Lockeretz W (ed.), Environmentally Sound Agriculture, Prager, New York pp. 351–374
Oni PI (1997) *Parkia biglobosa* (Jacq.) Benth in Nigeria: a resource assessment. PhD dissertation, School of Agricultural and Forest Sciences, University of Wales, Bangor pp 213 + appendix
Pullan RA (1974) Farmed parkland in West Africa. Savanna 3(2):119–151
Onyewotu LOZ, Ogingi MA, Stigter CJ (1994) A study of competitive effects between a Eucalyptus camaldulensis shelterbelt and an adjacent millet (Pennisetum typhoides) crop. Agriculture, Ecosystems and Environment 51(3): 281–286
Raebild A, Søndergård Larsen A, Svejgaard Jensen J, Ouedraogo M, De Groote S, Van Damme P, Bayala J, Ousmane Diallo B, Sanou H, Kalinganire A, Dahl Kjaer E (2011) Advances in domestication of indigenous fruit trees in the West African Sahel. New Forests 41(3):297–315
Ramsay DM, Wiersum KF. 1974. Problems of watershed management and development in the Upper Solo river basin. Conference on Ecologic guidelines for forest, land or Water Resources, Institute of Ecology, Bandung
Reynor WC, Fownes JH (1991) Indigenous agroforestry of Pohnpei. Parts 1 and 2. Agrofor Syst 16:139–157
Roothaert RL, Franzel S (2001) Farmers' preferences and use of local fodder trees and shrubs in Kenya. Agrofor Syst 52:239–252
Sanchez PA (1995) Science in agroforestry. Agrofor Syst 30:5–55

Singh D, Kohli RK (1992) Impact of Eucalyptus tereticornis Sm. shelterbelts on crops. Agroforest Syst 20: 253–266
Singh HP, Kohli RK, Batish DR (1998) Effect of Poplar (Populus deltoides) shelterbelt on the growth and yield of wheat in Punjab, India. Agroforest Syst 40: 207–213
Singh HP, Kohli RK, Batish DR (1999) Impact of Populus deltoides and Dalbergia sissoo shelterbelts on wheat - a comparative study. International Tree Crop Journal 10(1): 51–60
Singh RP, Ong CK, Saharan N (1989) Above and below ground interactions in alley cropping in semi-arid India. Agrofor Syst 9:259–274
Singh G, Mutha S, Bala N (2007) Effect of tree density on productivity of a *Prosopis cineraria* agroforestry system in North Western India. J Arid Environ 70(1):152–163
Sirois MC, Margolis HA, Camiré C (1998) Influence of remnant trees on nutrients and fallow biomass in slash and burn agroecosystems in Guinea. Agrofor Syst 40:227–246
Smith DM, Jarvis PG (1998) Physiological and environmental control of transpiration by trees in windbreaks. For Ecol Manag 105(1–3):159–173
Sonwa DJ, Nkongmeneck BA, Weise SF, Tchatat M, Adesina AA, Janssens MJJ (2007) Diversity of plants in cocoa agroforests in the humid forest zone of Southern Cameroon. Biodivers Conserv 16:2385–2400
Smith DM, Jarvis PG, Odongo JCW (1998) Management of windbreaks in the Sahel: the strategic implications of tree water use. Agrofor Syst 40(1):83–96
Spaan W, Bodnár F, Idoe O, de Graaff J (2004) Implementation of contour vegetation barriers under farmer conditions in Burkina Faso and Mali. Quart J Int Agric 43(1):21–38
Styger E, Rakotondramasy HM, Pfeffer MJ, Fernandes ECM, Bates DM (2007) Influence of slash-and-burn farming practices on fallow succession and land degradation in the rainforest region of Madagascar. Agr Ecosys Environ 119(3–4):257–269
Sun D, Dickinson GR (1994) A case study of shelterbelt effect on potato (Solanum tuberosum) yield on the Atherton Tablelands in tropical North Australia. Agroforest Syst 25: 141–151
Sun D, Dickinson GR (1997) Early growth of six native Australian tree species in windbreaks and their effect on potato growth in tropical northern Australia. Forest Ecol Manag 95(1): 21–34
Tejwani KG (1994) Agroforestry in India. Oxford and IBH Publishing Co Pvt Ltd, New Delhi
Thomson LAJ, Turnbull JW, Maslin BR (1994) The utilization of Australian species of Acacia, with particular reference to those of the subtropical dry zone. J Arid Environ 27(3):279–295
Topps JH (1992) Potential, composition and use of legume shrubs and trees as fodders for livestock in the tropics. J Agric Sci 118(1):1–8
Udawatta RP, Anderson SH (2008) CT-measured pore characteristics of surface soils influenced by agroforestry and grass buffers. Geoderma 145(1–3):381–389
Ujah JE, Adeoye KB (1984) Effects of shelterbelts in the Sudan savanna zone of Nigeria on microclimate and yield of millet. Agricultural and Forest Meteorology 33: 99–107
Vergara NT (1981) Integral agroforestry: a potential strategy for stabilizing shifting cultivation and sustaining productivity of the natural environment. Working paper. Environment and Policy Institute, East-West Center, Honolulu
Wezel A, Bender S (2003) Plant species diversity of homegardens of Cuba and its significance for household food supply. Agrofor Syst 57:39–49
Wiersum KF (1982) Tree gardening and taungya in Java: examples of agroforestry techniques in the humid tropics. Agrofor Syst 1:53–70
Wiersum KF (2006) Diversity and change in homegarden cultivation in Indonesia. In: Kumar BM, Nair PKR (eds) Tropical Homegardens: a Time-Tested Example of Sustainable Agroforestry. Springer, The Netherlands, pp 13–24
Wright B, Stuhr K (2002) Windbreaks: an agroforestry practice. USDA Agroforestry Note 25, USDA Forest Service National Agroforestry Center, University of Nebraska, Lincoln, NE. 4
Yuan Y, Bingner RL, Locke MA (2009) A review of effectiveness of vegetative buffers on sediment trapping in agricultural areas. Ecohydrology 2(3):321–336

Chapter 6
Participatory Domestication of New Crops using Agroforestry Techniques

Abstract Tropical forests and savannas provide many tree, shrub, liana and herbaceous species whose products (fruits, kernels, roots, leaves, bark and bark extracts) are commonly used for food, income generation, shelter building, materials, and medicine by local people. These indigenous species most often have multiple uses, and are virtually wild. Some of these multipurpose species have been identified in regional farmers' surveys as priority species for domestication. Participatory domestication, as implemented by the World Agroforestry Centre and its partners in Africa, is farmer-driven and market-oriented, and involves the selection, propagation and integration of these species in farmlands. In Latin America, tree domestication is participatory, and consists of farmer-driven tree selection, testing and adaptation of provenances, seed zone delimitation and transfer guidelines, and accelerating the delivery of high-value germplasm to farmers. In Oceania, tree domestication consists of the selection of a wild genotype or a seedling of a cultivated form, improvement of the plant's environment, and improvement of the crop's population composed of well-established selected seedlings. In Southeast Asia, tree domestication involves tree breeding, exploration and collection of populations of crops enrolled in the domestication program, development of propagation techniques, dissemination, multiplication and assessment of germplasm, facilitation of farmers' access to the market and market information, marketing of tree products, integration of high-value germplasm into land-use systems, dissemination of technical information, and empowerment of farmers with tree domestication techniques.

6.1 Introduction

Participatory domestication of multipurpose forest species with nutritive, medicinal, and commercial values has been an important focus of agroforestry research in the tropics since the 1990s. Drawing heavily on tree domestication activities implemented by the World Agroforestry Centre (or ICRAF) and its local partners in the tropics, in this chapter a framework on how to conduct domestication activities is proposed.

A. Atangana et al., *Tropical Agroforestry,* DOI 10.1007/978-94-007-7723-1_6,

6.1.1 History of the Participatory Domestication of Agroforestry Species

Slash-and-burn agriculture has been identified as one of the main causes of forest area reduction in the tropics. Agroforestry systems such as improved fallows and alley cropping were developed as alternatives for shifting cultivation techniques in the tropics. Despite efforts to implement these new agroforestry systems over the course of a decade, studies showed that tropical forests were becoming increasingly reduced (FAO 1996, 2006; Achard et al. 2002). The humid tropics of Africa, America and Asia registered massive forest losses between 1990 and 2005 (for more details, please see Chap. 2). Most of this tropical forest has been lost as a result of land use changes from forest to agriculture (WCFSD 1999). Moreover, the adoption of improved fallows and alley cropping agroforestry systems popularized in the 1980s was limited in the humid tropic rainforests. Constraints on adoption of these practices are listed in Chap. 4. The decline in global market prices of key cash crops like cocoa and coffee also negatively affected the management of these agroforests. Additionally, most farmers in the forest zones depend on these agroforests for food, income, medicine, and shelter during times of reduced availability from other sources. This factor has contributed to increased pressure on natural forests and forest resources. One of the actions that can be taken to address this urgent problem, according to WCFSD (World Commission on Forests and Sustainable Development), is to provide more extensive support to community-based agroforestry projects in order to reduce the exploitation of primary forests for subsistence products.

Poverty-reduction and forest-protection strategies could be achieved, in part, through the development and cultivation of marketable and under-utilized "new crops" from these forests (Leakey et al. 2005).

Studies have been conducted to identify forest products taking into consideration their use by farmers, their market potential, and opportunities to increase their production on farms. Numerous forest species traditionally overlooked by science offer the potential for several different products useful to farmers and of interest to markets and industry. Leakey and Newton (1994) argued for the need for a 'green revolution of woody plants', referring to the 'Cinderella species', plants widely used by farmers but overlooked by science. Work was subsequently undertaken to develop methods for prioritizing species to be improved (Jaenicke et al. 1995). These methods were based on several criteria: (1) the identification of user groups and their problems and needs, (2) the identification of desired products by these groups, (3) the introduction of technologies to provide products and needed services, and (4) the choice of species to adapt to the selected technology (Fig. 6.1).

A list of priority species for domestication using agroforestry practices has been established for the humid lowlands of Africa (Franzel et al. 1996), the semi-arid Sahel zone of Africa (ICRAF 1996), the Miombo woodlands (Southern Africa;

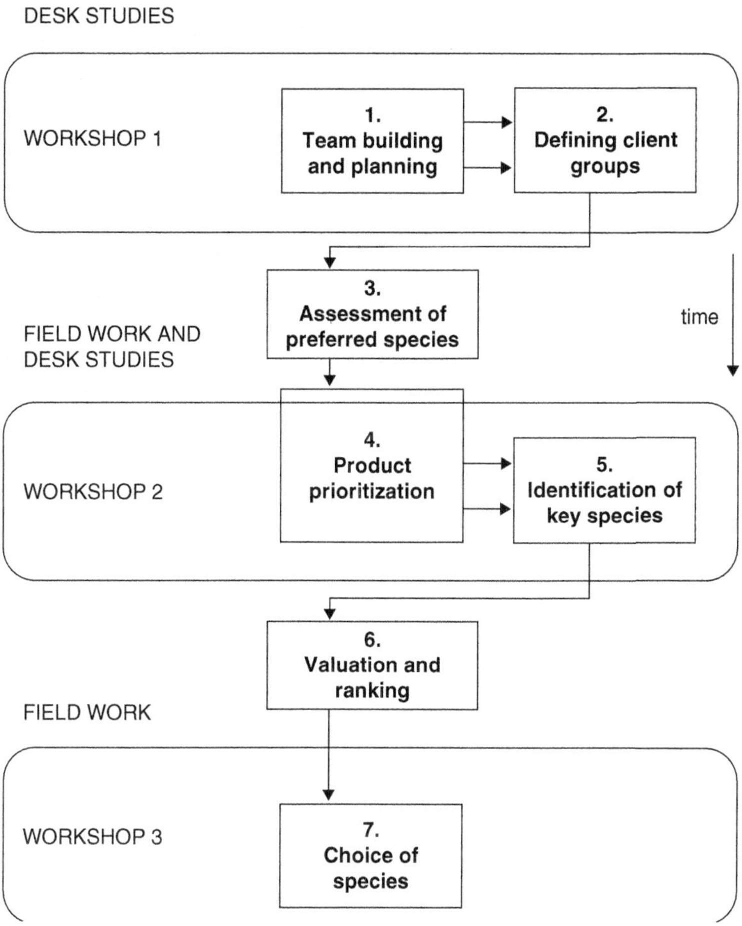

Fig. 6.1 The priority-setting process for indigenous fruit tree species. (Franzel et al. 2008)

Maghembe et al. 1998), and Latin America (Sotelo-Montes and Weber 1997). The domestication of indigenous trees producing products with high nutritive, medicinal and commercial values was used as a means to diversify and intensify agricultural ecosystems and create mature and diverse agro-ecosystems. Consequently, new initiatives in tropical forest tree improvement aiming at developing cultivars of trees with desired fruit, nut, and medicinal characteristics (Leakey and Newton 1994; Franzel et al. 1996; Simons and Leakey 2004) are being developed in Latin America, sub-Saharan Africa, and Southeast Asia. The goals of tree domestication are to help reduce deforestation as well as the integration of trees providing products of high nutritional and market values into farmland in order to increase and diversify income, contribute to poverty alleviation and food security, and improve human health in rural areas.

6.1.2 Participatory Domestication as Implemented by ICRAF and its Partners

Tree domestication can be simply defined as the selection and management of trees to increase their benefits to humankind. Benefits include timber and non-timber forest products, agroforestry tree products (i.e., products from trees cultivated on farms), including sap, gums, fruits, nuts, and ecological services such as soil improvement, shade, erosion control, and carbon sequestration. Domestication activities can include ethnobotany, market studies, conservation genetics, breeding programs, vegetative propagation, tree improvement, sustainable agriculture, interaction with environment, commercial adoption, assessment of the social environment, and economic impact (Leakey and Newton 1994). The priority-setting process for the domestication of indigenous fruit tree species is shown in Fig. 6.1. Five approaches can be used to domesticate forest trees (http://www.camcore.org/tree-domestication/):

- Eliminate undesirable individuals in natural populations
- Collect and plant seeds of individuals with desirable characteristics
- Vegetative propagation of individuals with desirable characteristics
- Direct pollination of individuals with desirable characteristics
- Conducting large-scale planned trials to evaluate complex traits

These practices are intended to increase the number of individuals with desirable characteristics in a population, but the rise in frequency of the occurrence of desirable traits is only observable after several generations. An example is the domestication of safou or African plum (*Dacryodes edulis* H.J. Lam, Burseraceae) in Cameroon and bush mango (*Irvingia gabonensis* (Aubry-Lecomte ex.O'Rorke) Baill., Irvingiaceae) in Nigeria. Farmers have grown, by direct seeding, trees with desired characteristics over several generations (Leakey et al. 2004). This has led to a greater observed frequency of *D. edulis* with large fruits in Cameroon (where safou is grown by farmers) than in Nigeria, where the species being grown is virtually wild. For *I. gabonensis*, it was just the opposite; it is semi-domesticated in Nigeria, but grows virtually wild in Cameroon. In Africa, there are around 3,000 wild tree species that could contribute to the health and welfare of rural communities. Through the participatory tree domestication program managed by the ICRAF in West and Central Africa, farmers and scientists have worked together to develop superior varieties of African plum, bush mango, kola nut (*Garcinia kola* Heckel, Clusiaceae) and several other species that are being planted on farmers' agricultural lands, thereby raising incomes, improving health and stimulating the rural economy (Pye-Smith 2010).

The domestication of agroforestry species as conceived and implemented by ICRAF is an iterative process involving the identification, production, management and adoption of high quality germplasm (Simons 2003). Tree domestication as executed by ICRAF is a farmer-driven and market-led process, which matches locally important trees to the needs of subsistence farmers, product markets, and agricultural environments (Simons and Leakey 2004). This process is based on the selection and vegetative propagation of individuals with desired characteristics. The process identifies priority species for domestication and traits to be improved using Participatory Rural Appraisal (PRA) tools (Franzel et al. 1996), develops descriptors of variation for traits of interest (Leakey et al. 2000), characterizes the intraspecific variation of desirable traits (Atangana et al. 2001, 2011; Wahuriu 1999; Leakey et al. 2005), aids in the selection of 'plus-trees' and the vegetative propagation of selected individuals using simple, robust and low-technology methods (Leakey et al. 1990; Atangana et al. 2006; Atangana and Khasa 2008), and integrates the new germplasm into existing land use systems (Tchoundjeu et al. 2006). Farmers are trained in vegetative propagation techniques (Leakey and Simons 1998; Tchoundjeu et al. 2006), and helped in the creation of rural nurseries while local resource centers supply the germplasm. ICRAF promotes diversity in mixed agricultural systems. It does not support the large-scale monocultural production of high-value trees, which is harmful to agroforestry systems. As farmers are encouraged to build entrepreneurial skills and get trained in agroforestry product processing, their access to markets is facilitated.

Tree domestication is participatory because farmers, as the first user of agroforestry products, are involved throughout the process, and market needs are considered. Characteristics that concern farmers in priority species for domestication include the reduction of the period before the plant begins to fruit, increased sizes of fruits and seeds, improved fruit flavors (juicier and less fiber content), and improved resistance to pests. Emphasis is placed on the seasonality of agroforestry products to allow year round production. Ideally, there should be a range of species whose products are harvested at different times during the year. Therefore, product diversification for various uses is also desirable.

6.1.2.1 Characterization of Phenotypic Variation and Tree Selection

Most candidate species for domestication are wild. Therefore, the first step towards the domestication of a forest tree species is the assessment of the phenotypic diversity of traits in the wild, focusing on those of interest for improvement (Atangana et al. 2011). The assessment involves the quantification of the variation between sites, as well as between and within trees, and is oriented towards farmers' preferences. The objective is to select 'plus-trees,' trees whose products are an improvement over the population mean. If fruit is the product of interest, phenotypic characterization of fruit should be carried out by sampling at least 24 fruits per tree (Leakey et al. 2000) from at least 30 trees per site, and measuring parameters of interest using robust, simple and easily transportable equipment (Leakey et al. 2000; Atangana et al. 2001). The equipment required to measure fruit parameters in the wild consists of an electronic balance, calipers, collection bags, a measuring tape (to measure tree diameter), and a GPS unit (to record tree locations). The use of small and transportable electronic scales is recommended for the estimation of fruit mass in the wild, and calipers graduated to 0.1 mm are recommended for the measurement of fruit size.

Fig. 6.2 Single node leafy stem cuttings in a non-mist poly propagator; this photo shows part of the experiment that investigated the amenability of *Allanblackia floribunda* to vegetative propagation (Atangana et al. 2006)

A distribution map of surveyed trees should be compiled to assist with any future data collection and to further tree domestication. Fruit collection should not occur without a prior meeting with the local population, and some farmers may serve as guides during fruit sampling. Chemical analyses of the fruits and seeds are sometimes needed in the characterization process. Characterization of fatty acid profiles in nuts and other fruit elements can be made using gas chromatography methods (Atangana et al. 2011), visco-analysis (Leakey et al. 2005a), or proteolysis (Giami et al. 1994). Data collected can be used for the assessment of variation between sites, trees, and fruits, preferably using a nested (hierarchical) design (Anegbeh et al. 2003; Leakey et al. 2005a, b; Atangana et al. 2011). For details on experimental designs in tropical agroforestry, refer to Chap. 14.

6.1.2.2 Vegetative Propagation of Agroforestry Species

An important element of the domestication of agroforestry species as implemented by ICRAF and its partners is the vegetative propagation of selected individuals. Vegetative propagation has several advantages over sexual reproduction of wild indigenous species as regards to tree domestication. Most often, there is a limited availability of seeds from desirable candidates for tree domestication; also, when seeds are available, the germination success is sometimes very poor, owing to seed dormancy or recalcitrant seeds. Further, vegetative propagation of trees better captures desirable characteristics (i.e., fruit size, fruit flesh sweetness, nut size, etc.) for tree improvement, and allows early fruiting. One of the methods used in vegetative propagation is rooting of cuttings. Here, propagation is accomplished by taking cuttings and rooting them using non-mist (i.e., without spraying mist), simple, and robust propagators (Fig. 6.2; Tchoundjeu 1989; Leakey et al. 1990). Leafy uninodal cuttings are taken from young seedlings or coppicing stumps (Tchoundjeu et al. 2004; Ngo Mpeck and Atangana 2007; Atangana and Khasa 2008; Atangana et al. 2006; Ngo Mpeck et al. 2009).

The non-mist poly propagators used for rooting cuttings can be constructed following a design of Howland (1975), modified by Leakey et al. (1990). These

Fig. 6.3 Different substrate layers of a non-mist poly propagator (Source: ICRAF-Cameroon)

propagators consist of a wooden frame approximately 3 × 1 × 1 m in size, enclosed in a clear polythene sheet so that the base of the propagator is watertight (Fig. 6.3). Access to the propagator is provided by a fitted lid, which is also covered with a polythene sheet and is airtight. The base of the propagator is covered with a thin layer of fine river sand to prevent the polythene from being punctured by the stones, which is followed by successive layers of small stones and gravel, and the rooting medium to a depth of 10 cm on top of the gravel (Fig. 6.3). Relative humidity in the propagator is kept high, close to the values found in tropical rainforests, in order to maintain high water content in the cuttings, and to mimic the natural environment of species to be rooted (Newton and Jones 1993).

Rooting of cuttings is influenced by the microclimate of the propagation medium (Mesén et al. 1997). The development of a practical rooting protocol for leafy stem cuttings of a previously unstudied species involves the identification of factors influencing rooting of cuttings. Factors affecting rooting of leafy stem cuttings include the propagation environment (i.e., within the propagator), auxin application, leaf area supported on the cutting (e.g., 25, 50 or 75 cm^2), cutting length and

Fig. 6.4 Rooted single node leafy stem cuttings of *Allanblackia floribunda* (Atangana et al. 2006)

diameter, origin and stockplant management. Genetic origin can also influence the morphology and physiology of the cuttings (Dick et al. 2004; Leakey 2004). Of these factors, auxin application on cuttings prior to insertion into the propagators was reported as the most important for the success of the propagation (Hartmann et al. 2002; Leakey 2004). Other factors that greatly impact rooting are trimming of leaf area to optimize for the cutting size, and the determination of optimal cutting length (Tchoundjeu and Leakey 1996). The most widely used auxins are indole-3-acetic acid (IAA), indole-3-butyric acid (IBA), 4-chloro-indole-3 acetic acid (4-Cl-IAA), and phenylacetic acid (PAA; Slovin et al. 1999; Hartmann et al. 2002; Leakey 2004). IBA most frequently yielded the best results for promoting the rooting of cuttings (Nordström et al. 1991; Leakey 2004). Some species, such as *Allanblackia floribunda* Oliv. (Clusiaceae) (Fig. 6.4) and *Baillonella toxisperma* Pierre (Sapotaceae), are insensitive to the application of hormones (Atangana et al. 2006; Ngo Mpeck and Atangana 2007). When the rooting of stem cuttings is responsive to hormone application, it is necessary to determine the concentration required to apply to the base of the cutting, and the hormone can either be applied in a powder form (e.g., Seradix) or in the liquid form using a syringe.

Other factors influencing the rooting of leafy stem cuttings are the type of rooting medium used, node position and the physiological and ontological aging of the cutting (Leakey 2004). Several factors may simultaneously affect the rooting of leafy stem cuttings (Dick and Dewar 1992; Atangana et al. 2006).

The rooting period varies among species, and can range from 2 to 3 weeks to more than 10 weeks (Atangana et al. 2006; Ngo Mpeck and Atangana 2007; Atangana and Khasa 2008). A non-mist poly-propagator used for the mass propagation of plants may contain up to 300 cuttings per batch. Rooting of cuttings using poly-propagators requires the following material:

- Non-mist poly propagators, similar to the one designed by Leakey et al. (1990)
- Young plants or coppicing stumps. Cuttings from seedlings and coppicing stumps root better than cuttings from mature trees
- A knapsack sprayer to mist cuttings with water when necessary

- Sharp scissors (to cut the cuttings)
- Watering cans
- Syringes (if auxins are to be applied as in the liquid form to the base of the cutting)
- Polythene bags for potting the rooted cuttings
- A nursery bed

Daily monitoring of cuttings in propagators is required. At least once a day, preferably early in the morning, the propagators should be inspected, during which the water level in the water table within the propagator is checked, and the propagator is cleaned of dew on the inner walls. Knowledge of rooting protocol (which substrate to use, what leaf area level per cutting yields the best result and how to apply hormones) is necessary prior to cutting, and insertion of the cuttings into the propagators. If the species is easy to root, an assessment of rooting success should be done weekly after the first two weeks by lifting the cuttings out of the rooting medium. Usually, a cutting is considered to be rooted when it has at least one root 10 mm or longer (Fig. 6.4). Rooted cuttings are potted in polythene bags filled with soil taken from the base of the coppicing stumps that provided the cuttings, and left for rehabilitation in a propagator that is regularly left open. The opening time of the rehabilitation propagator increases gradually from a couple of hours a day to a couple of days a week. This serves to gradually accustom the cuttings to the environment outside of the propagator. After an acclimatization (rehabilitation) period, the young plants are moved to the nursery, and watered when necessary, before being transferred to a farm.

Marcotting and grafting are also used in tree domestication to capture traits of interest in mature trees. Marcotting or air layering is done on stems of trees with desirable characteristics for domestication, and consists of removing stem bark around the portion of the stem to be rooted using a sharp knife. The width of the stem portion on which bark is removed may vary from 2–6 cm. Then, a rooting medium consisting of moist soil taken from the base of the tree on which marcotting is placed around the debarked stem and wrapped with a transparent plastic sheet so as to easily see later if roots have developed. Marcotting is done on mature trees with known fruit/nut characteristics, whereas rooting of cuttings yields good rooting percentage (i.e., more than 90 % rooting success) when cuttings are taken from young seedlings or coppicing stumps. However, the number of marcot plants obtained from a given tree is limited, as a maximum of ten marcots is set per tree, and because not all the marcots set develop roots; also, not all the young marcot plants survive in nursery (Tchoundjeu et al. 2002, 2006). Marcotting is also used as an alternative to rooting of leafy stem cutting for species with poor rooting success of leafy stem cuttings, as was the case of *D. edulis* (Tchoundjeu et al. 2002).

Grafting is an old horticultural technique, which consists of inserting tissues (i.e., the graft) from one plant (which most often has proven desirable characteristics for fruit and nut production, i.e., fruit size, flesh fruit sweetness, number of fruits, nut size, etc.) into those of another plant (the rootstock, which has proven desirable characteristics for rusticity) so that the two sets of tissues may join together. Grafting has the advantage of massively reproducing trees that have proven desirable characteristics for domestication, as many grafts can be taken from mature trees. However, incompatibility between graft and rootstock may occur. Side-tongue grafting (Hartmann et al. 2002) approach has been effective for the vegetative propagation of *A. floribunda* (80 % successful union between grafts and rootstock by nine weeks after grafting; Asaah et al. 2011), a reputed hard-to-root species (68.7 % rooting success of leafy stem cuttings; Atangana and Khasa 2008).

6.1.2.3 Germplasm Management and Integration in Land use Systems

The main objective in germplasm management is to identify a place suitable for seedling planting. Usually, farmers choose the site of planting, which can be a homegarden, perennial crop farm (coffee, cocoa, or rubber farms), parkland, or food crop farm/fallow. The development of diverse and complex agro-ecosystems, where several species of domesticated trees are integrated, can constitute an interesting alternative to shifting cultivation (Leakey and Tchoundjeu 2001). This is the case because farmlands cultivated with high value trees are permanent systems no longer subjected to slash-and-burn practices. Another option is the introduction of high-value indigenous tree species into community forests. The number of community forests in the Congo Basin is increasing; 200 community forests with a valid license exist in Cameroon in 2011, and more and more management committees are enriching their community forests with improved germplasm of indigenous trees.

Plant management mostly consists of techniques and practices intended to facilitate plant growth, including weeding, mulching, pruning, and pest management practices.

6.1.2.4 Genetic Resources Conservation

A key point in the domestication of agroforestry species is the conservation of genetic resources during tree improvement. Preserving genetic resources in a species should be based on a detailed inventory of the genetic diversity of that species. This inventory is done by examining species undergoing breeding in common gardens. Though numerous species are under domestication in the humid tropics of West and Central Africa, preservation efforts of genetic resources are restricted to the establishment of gene banks in Cameroon and Nigeria for only *I. gabonensis*. However, collections of germplasm of safou and *Ricinodendron heudelotii* (Baill.) Pierre ex Pax (Euphorbiaceae) have been established at the Minkoameyos research station, in Nkolbisson, Yaoundé. Also, provenance trials of *Adansonia digitata* were established in 1989 at two sites, Gonsé and Djibo in Burkina Faso, and included accessions from Burkina Faso, Côte-d'Ivoire, Senegal, and Kenya (Raebild et al. 2010). Only the site at Djibo is still operational today. Other collections of *A. digitata* provenances began in 2006 and 2007 with material from Burkina Faso, Mali and Niger (12 provenances selected), Benin, Senegal and Togo (5 provenances), Kenya,

Malawi, Mozambique, Tanzania, and Sudan (12 provenances; Raebild et al. 2010). *Parkia biglobosa* seed sources were collected in West Africa (Teklehaimanot 1997), and two trials in Burkina Faso, as well as one in Nigeria, were established. In the same period, seeds of *P. biglobosa* were collected from Mali and Burkina Faso, and trials established in these two countries (Raebild et al. 2010). Nine provenances of *Tamarindus indica* were collected, and established in Burkina Faso, Mali and Niger (Raebild et al. 2010). *Vitellaria paradoxa* seeds of five provenances from Burkina Faso, Mali and Senegal were collected from trees selected on the basis of desirable characteristics, and established in Burkina Faso (Bayala et al. 2009).

In Southern and Eastern Africa, seeds of *Uapaca kirkiana* Müll. Arg. (Phyllanthaceae) were collected from 26 provenances from Malawi, Tanzania, Zambia, Mozambique and Zimbabwe, and *Sclerocarya birrea* from the same areas, as well as from Mali, Namibia and Kenya. Multi-location trials were established with 12–16 provenances per country (Kwesiga et al. 2000; Akinnifesi et al. 2004b).

6.1.3 Priority Species for Domestication

The first step in tree domestication consists in conducting a priority-setting process for identification of priority species and desirable characters for tree improvement (Franzel et al. 1996, 2008). Priority species for domestication are listed by geographic area. Regional surveys of agroforestry species prioritization have been documented for the Sahel, southern Africa and West Africa (Jaenicke et al. 1995; Sigaud et al. 1998; Maghembe et al. 1998), and for individual countries including Bangladesh, Brazil, Ghana, India, Indonesia, Peru, Philippines, and Sri Lanka (Sotelo and Weber 1997; Lovett and Haq 2000).

6.1.3.1 Examples of Tree Species Under Domestication in the Humid Lowlands of West and Central Africa

Priority species for domestication in the humid lowlands of Africa include *I. gabonensis*, *Irvingia wombolu* Vermoesen, *D. edulis*, *Chrysophyllum albidum* G. Don (Sapotaceae), *Ricinodendron heudelotii*, *Garcinia kola*, and *Cola* spp. and *Coula edulis* Baill. (Olacaceae; Franzel et al. 1996). Among those species, *Irvingia* spp, *D. edulis*, *R. heudelotii*, *C. albidum*, and *Coula edulis* are a food source, providing either fruits, nuts or seeds used as soup-thickeners, whereas *G. kola*, *Cola acuminata* Schott & Endl. (Malvaceae) and *Cola anomala* K. Schum. (Malvaceae) are sources of various stimulants. Other priority species include *Prunus africana* (Hook.f.) Kalkman (Rosaceae), *Pausinystalia yohimbe* (K.Schum.) Pierre ex Beille (Rubiaceae), and *Annickia chlorantha* (Oliv.) Setten & Maas (Annonaceae), whose barks and seeds are used for medicinal purposes.

However, the domestication of agroforestry species and priority-setting in particular is a dynamic process. For example, *A. floribunda* and Eru or Okok (*Gnetum africanum* Welw. *Gnetum bucholzianum* Engl., Gnetaceae) were not identified as top priority species for domestication in the humid lowlands of West and Central Africa when the domestication program started in the mid-1990s (Franzel et al. 1996), but were added later because of their importance for local people and for trade. *Allanblackia floribunda*, has seeds that are very rich in fats, consisting mainly of stearic and oleic acids (Atangana et al. 2011), and used in the food and cosmetic industry. *Gnetum africanum* and *G. buccholzianum,* are vines whose leaves are commonly consumed as vegetables in West and Central Africa, and extensively traded.

Tree domestication activities in the humid lowlands of West and Central Africa are underway in Cameroon, Nigeria, Democratic Republic of Congo (DRC), Equatorial Guinea and Gabon. Priority species for domestication vary between these countries. *I. gabonensis* is highly valued in Gabon, Equatorial Guinea, Cameroon and DRC, whereas *I. wombolu* is preferred in Nigeria and Southwest Cameroon. *Chrysophyllum albidum* is highly valued in Nigeria, while *G. africanum* is widely consumed and traded in Cameroon, DRC and Nigeria. The following section describes domestication processes for a number of priority species in the humid lowlands of West and Central Africa.

Irvingia gabonensis/Irvingia wombolu (Bush mango)

The genus *Irvingia* comprises seven species, of which six occur in tropical Africa and one in South-East Asia. *I. gabonensis* and *I. wombolu* are among the six species found in tropical Africa. Locally known as 'ogbono', 'ugiri' (Nigeria), 'bush mango', 'andok' or 'dika nut', these Irvingiaceae were often taken to be the same species, before varietal or taxonomical delineation was made by Okafor (1974), and subsequently revised by Harris (1996). Random amplified polymorphic DNA (RAPD) analyses indicated that the two species are genetically distinct with significant genetic integrity in the two bush mango species (Lowe et al. 2000). Both species produce fruits in the shape of a small mango (Fig. 6.5). The flesh from *I. gabonensis* fruit is edible and an excellent source of vitamin C (Leakey 1999). The *I. wombolu* fruit is not edible, as its flesh is very fibrous and bitter. *I. gabonensis* grows naturally from Nigeria to Congo, while *I. wombolu* is naturally distributed from Senegal to Uganda (Harris 1996). The seeds extracted from fruits of both species are used as a food-thickening agent. The food-thickening properties and fat and protein contents of the Dika nut were assessed by Leakey et al. (2005a). Important chemical food properties, such as fat content, viscosity, and drawability, of the bush mango, were found to vary per tree, allowing the selection of 'ideotypes' for domestication (Leakey et al. 2005a). Kernel fat content was not found to vary with viscosity or drawability (Leakey et al. 2005a). Various methods exist for the preservation of bush mango kernels (Tchoundjeu et al. 2005), the most common involving roasting and grinding the kernels, and leaving the resulting cake to dry for a few hours. Once solid, the cake is ready to use. The oil that drips from the cake is used for cooking (Tchoundjeu et al. 2005). Another method consists of sun drying the kernels for one to two days, and keeping them in a container. In Cameroon, the trade

Fig. 6.5 *Irvingia gabonensis* fruits (Source: Ann Degrande)

in these bush mango kernels to Gabon, Nigeria, Equatorial Guinea and the Central African Republic has been valued at US$260,000 per annum (Ndoye et al. 1997).

Tree domestication of *I. gabonensis* began with studies on vegetative propagation of the species using the rooting of cuttings (Shiembo et al. 1996), followed by the characterization of phenotypic variation (Leakey et al. 2000; Atangana et al. 2001). 'Plus trees' were selected for domestication (Atangana et al. 2002), and studies undertaken on marcotting (air-layering) and grafting in Onne, Nigeria, and Yaoundé, Cameroon, by the ICRAF. Germplasm from marcots and grafted plants was established in these localities. Preliminary assessments of these germplasm collections indicated that propagated plants bore fruit 3–4 years after planting. Using Random Amplified Polymorphic DNA (RAPDs), Lowe et al. (2000) assessed the genetic variation in *I. gabonensis* and *I. wombolu*. This study revealed significant genetic integrity in the two species, and identified 'hot spots' of genetic diversity clustered in southern Nigeria and southern Cameroon for *I. wombolu*, and in southern Nigeria, southern Cameroon, and central Gabon for *I. gabonensis*. Continuing *I. gabonensis* tree domestication steps include the establishment of multi-location provenance trials, assessment of the genetic control of the traits of interest for domestication, selection of 'elite trees' for improvement, and a description of the mating system in the species. Provenance trials in agroforestry systems should also be designed so that interactions between bush mango and main crops in the region could be investigated.

Dacryodes edulis (Safou or African Plum)

Locally known as safou, *D. edulis* (Burseraceae) is an oleiferous fruit tree whose natural distribution range spans from Central Africa to Sierra Leone, Uganda and Angola (Troupin 1950). The species is thought to be indigenous to the Gulf of

Guinea, specifically from Southern Nigeria and Cameroon (Vivien and Faure 1996). The flesh of safou fruits is part of the staple diet in the Congo Basin. Fruits are eaten roasted or boiled often with cassava, plantain or bread, and are also a source of edible oil (Fonteh 1998). Safou is a dioecious species with an allogamous reproductive system (Kengue et al. 2002), and is insect-pollinated. The pulp of the safou is rich in amino acids and ascorbic acids (Omoti and Okiy 1987; Achinewhu 1983). Flesh from safou fruit is rich in fat, and its properties were reviewed by Leakey (1999). The fatty acid profiles of safou are similar to that of *Elaeis guineensis* Jacq. nuts (Leakey 1999).

Dacryodes edulis fruits are widely traded in the Western and Central Africa, and exported to Europe (Tabuna 1999). The annual value of the safou trade in Cameroon was estimated at over US$ 7 million (Awono et al. 2002), and the trade value of safou in Europe and North America was estimated at another US$ 2.2 million (Awono et al. 2002). *D. edulis* is widely distributed within the Gulf of Guinea, where it is thought to originate. The species is cultivated in Cameroon, where farmers have selectively bred the plant for generations in order to improve the fruit characteristics (Leakey et al. 2004). Phenotypic variation in safou was done by Leakey et al. (2002), and a germplasm collection from marcots was established at Minkoameyos, Yaoundé (Cameroon) by ICRAF.

Germplasm collection was carried out with the help of local farmers to identify and collect seeds and marcots from superior trees (i.e., trees that produce fruits with desired characteristics for improvement). Collection efforts were centered in four sub-regions from the Central and Western provinces of Cameroon (Tchoundjeu et al. 2002). Marcots were collected from these villages (10 marcots per site) in 1998 and planted in demonstration plots in eight pilot villages in Cameroon and two in Southeast Nigeria (Tchoundjeu et al. 2002). In 1995, provenance trials involving 20 accessions from the humid lowlands and western highlands of Cameroon were established in Barombi-Kang and Minkoameyos research stations in Cameroon by IRAD (Institut de Recherche Agricole pour le Développement). Kengue and Singa (1998) reported variations between accessions and between trees in these collections. A range-wide seed collection was carried out with farmers in Cameroon, Equatorial Guinea, Gabon and Nigeria. Farmers from each selected village identified twenty to thirty trees with desirable fruit characteristics for the establishment of live gene banks in these countries (Tchoundjeu et al. 2002). Village nurseries were established for the propagation of selected 'plus trees' (Tchoundjeu et al. 2002), and research on post-harvest processing, market development and integration of high-value germplasm is underway.

Future goals in *D. edulis* domestication might include the assessment of heritability values in fruit traits, fine-tuning post harvesting methods, and an investigation on pest management, as syrphid flies (Syrphidae) can cause the abortion of flowers in the species. Interactions between safou and crops should also be investigated, to determine optimal spacing for safou-based agroforestry systems.

Fig. 6.6 Ndjansang kernels (Source: Ann Degrande)

Ricinodendron heudelotii (Baill.) Pierre ex Pax (ndjansang)

Ricinodendron heudelotii is commonly called 'ndjansang' in the Congo Basin. Kernels extracted from *R. heudelotii* fruits (Fig. 6.6) are used as condiments and the oil extracted from these seeds is used in cooking. Ndjansang is a large, deciduous, dioecious fast-growing tree, usually averaging 20–30 m in height, but can reach up to 50 m. The species is endemic to tropical Africa (Good 1964), and occurs naturally in Senegal, Guinea Conakry, Sierra Leone, Liberia, Ivory Coast, Ghana, Benin, Nigeria, Fernando Po, Cameroon, Central African Republic, Sudan, Gabon, Congo, Democratic Republic of the Congo, Uganda, Kenya, Angola, Zambia, Tanzania and Madagascar (Fig. 6.7). Fruiting occurs in September and October. The fruit is a somewhat plum-like, indehiscent yellow-green capsule, approximately 3.5–5 cm long and 2.5–4 cm wide (Ngo Mpeck et al. 2003). It is generally spherical, with 1, 2 or 3 seeded lobes (Fondoun et al. 1999; Ngo Mpeck et al. 2003). These seeds are also widely traded in the Central African sub region and Europe (Tabuna 1999). The trade value of *R. heudelotii* kernels in New-Bell market, Douala (Cameroon) was estimated at US$248,700 in 1998, and US$ 464,235 in 1999 (Ngono and Ndoye 2004).

The paste of the ground kernels is sometimes used as a thickening agent for soups, and incorporated in to baby cereals and cakes, due to their high oil retention capacity (Leakey 1999). The physicochemical properties of the dry kernels and the defatted oil extracted from the kernels were reviewed by Tchoundjeu and Atangana (2006). These kernels are rich in oils (47.4–55.3 % fatty acid), crude protein, and fibres. The fat content of ndjansang kernels vary with the geographic origin of the tree (Dandjouma et al. 2000; Tiki Manga et al. 2000).

Domestication activities in ndjansang have been restricted to the assessment of phenotypic variation in fruit and seed traits (Ngo Mpeck et al. 2003), and the vegetative propagation of the species using rooting of single-node leafy stem cuttings

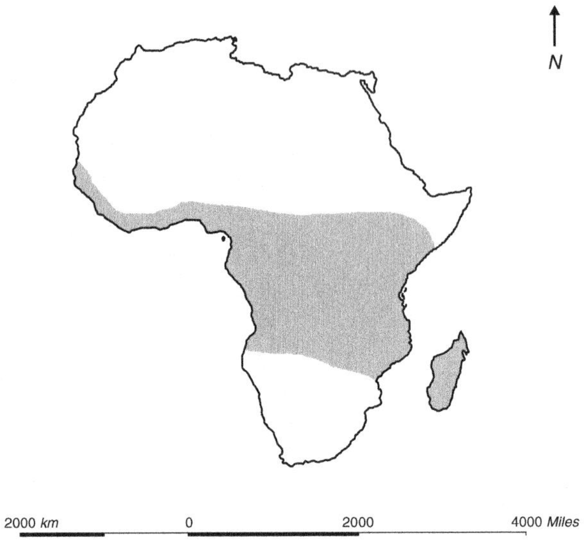

Fig. 6.7 A map of the distribution of Ndjanssang in Africa (Source: Tchoundjeu and Atangana 2006)

(Shiembo et al. 1997). On-station experiments on vegetation propagation are underway. No germplasm collection has been made yet, although trees are being cultivated at Minkoameyos research station, Yaoundé. Major constraints in ndjansang tree domestication include the difficulty in distinguishing between male and female trees, a lack of characterization data on a wide range of genetic resources, a lack of knowledge on flowering phenology and pollination mechanisms, and high seed dormancy (Tchoundjeu and Atangana 2006). The germination rate of seeds is less than 4 %; however, this can be overcome by hand scarification (Mapongmetsem et al. 1999). Interactions between ndjansang trees and the main cultivated food and cash crops in the Congo Basin also need to be investigated.

Chrysophyllum albidum (African star apple)

African or white star apple (*Chrysophyllum albidum*) is one of the 150 edible woody species covering 103 genera and 48 families in the Nigerian forest zone (Okafor 1980). African star apple fruits (Fig. 6.8) are widely consumed in Nigeria, and are a good source of Vitamin C (Achinewhu 1983). The juice from the fruit can potentially be used in the manufacture of soft drinks and wine (Ajewole and Adeyeye 1991). The nutritional value and mineral contents of the African star apple were reported by Nwadinigwe (1982). The moisture, ash, crude fiber, protein, sugar, starch, oil and ascorbic acid contents of African star apple are given in Table 6.1. The mineral elements in the fruit portions of African star apple are shown in Table 6.2. The major fatty acids of the seed oil are oleic, linoleic and palmitic, and unsaturated acids constitute 74 % of the fatty acid content (Ajewole and Adeyeye 1991).

Fig. 6.8 *Chrysophyllum albidum* fruits. (Source: ICRAF-Cameroon)

Fig. 6.9 *Garcinia kola* fruits (Source: Ann Degrande)

Very little has been done for the domestication of this species, despite its importance. Information is needed on the plant's reproductive biology, genetic variation in fruit and seed traits, propagation, and germplasm management.

Garcinia kola (Bitter cola)

The seeds of the *Garcinia kola* plant (Fig. 6.9) are eaten as a stimulant in sub-Saharan Africa and may have potential use in the brewing industry as an alternative to hops (Aniche and Uwakwe 1990; Ogu and Agu 1995). Owing to the bitterness of the plant, *G. kola* is locally called 'Bitter cola', and it is widely traded. In South Cameroon, Bitter kola can generate up to US$ 1,167.6 annual revenue per household (Fondoun and Tiki-Manga 2000). *Garcinia kola* sticks are also chewed for dental care in West and Central Africa. Chewing bitter kola relieves coughs, hoarseness,

Table 6.1 Percentage/unit weight of moisture, ash, crude fiber, oil, protein, starch, total sugar, and content of ascorbic acid in fresh *C. albidum* fruit (Nwadinigwe 1982)

Parameter		Fresh fruits, sample A							
		Determination (percentage/unit weight)							
		1	2	3	4	5	6	7	Average
Moisture	Peel	55.89	55.87	55.87	55.87	55.85	55.86	55.88	55.87
	Pulp	72.55	72.12	72.34	72.34	72.55	72.12	72.34	72.34
Ash	Peel	2.16	2.20	2.17	2.18	2.20	2.16	2.17	2.18
	Pulp	2.85	2.65	2.71	2.74	2.85	2.71	2.65	2.74
Crude fiber	Peel	13.13	13.20	13.11	13.15	13.19	13.11	13.15	13.15
	Pulp	14.92	14.80	14.96	14.89	14.80	14.96	14.92	14.89
Oil	Peel	6.05	5.23	5.91	5.73	6.05	5.23	5.91	5.73
	Pulp	28.73	27.72	28.66	28.37	28.66	28.73	27.72	28.37
Protein	Peel	8.76	8.33	9.27	8.79	8.79	8.77	8.81	8.79
	Pulp	11.60	11.38	11.92	11.63	11.38	11.61	11.91	11.63
Starch	Pulp	14.75	14.80	14.62	14.72	14.75	14.72	14.69	14.72
Total sugar	Pulp	19.13	19.14	19.24	19.17	19.23	18.17	19.25	19.19
Ascorbic acid (mg 100 ml^{-1})	Pulp	96.39	96.39	96.40	96.39	96.38	96.38	96.41	96.39

Table 6.2 Mineral elements in the various parts of *C. albidum* fruit (Nwadinigwe 1982)

	Amounts in various parts of the fruit (100 g^{-1})					
	In raw (fresh) fruit			In ash		
Elements detected	Peel	Pulp	Seed	Peel	Pulp	Seed
Al	0.3350	0.1918	0.1606	16.19	18.62	19.83
Ca	0.0540	0.0269	0.0157	2.62	2.61	1.94
Fe	0.0044	0.0020	0.0028	0.21	0.19	0.34
K	0.0035	0.0002	0.0009	0.17	0.02	0.11
Mg	0.0654	0.0364	0.0565	3.16	3.53	6.97
P	0.0697	0.0446	0.0294	3.37	4.33	3.63
Zn	0.0064	0.0034	0.0034	3.11	3.27	3.27
S	0.0100	0.0083	0.0045	4.81	8.02	5.61
Mn[a]						

[a] Not quantified

bronchial and throat troubles. The fruit pulp is edible and rich in nutrients (Leakey 1999). Bitter Kola has been identified as a potent antibiotic and aphrodisiac, which could be effective in the treatment of many diseases including malaria, dysentery, osteoarthritis, and as an antidote against poisoning. *Garcinia kola* is native to the evergreen, wet, and moist semi-deciduous forest zones of West and Central Africa.

Domestication of this species is still in its early stages. Air-layering (marcotting) and rooting of single-node leafy stem cuttings of *G. kola* have proved difficult, and

Fig. 6.10 Children with *Gnetum* leaves (Source: Catherine Momha)

the plant is instead propagated using grafting techniques (Tchoundjeu et al. 2006). However, methods have been developed to break seed dormancy in *G. kola*, including seed coat removal followed by soaking in cold water (Eyog-Matip et al. 2007). De-coating or soaking of *G. kola* seeds in 70 % ethanol also significantly increases germination to 92 %, as well as rate of germination (Agyili et al. 2007).

Gnetum africanum/Gnetum buccholzianum (Eru or Fumbua)

Gnetum africanum (Gnetaceae) and related species such as *G. buccholzianum* are lianas naturally distributed in the Gulf of Guinea, from Southeast Nigeria to Angola (Lowe 1984). They are locally called 'Okok', 'Eru' and 'Koko' in Cameroon, 'Fumbua' in Democratic Republic of Congo, and 'Okasi' in West Africa. Leaves (Fig. 6.10) from these dioecious vines are very rich in proteins and minerals (Mialoundama and Paulet 1985), and are widely used as a vegetable. The leaves are traded in sub-Saharan Africa, Europe, and North America. Between 1985 and 1994, 5,296.42 tons of Eru were exported from Cameroon (Bokwe and Ngantoum 1994), and data from Yaoundé Nsimalen airport indicated that over 7.6 tons of Eru is exported annually through that airport alone (Asaha et al. 2000). Nkefor (2000) and Asaha et al. (2000) revealed that 3,600 tons of Eru is shipped every year from the Southwest province of Cameroon to Nigeria, and exported to Europe and North America. The commercialization of *Gnetum* leaves harvested in the major production areas in the Southwest province of Cameroon and in the humid forest zone of Nigeria involved at least 2,250 people across the chain value from the harvester to the consumer in a survey carried out between March 2009 and February 2010 (Ingram et al. 2012). The study revealed that *Gnetum* leaves contribute to 62 % of a harvesters' annual income (1,125 US$; Ingram et al. 2012).

Ndam et al. (2001) reported domestication activities of Eru that were carried out in Southeast Cameroon by the Limbe Botanic Garden. These activities include vegetative propagation using rooted cuttings, cultivation trials, and the introduction of Eru plants from rooted cuttings into farmlands. The Botanic Garden reported a 94 % success rate for rooting, the rooting period lasting for about 3.5 weeks. A previous study had reported successful rooting of Eru cuttings using non-mist poly propagators (Shiembo et al. 1996). In the cultivation trials, Eru performed well in the 50 % light environments, with a mean biomass production of 100–200 g per plant. The substrate that produced the highest biomass was a volcanic soil. On-farm survival rates range from 94 to 98 %. The first harvest occurred 12 months after planting, with further harvesting being done 4–6 months later (Ndam et al. 2001). A gene bank has been established in the Limbe Botanic Garden in Cameroon, using cuttings of *Gnetum* collected from 19 sites from the Southwest, Littoral, Centre, and South provinces of Cameroon (Ndam et al. 2001).

Cola acuminata, C. anomala and *C. nitida* (Kola nut or Cola nuts)

Cola nuts (Fig. 6.11) are widely consumed as stimulants in West and Central Africa, and are used in various ceremonies. The nuts are also widely traded. In 1985, Nkongmeneck estimated the size of the *Cola acuminata* market at 20,400 tons. One decade later, *C. acuminata* nuts were identified as the second most important Non-Timber Forest Product (NTFP) traded in the markets of the humid forest zone of Cameroon. Sales of *Cola* nuts in 1995 were valued at 43,432,200 cfa (US$ 1 ~ 460.68 cfa; Ndoye et al. 1997). The genus *Cola* Schott and Endl. (Sterculiaceae) comprises about 90 species. Three of the species, namely *Cola acuminata, Cola nitida* and *Cola anomala*, are of economic importance in the sub-Saharan region. Kola nuts are traditionally used as a caffeine stimulant and have also been used for their euphoric qualities, and thus, have been used as a common additive to American and European soft drinks. *Cola acuminata* and *C. nitida* occur mainly in the evergreen and semi-deciduous lowland forest zones, whereas *C. anomala* is mostly found in the montane forest zone of West Cameroon. *Cola* tree domestication activities in Cameroon have focused on vegetative propagation of selected trees using air-layering, development of storage methods for the nuts, and farmer enterprise development. The biggest problem in Cola nut production is parasites. Two weevils, *Balanogastris kolae* and *Saphrorhinus spp.*, attack the nuts both prior to harvest and during storage.

Fig. 6.11 *Cola acuminata* nuts (Source: Ann Degrande)

In Nigeria, exploration and collection of many accessions of *Cola* species has been ongoing since the 1960s, and gene banks have been established at the Cocoa Research Institute of Nigeria. During the flowering and fruiting seasons between 1999 and 2002, germplasm characterization and morphological assessments were carried out (Adebola and Morakinyo 2006). The evaluation of twenty-seven quantitative and thirty-three qualitative morphological traits revealed enormous variability among the plants studied, and indicated that wild *Cola* species form a special group with many exploitable agronomic traits (Adebola and Morakinyo 2006). The establishment of multi-location trials is needed for the selection of a suitable breeding population within the genus. Interspecific hybridization should also be investigated, as well as gene flow between breeding and wild populations.

Prunus Africana (African cherry)

Prunus africana (Rosaceae), also called African cherry, is an evergreen tree widespread in montane forests of West, Central, East, and Southern Africa, and Madagascar. African cherry trees are found in isolated populations between 700 and 3,000 m elevation in sub-Saharan Africa, namely in Angola, Burundi, Cameroon, Equatorial Guinea, Ethiopia, Mozambique, DRC, Uganda, Rwanda, São Tomé e Príncipe, South Africa, Sudan, Tanzania, Swaziland, Zambia, Zimbabwe, and Madagascar. Bark extracts are used for the treatment of prostate gland hypertrophy and benign prostatic hyperplasia (Bombardelli and Morazzoni 1997). Consequently, the raw bark is shipped mainly from Cameroon, Equatorial Guinea, Kenya, and Madagascar to Europe. Trade of *P. africana* bark obtained from natural populations generates nearly US$ 250 million in Cameroon, the largest exporter in the world (Cunningham 1995). Most bark collection methods are destructive and unsustainable. Frequently, the wild trees are completely stripped of their bark, or are felled and then stripped. Concerns were raised about the conservation of genetic resources in the species. In 1995, CITES (Convention on International Trade in Endangered Species) listed *P. africana* as an endangered species. Consequently, trade of *P. africana* barks has been restricted in Cameroon since 1995, reducing the volume of bark exported from Cameroon annually.

Due to concerns over the preservation of the species' genetic resources, it would be imperative to conduct an assessment of genetic diversity in the species over its whole distribution range, including gene flow and mating system descriptions using informative molecular markers. Efforts by Dawson and Powell (1999) and Muchugi et al. (2006) to investigate genetic variation in the species using RAPD data were hampered by a lack of complete genotypic information resulting from dominance nature of RAPD markers (Lynch and Milligan 1994). In Cameroon, Avana et al. (2004) used microsatellites and Amplified-Fragment Length Polymorphisms (in addition to RAPD) to assess the genetic diversity of a few samples of *P. africana.* A thorough assessment of the genetic diversity in the species using highly informative markers such as microsatellite markers is required, and should cover the entire natural distribution range. An investigation on the genetic control of the chemical constituents of *P. africana* bark is also needed in order to set up breeding programs.

Factors influencing the rooting of *P. africana* leafy stem cuttings have been investigated by Tchoundjeu et al. (2002), and rooting percentages of up to 80% have been obtained. Cultivation of the species has been promoted, and enrichment plantations were established in Ntingue (Dschang, Cameroon) by the National Office of Forest Development of Cameroon in 1972. Other plantations were established by the Forest Department of Kenya for timber production (Dawson et al. 2000), and in Madagascar. In Cameroon, a 2 ha plantation has been established near Buea, where harvested materials carry a 2% regeneration tax on the value of the raw material and a transformation tax, both payable to the Forestry Department. The establishment of gene banks and germplasm collections is imperative to better service future development of this species.

Pausinystalia johimbe (Yohimbe)

Pausinystalia johimbe, also called Yohimbe, is a medicinal tree native to the Gulf of Guinea, and is found in evergreen forests from Southern Nigeria to Mayombe, Congo (Vivien and Faure 1985). Bark extract contains an alkaloid (Yohimbine) used to treat cardiac disease and male impotence (Sunderland et al. 1997). Consequently, *P. johimbe* bark is used directly as an aphrodisiac. The bark is widely traded. In Cameroon, trade of *P. johimbe* bark generates US$ 640,000 annually. As the species is not yet cultivated, bark is obtained from trees from natural populations using unsustainable methods, usually by felling the tree and stripping off the bark. *Pausinystalia johimbe* has small winged seeds, which are difficult to collect in wild stands, and exhibit a poor survival and growth rate after germination. However, the species has been successfully propagated using single-node leafy stem cuttings from coppicing stumps (Tchoundjeu et al. 2004). Domestication activities in the species are still in their infancy, and no germplasm collection has been reported as of yet.

Annickia chlorantha (African Whitewood)

Annickia chlorantha, colloquially known as fever bark, African whitewood, African yellow bark, and yellow stick, is a medium-sized forest tree naturally distributed from Nigeria to Angola and DRC (Le Thomas 1969). The primary economic value of this species comes from its bark, which is an antimalarial used in Cameroon (Elesha et al. 1999). The bark of *A. chlorantha* is the most commonly sold bark in the markets in Cameroon (Facheux et al. 2003). The pharmaceutical properties of *A. chlorantha* bark continue to be characterized; a bark extract, protoberberin, was used to produce a drug that combats human viral hepatitis (Virtanen et al. 1989).

Domestication activities in *A. chlorantha* have been limited to the development of vegetative propagation methods using single-node leafy stem cuttings (Ngo Mpeck et al. 2009). Due to the rarity of the species (Versteegh and Sosef 2007) and the unsustainable methods used to exploit the species for its bark, an assessment of the genetic diversity, followed by the establishment of gene banks, is imperative.

Allanblackia floribunda Oliv. (Tallow Tree)

Allanblackia floribunda, or tallow-tree, is a medium-sized forest tree species about 30 m tall (Fig. 6.12), that grows naturally in evergreen lowland and deciduous forests ranging from Benin to DRC and North Angola (Bamps 1969). The species is valued for the fat that can be extracted from its fruit seeds. Seeds from the tallow-tree are rich in a white fat consisting mostly of stearic and oleic acids (44.16–66.12 %, and 24.95–48.42 %, respectively; Atangana et al. 2011). The fat content of the seeds is between 67 and 73 % (Foma and Abdala 1985). The fat profile makes *A. floribunda* seed ideal for margarine production. Seeds from the tallow-tree are purchased by Unilever PLC, which has created a guaranteed market for the product. The market is expected to grow to over US$ 100 million in Africa.

Phenotypic characterization of fruit and seed traits has been carried out (Atangana et al. 2011), and methods for vegetative propagation using single-node leafy stem cuttings have been developed (Atangana et al. 2006). Grafting methods, and seed germination rates in a closely related species (*A. parviflora* A. Chev.) were reported and, although rates are very low (approximately 35 %), the strategies may be useful for tallow-tree domestication (Asaah et al. 2011; Ofori et al. 2011). Molecular genetic markers (microsatellites) were developed for tallow trees and genetic diversity was assessed in wild stand populations in Cameroon (Atangana et al.

Fig. 6.12 A fruiting tree of *Allanblackia floribunda*. (Source: Alain Atangana)

2010). Microsatellite markers were also developed for *A. stuhlmannii* Engl. and genetic diversity assessed (Russell et al. 2009). Future steps in the domestication of tallow trees would include the formation of a rangewide germplasm collection, the establishment of multi-location provenance trials, and the selection of 'elite trees' (i.e., trees that have proven genetic superiority in desirable traits for improvement). A description of the mating system in the species is also needed.

6.1.3.2 Examples of Priority Species for Domestication in the Sahel

Priority species for domestication in the Sahel and East and Central Africa drylands include *Ziziphus mauritiana* (desert apple or jujube), *Faidherbia albida*, *Balanites aegyptiaca*, *Detarium microcarpum*, *Tamarindus indica* (tamarind), *Parkia biglobosa*, *Adansonia digitata* (baobab) and *Sclerocarya birrea* (marula; Chikamai et al. 2004; ICRAF 1996). In 1983, some species were identified as having the potential for introduction into the savannah agroforestry systems of Senegal, those species include *A. digitata*, *B. aegyptiaca*, *Z. mauritiana*, *Boscia senegalensis* (Pers.) Lam. ex Poiret, *Senna obtusifolia* (L.) H.S.Irwin & Barneby and *Sclerocarya birrea* (Becker 1983). Marula produces edible fruit, commonly eaten in Niger (Glew et al. 2004).

Table 6.3 Nutritional composition of the fruit and kernels of *Adansonia digitata, digitata, Sclerocarya birrea, Tamarindus indica* and *Ziziphus mauritiana* (Maundu et al. 1999)

Species	Plant part	Vitamin C (mg)	P (mg)	Ca (mg)	Fe (mg)	K (mg)	Protein (g)	Fat (g)
Baobab	Fruit	20.9	76.2	335	2.65	2409	2.7	0.2
	Kernel		5.12	273	6.55	1275	33.7	30.6
	Pulp	270	118	284	7.4		2.2	0.8
Marula	Flesh	194	11.5	20.1	0.5	317	0.5	0.4
	Fruit raw	68	19	6.0	0.1		0.5	0.1
	Kernel		808	118	4.87	601	28.3	57.3
Tamarind	Fruit raw	8.0	97	60.0			2	0.2
	Fruit dried	9.0	190	166	2.2		5	0.6
	Fruit pulp						3.1	0.4
	Seeds						16	5.5
Desert apple	Fruit dried	24.0	210	56	3		4.3	0.1

Four types of *A. digitata* are found in Sahelian Africa, and species can be distinguished by the color of the bark or leaves. The leaves of 'dark leaf' trees are popular because the leaves can be eaten as vegetables, and the fruit of the 'black bark' and 'red bark' trees are edible. Baobab leaves are an excellent source of calcium, iron, potassium, manganese, phosphorus and zinc (Yazzie et al. 1994), and the fruit are an excellent source of vitamin C (Ibiyemi et al. 1988).

Jujube fruits are eaten fresh, boiled (with rice) or used in baking in the region, and the fruit are rich in sugar and vitamin C (Geurts 1982; Becker 1983). Some varieties of Jujube are cultivated. Jujube fruits are traded in India and Pakistan. The seeds extracted from *V. paradoxa* fruit are rich in fat (5 kg of fat is produced from about 10 kg of seeds) that is commonly used in cooking and by the cosmetic and soap industries. The nutritional composition of *A. digitata, S. birrea, T. indica* and *Z. mauritiana* fruits and seeds from Kenya were reported by Maundu et al. (1999). The fruits of these species are rich in vitamin C, Ca and Fe (Table 6.3). The fatty acid profile and properties of this fat were reviewed by Leakey (1999). The seeds of Néré (*Parkia biglobosa*) are fermented to get 'Soumbala' or 'Dawawa', a paste that is widely consumed in West Africa, and the endosperm of Tamarind seeds produces a gum which, when purified and refined, is used as a stabilizer for food. Uses and properties of Tamarind fruits and seeds have been reviewed by Leakey (1999).

The status of ongoing domestication efforts of Baobab, Néré, Tamarind, Jujube, and *V. paradoxa* in the African Sahel was reported by Raebild et al. (2010). Germplasm collection has been less intense in Central West Africa than elsewhere in the natural distribution ranges. The genetic parameters of fruit traits and mating systems are unknown in all species under domestication, except for Tamarind (Diallo et al. 2008). However, the characterization of genotypes and morphotypes is well underway (Assogbadjo et al. 2006, 2009; Sanou et al. 2006), and the genetic diversity of *V. paradoxa* and Tamarind has been assessed (Bouvet et al. 2004; Sanou et al. 2005; Fontaine et al. 2008; Diallo et al. 2007). Vegetative propagation of these species using simple and robust methods is possible for all species, except *P. biglobosa* (Teklehaimanot et al. 1996; Danthu and Soloviev 2000; Danthu et al. 2004; Sanou et al. 2004).

6.1.3.3 Domestication of Priority Species in Southern Africa

Since 1989, ICRAF has been conducting research on the domestication and commercialization of indigenous fruit trees from five countries in Southern Africa (Maghembe et al. 1998; Akinnifesi et al. 2004a). The program's philosophy is to strengthen local farmers' desire to cultivate indigenous fruit trees, increase the ways in which these species will promote food and nutritional security, increase and diversify household incomes, create jobs, and diversify farming systems in rural areas. Priority species for domestication in the Miombo woodlands of Southern Africa include *S. birrea, Uapaca kirkiana, Vangueria infausta* Burch., *A. digitata, Syzygium cordatum* (Hochst.), *Parinari curatellifolia* Planch ex Benth., *Strychnos cocculoides* Baker, *Flacourtia indica* (Burm. F.) Merr., *Syzygium guineense* Wall., *Azanza garckeana* (F. Holm.) Exell et Hillc., *T. indica* and *Vitex* spp. (Maghembe et al. 1995, 1998; Mateke 2000, 2003; Kwesiga et al. 2000; Akinnifesi et al. 2004a, b). Fruits from Marula are used in the manufacture of Amarula liquor, traded worldwide. Surveys of local farmers have identified the traits most important for improvement, namely taste, fruit size, early fruiting, and reduction of tree height.

Four priority tree species of the Miombo woodlands are currently under domestication in Southern Africa, i.e. *U. kirkiana, S. cocculoides, P. curatellifolia* and *S. birrea* (Akinnifesi et al. 2008). The process is participatory, as farmers are involved in all stages of domestication, product development, and commercialization. The phenotypic variation in fruit characteristics was investigated by Thiong'o et al. (2002) and Leakey et al. (2005b, c) for *S. birrea*, and by Akinnifesi et al. (2004b) for *U. kirkiana*. Nurseries have been established and germination protocols developed (Maghembe 1995; Mkonda et al. 2003). Grafting of some of these priority species, namely baobab, *U. kirkiana, S. cocculoides, S. birrea, V. infausta*, and *P. curatellifolia,* has been successful (Mhango and Akinnifesi 2001; Akinnifesi et al. 2004a). Multi-location provenance trials have been established for *U. kirkiana* and *S. birrea* and farmers have been trained in tree domestication techniques (Akinnifesi et al. 2008).

6.1.3.4 Domestication of Priority Species in Latin America

Sotelo Montes and Weber (1997) surveyed famers in the Peruvian Amazon, and identified 23 high priority species for domestication, out of the 150 tree species farmers would like to cultivate. Domestication activities were implemented by ICRAF and the Peruvian National Institute for Agricultural Research and Extension (INIEA). Activities started with four species that had significant importance in the farm economy (Labarta and Weber 1998). Those four species are *Bactris gasipaes* Kunth. (also known as peach palm), *Calycophyllum spruceanum* Benth.

(Mulateiro or Bayabochi), *Guazuma crinita* Mart., and *Inga edulis* Mart. Participatory domestication in the Peruvian Amazon began with the documentation of farmers' knowledge about intraspecific variation, taking into account potential differences in perceptions. Women have identified six varieties of *I. edulis*, based on pod size and the size, shape and color of the leaves (Potters 1997). The local knowledge provided a basis for testable research hypotheses that could be used to accelerate the delivery of improved planting materials to farmers. For example, farmers told the researchers that peach palm fruits with red, waxy coats have higher oil content than those with red or yellow, non-waxy coats (Weber et al. 2001). A starchy fruit is more suitable for producing flour, while an oily fruit can be used to produce cooking oil.

The domestication strategy in the Peruvian Amazon involves participatory and farmer-driven tree selection, testing adaptation of seed sources (for seed zone delimitation and transfer guidelines), and accelerating the delivery of high-value germplasm to farmers. In 1996, 11 natural provenances of both *C. spruceanum* and *G. crinita* were identified in the Peruvian Amazon, following a systematic collection strategy. In order to ensure a representative sample of the variation within natural populations, samples from 35 trees were collected at random within each population, with a minimum distance of 100 m between trees (Dawson and Were 1997; Weber et al. 2001). Seeds were then collected from each of these provenances (Weber et al. 2001). On-farm trials were established to identify the most promising provenances for different products under various rainfall conditions (Weber et al. 2001). Preliminary results indicated the potential gains that farmers could obtain in domesticating some provenances, and the use of molecular techniques allowed identification of the most diverse provenances of *C. spruceanum* (Russell et al. 1999). Recommendations were made that farming communities manage the provenances for in situ conservation and seed production (Sotelo Montes et al. 2000; Weber et al. 2001).

To test tree adaptation, the progeny of 200 *C. spruceanum* and *G. crinita* trees selected by farmers were established in progeny trials on 15 farms in the lower, middle, and upper parts of the Aguaytia watershed, representative of many other watersheds in the Peruvian Amazon in 2000/2001 (Weber et al. 2001). These provenance trials helped identify the best seed sources, while progeny tests help to identify the best mother trees within a selected seed source. Collection of germplasm from the best mother trees was performed to establish seedling or clonal seed orchards and the high-quality seed produced from these orchards was disseminated to farmers (Weber et al. 2001). For germplasm delivery, key farmers were identified and involved in the multiplication and dissemination of germplasm.

Genetic differentiation among domesticated populations of peach palm along the Paranapura River (56 plants from four populations) and Cuiparillo River (145 plants from 12 populations) was assessed by Adin et al. (2004) as a germplasm management activity. No relation was found between any genetic differentiation and the geographic location of populations. The authors speculated that the exchange of materials by farmers and commercial traders may be responsible for most of the gene flow among the populations studied. The ICRAF-INIEA participatory program of peach palm improvement in the Peruvian Amazon focused on genetic improvement, genetic conservation, and seed production for community development (Cornelius et al. 2006). Genetic conservation and timely germplasm delivery were emphasized, rather than genetic gain (Cornelius et al. 2006). Trade-off analysis between genetic gains obtained from tree improvement *versus* the conservation of genetic resources in peach palm suggested that, with careful management, genetic diversity can be effectively conserved through 20 generations of improvement (Cornelius et al. 2006).

Ensuring long-term conservation of genetic resources of a species under domestication could be achieved through a good collection practice of seed from already cultivated species. Pre-existing examples include an analysis of *I. edulis* from natural and planted stands at five sites in the Peruvian Amazon (Dawson et al. 2008). The study found that the genetic material of *I. edulis* is primarily of non-local origin, indicating that conservation based on new and wide-scale infusions from local wild stands into farm stands may be inappropriate in the region. An analysis of Simple Sequence Repeat loci comparing genetic diversity of planted and unplanted *I. edulis* in the Peruvian Amazon indicated that farmers reduced genetic diversity when they domesticate tropical trees, although allelic variation in planted stands was still high (Hollingsworth et al. 2005). Therefore, strategies to conserve genetic resources throughout the domestication of multipurpose tree species in the Amazon should be developed and implemented. O'Neill et al. (2001) speculated that the implementation of improved seed collection systems and simple seed transfer guidelines would better address issues of reducing the collection of poor quality seed with low genetic diversity, and avoid maladapted plantings. The authors also suggested the strategic identification and design of *in situ* conservation areas that would help to ensure the viability of conserved populations. However, this would require the forfeiture of significant revenue from timber concessions.

6.1.3.5 Domestication of Priority Species in India, Southeast Asia, and Oceania

Roshetko and Verbist (2000) reported that smallholder tree production systems, with emphasis on both indigenous and exotic timber and fruit species, are priorities for domestication in Southeast Asia. Species of interest for domestication include *Vitex pubescens* Vahl (planted in SE Asia for charcoal), jackfruit (*Artocarpus heterophyllus* Lam.), *Durio zibethinus* L. (durian), *Gnetum gnemon* L. (melinjo), *Gliricidia sepium, Leucaena leucocephala, Paraserianthes* and *Parkia speciosa* Hassk. (petai) and *Litchi sinensis* Sonn.

Durian is cultivated throughout Southeast Asia and is well-known for its remarkable pungency. The aril is eaten either fresh or fermented, with salt or sugar. The Durian rind contains polysaccharides with pharmaceutical properties (Hokputsa et al. 2004). Durian trees are also important in financial systems in West Java (Dury et al. 1996). The flavor and sensory characteristics of Durian vary between cultivars (Voon et al. 2007). This variation can be useful in tree domestication. Durian is subject to breeding by scientists in government institutions in the Malay Peninsula (Natanchai 1994). Tree domestication includes tree breeding and involves many other activities, such as exploration and collection of populations, development of propagation techniques, dissemination, multiplication and assessment of germplasm, facilitation of farmers' access to markets and market information, marketing of tree products, integration of high-value germplasm in land-use systems, dissemination of technical information, and instruction of farmers in tree domestication techniques.

In Oceania, indigenous species with economic potential include *Artocarpus altilis* (Parkinson) Fosberg, *Barringtonia edulis* Seem., *Barringtonia novae-hiberniae* Lauterbach, *Barringtonia procera* (Miers) R. Knuth, *Burckella fijiensis* (Hemsl.) A.C.Sm. & S.P.Darwin, *Burckella obovata* (G.Forst.) Pierre, *Burckella* spp., *Cassidispermum megahilum* Hemsl, *Canarium harveyi* Seem, *Canarium indicum* L., *Dracontomelon vitiense* Engl., *Inocarpus fagifer* (Parkinson ex Zollinger) Fosberg, *Morinda citrifolia* L., *Spondias dulcis* L., *Syzygium malaccense* (L.) Merr. & L.M.Perry, and *Terminalia catappa* L. (Lebot et al. 2008). Tree domestication in Oceania consists of (Lebot et al. 2008):

- Selection of a wild genotype or a seedling from a cultivated form
- Improvement of the environment
- Improvement of the population composed of well-established selected seedlings

Numerous indigenous genera from Oceania are used for forage, including *Aceratum*, *Burckella*, *Corynocarpus*, *Ficus*, *Garuga* and *Garcina* (Lebot et al. 2008). Few indigenous species are cultivated, mostly around homesteads, including *Artocarpus altilis*, *Morinda citrifolia* and *Gnetum gnemon* (Lebot et al. 2008). These species are consumed as staple foods.

Canarium indicum has been domesticated by farmers through selection and propagation of trees with desired fruit characteristics (Tio Nevenimo et al. 2007). Tio Nevenimo et al. (2007) further suggested how domestication and commercialization of Canarium fruits could be advanced to improve the livelihoods of rural populations in Oceania. Because the tree nuts offer the best market prospects, various suggestions were made to improve the supply of high quality *C. indicum* kernels. The size and quality of the available resources can be increased by promoting planting in home gardens and integration in agroforestry systems with cocoa and other cash crops to provide shade and a wider range of products. Improvements to the quality and uniformity of the products can be accomplished through the domestication of the species as a crop so that these plantings can increasingly be made with a selected cultivar. Furthermore, the market for *Canarium* kernels still requires development. Tree-to-tree and continuous variations exist in the nutritional and medicinal properties of *C. indicum*. Opportunities exist for multiple-trait selection in cultivar development in the species as part of its tree domestication program (Leakey et al. 2008). The assessment of genetic resources using molecular markers in Australian tree species undergoing domestication was reviewed by Moran et al. (2000). The domestication of *Acacia mangium* resulted in a high proportion of the genetic resources being included in breeding programs. *Acacia aulacocarpa* Benth. domestication, on the other hand, involved a significant fraction of the genetic resources not being incorporated into the baseline populations (Moran et al. 2000). Very limited loss of genetic diversity was found in *Eucalyptus sieberi* L.A.S.Johnson.

In India, farmers have traditionally cultivated indigenous fruit trees around homesteads, including: Mango (*Mangifera indica*), *Artocarpus heterophyllus*, *Phyllanthus emblica* L., *Aegle marmelos* (L.) Corr.Serr., *Annona squamosa* L., *Syzygium cuminii*, *Tamarindus indica* and *Carissa congesta* Wight (Kumar 2008). The selection of individuals to be propagated is done on the basis of desirable fruit characteristics such as fruit size and flesh sweetness, and germplasm are multiplied using vegetative means and distributed among farmers (Muthulakshmi et al. 2005; Nazeem et al. 1984; Puri and Swamy 1999; Nair et al. 2005; Tewari and Bajpai 2005). A strategy for the domestication of these species needs to be developed (Kumar 2008).

Bibliography

Achard F, Eva HD, Stibig H-J, Mayaux P, Gallego J, Richards T, Malingreau J-P (2002) Determination of deforestation rates of the world's humid tropical forests. Science 297:999–1002

Achinewhu SC (1983) Ascorbic acid content of some Nigerian local fruits and vegetables. Plant Foods Human Nutrition 33:261–266

Adebola PO, Morakinyo JA (2006) Evaluation of morpho-agronomic variability of wild and cultivated kola (*Cola* species Schott et Endl.) in South Western Nigeria. Genetic Res Crop Evol 53:687–694

Adin A, Weber JC, Sotelo-Montes C, Vidaurre H, Vosman B, Smulders MJM (2004) Genetic differentiation and trade among populations of peach palm (Bactris gasipaes Kunth) in the Peruvian Amazon-implications for genetic resources management. Theoret Appl Genet 108:1564–1573

Agyili J, Moctar Sacande, Koffi E, Peprah T (2007) Improving the collection and germination of West African *Garcinia kola* Heckel seeds. New Forests 34:269–279

Ajewole K, Adeyeye A (1991) Seed oil of white star apple (*Chrysophyllum albidum*)-Physicochemical characteristics and fatty acid composition. J Sci Food Agr 54:313–315

Akinnifesi FK, Kwesiga FR, Mhango J, Mkonda A, Chilanga T, Swai R (2004a) Domesticating priority miombo indigenous fruit trees as a promising livelihood option for smallholder farmers in southern Africa. Acta Horticulturae 632:15–30

Akinnifesi FK, Chilanga T, Mkonda A, Kwesiga F (2004b) Domestication of *Uapaca kirkiana* in southern Africa: a preliminary evaluation of provenances in Malawi and Zambia. In: Rao MR, Kwesiga FR (eds) Proceedings of Regional Agroforestry Conference on Agroforestry Impacts on Livelihooods in Southern Africa: Putting Research into Practice: World Agroforestry Centre: Nairobi, Kenya, pp 85–92

Akinnifesi FK, Kwesiga F, Mhango J, Chilanga T, Mkonda A, Kadu CAC, Kadzere I, Mithofer D, Saka JDK, Sileshi G, Ramadhani T, Dhliwayo P (2006) Towards the development of miombo fruit trees as commetcial tree crops in southern Africa. Forests Trees Livelihoods 16:103–121

Akinnifesi FK, Leakey RRB, Ajayi O, Sileshi G, Tchoundjeu Z, Matakala P (2008) Indigenous fruit tree in the tropics:domestication, utilization and commercialization. CABI

Anegbeh PO, Usoro C, Ukafor V, Tchoundjeu Z, Leakey RRB, Schreckenberg K (2003) Domestication of *Irvingia gabonensis*: 3. Phenotypic variation of fruits and kernels in Nigeria. Agrofor Syst 58:213–218

Aniche GN, Uwakwe GU (1990) Potential use of Garcinia kola as substitute in lager beer brewing. World J Microbiol Bio Tech 6:323–327

Asaah E, Tchoundjeu Z, Ngahane W, Tsobeng A, Kouodiekong L, Jamnadass R, Simons A (2011) *Allanblackia floribunda*: a new oil tree crop for Africa: amenability to grafting. New Forests 41:389–398

Asaha S, Tonye Mahop M, Ndam N, Blackmore P (2000) State of knowledge study on *Gnetum africanum* Welw. and *Gnetum buchholzianum* Engl. A report commissioned by the Central African Regional Program for Environment. Limbe Botanic Garden Library (11–15), pp 26 + 13

Assogbadjo AE, Kyndt T, Sinsin B, Gheysen G, Van Damme P (2006) Patterns of genetic and morphometric diversity in baobab (Adansonia digitata) populations across different climatic zones of Benin (West Africa). Annal Botany 97:819–830

Assogbajo AE, Kyndt T, Chadare FJ, Sinsin B, Gheysen G, Eyog-Matig O, Van Damme P (2009) Genetic fingerprinting using AFLP cannot distinguish traditionally classified baobab morphotypes. Agrofor Syst 75:157–165

Atangana AR (2010) Phenotypic diversity in fruit and seed traits, and neutral genetic diversity in *Allanblackia floribunda*. Ph.D. dissertation, Université Laval, Québec, Canada

Atangana AR, Khasa DP (2008) Preliminary survey of clonal variation in rooting of *Allanblackia floribunda* leafy stem cuttings. Canadian J Forest Res 38:10–15

Atangana AR, Tchoundjeu Z, Fondoun J-M, Asaah E, Ndoumbe M, Leakey RRB (2001) Domestication of Irvingia gabonensis: 1. Phenotypic variation in fruits and kernels in two populations from Cameroon. Agrofor Syst 53:55–64

Atangana AR, Ukafor V, Anegbeh P, Asaah E, Tchoundjeu Z, Fondoun J-M, Ndoumbe M, Leakey RRB (2002) Domestication of *Irvingia gabonensis*: 2. The selection of multiple traits for potential cultivars from Cameroon and Nigeria. Agrofor Syst 55(3):225–229

Atangana AR, Tchoundjeu Z, Asaah EK, Simons AJ, Khasa DP (2006) Domestication of Allanblackia floribunda: amenability to vegetative propagation. Forest Ecol Manag 237:246–251. doi:10.1016/j.foreco.2006.09.081

Atangana AR, Beaulieu J, Khasa DP (2010) Wild genetic diversity preservation in a small-sized first generation breeding population of Allanblackia floribunda (Clusiaceae). Tree Genet Genom 6:127–136

Atangana AR, van der Vlis E, Khasa DP, van Houten D, Beaulieu J, Hendrickx H (2011) Tree-to-tree variation in stearic and oleic acid content in seed fat from Allanblackia floribunda from wild stands: potential for tree breeding. Food Chem 126:1579–1585. doi:10.1016/j.foodchem.2010.12.023

Avana ML, Tchoundjeu Z, Bell JM, Vaillant A, Chevallier M-H (2004) Diversité génétique du *Prunus africana* (Hook. F.) Kalkman au Cameroun. Bois et Forêts des Tropiques 282:41–49

Awono A, Ndoye O, Schreckenberg K, Tabuna H, Isseri F, Temple L (2002) Production and marketing of Safou (*Dacryodes edulis*) in Cameroon and internationally: Market development issues. Forest Trees Livelihoods 12:125–147

Bamps P (1969) Notes sur les Guttiferae d'Afrique tropicale. Bulletin du Jardin botanique national de Belgique 39(4):345–372

Bayala J, Ouédraogo SJ, Ong CK (2009) Early growth performance and water use of planted West African provenances of Vitellaria paradoxa C.F. Gaertn (karite') in Gonse, Burkina Faso. Agrofor Syst 75:117–127

Becker B (1983) The contribution of wild plants to human nutrition in the Ferlo (Northern Senegal). Agrofor Syst 1:257–267

Bokwe A, Ngatoum D (1994) Rapport de la mission effectuee autour du Mont Cameroun (Province du Sud-Ouest) relative au recensement de certaines espèces des produits forestiers secondaires en voie de disparition, pp 2–6

Bouvet JM, Fontaine C, Sanou H, Cardi C (2004) An analysis of the pattern of genetic variation in Vitellaria paradoxa using RAPD markers. Agrofor Syst 60:61–69

Chikamai B, Eyog-Matip O, Mbogga M (2004) Review and appraisal on the status of indigenous fruits in Eastern Africa. A report prepared for the IPGRI-SAFORGEN in the framework of AFREA/FORNESSA, Kenya Forestri Institute, Nairobi, Kenya, p 199

Cornelius JP, Clement CR, Weber JC, Sotelo-Montes C, van Leeuwen J, Ugarte-Guerra LJ, Ricse-Trembladera A, Arevalo-Lopez L (2006) The trade-off between genetic gain and conservation in a participatory improvement programme: the case of peach palm (Bactris gasipaes Kunth). Forests Trees Livelihoods 16:17–34

Cunningham AB (1995) An economic evaluation of *Prunus africana* cultivation for medicinal bark: a case study from Cameroon. WWF/UNESCO/Kew people and Plants initiative, unpublished trip report; April 1995, pp 3–26

Dandjouma AK, Tchiegang C, Ndjouenkeu R, Kapseu C, Mbofung CMF (2000) Caracteristiques morphologiques des amandes et physico-chimiques des huiles de *Ricinodendron heudelotii* (Bail.) du Cameroun. In: Kengue J, Kapseu C, Kayem GJ (eds) Third International Workshop on the Improvement of Safou and other Non-Conventional Oil Crops. Presses Universitaires d'Afrique, Yaoundé, pp 288–299

Danthu P, Soloviev P (2000) Propagation par greffage de trois espèces forestie`res fruitières des zones tropicales sèches: *Adansonia digitata, Balanites aegyptiaca et Tamarindus indica*. Le Flamboyant 53:22–24

Danthu P, Touré MA, Soloviev P, Sagna P (2004) Vegetative propagation of *Ziziphus mauritiana* var. Gola by micrografting and its potential for dissemination in the Sahelian Zone. Agrofor Syst 60:247–253

Dawson IK, Were J (1997) Collecting germplasm from trees: some guidelines. Agroforestr Today 9(2):6–9

Dawson IK, Powell W (1999) Genetic variation in the Afromontane tree Prunus africana, an endangered medicinal species. Mol Ecol 8:151–156

Dawson I, Were J, Lengkeek A (2000) Conservation of *Prunus Africana*, an over-exploited African medicinal tree. Forest Genet Res 28

Dawson IK, Hollingsworth PM, Doyle JJ, Kresovich S, Weber JC, Sotelo Montes C, Pennington TD, Pennington RT (2008) Origins and genetic diversity conservation of tropical trees in agroforestry systems: a case study from the Peruvian Amazon. Conserv Genet 9:361–372

Diallo BO, Joly HI, McKey D, Hossaert-McKey M, Chevallier MH (2007) Genetic diversity of *Tamarindus indica* populations: any clues on the origin from its current distribution? African J Biotechnol 6:853–860

Diallo BO, McKey D, Chevallier MH, Joly HI, Hossaert-Mckey M (2008) Breeding system and pollination biology of the semidomesticated fruit tree, *Tamarindus indica* L. (Leguminosae: Caesalpinioideae): Implications for fruit production, selective breeding, and conservation of genetic resources. African J Biotechnol 7:4068–4075

Dick JMP, Dewar R (1992) A mechanistic model of carbohydrate dynamics during adventitious root development in leafy cuttings. Ann Bot (Lond) 70:371–377

Dick JMCP, Leakey RRB, McBeath C, Harvey F, Smith RI, Woods C (2004) Influence of nutrient application rate on the growth and rooting potential of the West African hardwood *Triplochiton scleroxylon*. Tree Physiol 24:35–44

Dury S, Vilcosqui L, Mary F (1996) Durian trees (*Durio zibethinus* Murr.) in Javanese homegardens: their importance in informal financial systems. Agrofor Syst 33:215–230

Elesha SO, Onabanjo AO, Agbaje EO (1999) Chronic toxicity of *Enantia chlorantha* in animal model. Nig Q J Hosp Med 2:143–145

Eyog-Matip O, Aoudji AKN, Linsoussi C (2007) *Garcinia kola* Heckel seeds dormancy-breaking. Appl Ecol Environ Res 5(1):63–71

Facheux C, Asaah E, Ngo Mpeck M-L, Tchoundjeu Z (2003) Studying markets to identify medicinal species for domestication: the case of *Annickia chlorantha* in Cameroon. Herbal Gram 60:38–46

FAO (1996) Report on the state of the World's Plants Genetic Resource for food and Agriculture. FAO, Rome, Italy, p 511

FAO (2006) Evaluation des ressources forestières mondiales 2005: Progrès vers la gestion forestière durable. FAO, Rome, Italie, p 320

Fondoun JM, Tiki-Manga T (2000) Farmers indigenous practices for conserving *Garcinia kola* and *Gnetum africanum* in southern Cameroon. Agrofor Syst 48:289–302

Fondoun JM, Tiki Manga T, Kengue J (1999) Ricinodendron heudelotii (Djansang): ethnobotany and importance for forest dwellers in Southern Cameroon. Plant Genet Res Newslett 117: 1–11

Fontaine C, Lovett PN, Sanou H, Maley J, Bouvet JM (2004) Genetic diversity of the shea tree (*Vitellaria paradoxa* C.F. Gaertn), detected by RAPD and chloroplast microsatellite markers. Heredity 93:639–648

Fonteh A (1998) The lipid content of the African pear and black olives from some agroecological zones in Cameroon. In: Kapseu C, Kayem GJ (eds) Proceedings of the 2nd International workshop on African pear improvement and other new sources of vegetable oils. ENSAI, Presses Universitaires de Yaoundé, pp 185–192

Franzel S, Jaenicke H, Janssen W (1996) Choosing the right trees: Setting priorities for multipurpose tree improvement, research report, 8. ISNAR, The Hague, The Netherlands, p 87

Franzel S, Akinnifesi FK, Ham C (2008) Setting priorities among indigenous fruit tree species in Africa: examples from southern, eastern and western Africa regions. In: Akinnifesi FK, Leakey RRB, Ajayi O (eds) Indigenous fruit trees in the tropics: domestication, utilization and commercialization. CABI, pp 1–27

Geurts IF (1982) The Indian jujube or ber (*Zizyphus mauritiana* Lamk.) Aspects related to germplasm conservation. A Preliminary Report, Royal Tropical Institute, Amsterdam, The Netherlands

Giami SY, Okonkwo VI, Akusu MO (1994) Chemical composition and functional properties of raw, heat-treated and partially protelolysed wild mango (*Irvingia gabonensis*) seed flour. Food Chem 49(3):237–243

Glew RS, VanderJagt DJ, Huang Y-S, Chuang L-T, Bosse R, Glew RH (2004) Nutritional analysis of the edible pit of *Sclerocarya birrea* in the Republic of Niger (*daniya*, Hausa). J Food Compos Anal 17(1):99–111

Good RL (964) The geography of the flowering plants, 3rd edn, vol 1. Wiley, London, p 518

Harris DJ (1996) A revision of the Irvingiaceae in Africa. Bulletin du Jardin Botanique National de Belgique 65:143–196

Hartmann, HT, Kester DE, Davies FT Jr, Geneve RL (2002) Plant Propagation: Principles and Practices, 7th edn. Prentice Hall, New Jersey, p 880 with indexes (Upper Saddle river)

Hokputsa S, Gerddit W, Pongsamart S, Inngjerdingen K, Heinze T, Koschella A, Harding SE, Paulsen BS (2004) Water-soluble polysaccharides with pharmaceutical importance from Durian rinds (Durio zibethinus Murr.): isolation, fractionation, characterisation and bioactivity. Carbohydrate Polymers 56(4):471–481

Hollingsworth PM, Dawson IK, Goodall-Copestake WP, Richardson JE, Weber JC, Sotelo-Montes C, Pennington RT (2005) Do farmers reduce genetic diversity when they domesticate tropical trees? A case study from Amazonia. Mol Ecol 14:497–501

Howland P (1975) Vegetative propagation methods for *Triplochiton scleroxylon* K. Schum. Proc. symp. On variation and breeding systems of *Triplochiton scleroxylon K. Schum*. Federal Department of Forest Research, Ibadan, Nigeria, pp 99–109

Ibiyemi SA, Abiodun A, Akanji SA (1988) *Andasonia [sic] digitata Bombax* and *Parkia filicoideae Welw*: Fruit pulp for the soft drink industry. Food Chem 28:111–116

Ingram V, Ndumbe LN, Ewane ME (2012) Small scale, high value: *Gnetum africanum* and *buccholzianum* value chains in Cameroon. Small-Scale Forestry. doi:10.1007/s11842-012-9200-8

Jaenicke H, Franzel S, Boland DJ (1995) Towards a method to set priorities amongst species for tree improvement research—a case study from West and Central Africa. J Trop Forest Sci 7(3):490–506

Kengue J, Singa EM (1998) Preliminary characterization of some Plum accessions in Barombi-Kang germplasm collection. In: Kapseu C, Kayem GJ (eds) Proceedings of the 2nd International workshop on African pear improvement and other new sources of vegetable oils. ENSAI, Presses Universitaires de Yaoundé, pp 113–122

Kengue J, Tchuenguem Fohouo FN, Adewusi HG (2002) Towards the improvement of safou (*Dacryodes edulis*): population and reproductive biology. Forests Trees Livelihoods 12(1/2):73–84

Kumar BM (2008) Homegarden-based indigenous fruit tree production in peninsular India. In: Akinnifesi FK, Leakey RRB, Ajayi O, Sileshi G, Tchoundjeu Z, Matakala P (eds) Indigenous fruit tree in the tropics: domestication, utilization and commercialization. CABI, pp 84–99

Kwesiga F, Akinnifesi FK, Ramadhani T, Kadzere I, Saka J (2000) Domestication of Indigenous fruit trees of the miombo in southern Africa. In: Shumba EM, Lusepani E, Hangula R (eds) The Domestication and Commercialization of Indigenous Fruit Trees in the SADC Region. SADC Tree Seed Centre Network

Labarta Chavarri R, Weber JC (1998) Valorizacion economica de bienes tangibles de cinco especies arboreas agroforestales en la Cuenca Amazonica Peruana. Revista Forestal Centroamericana 23:12–21

Leakey RRB (1999) Potential for novel food products from agroforestry trees. Food Chem 66:1–14

Leakey RRB (2004) Physiology of vegetative reproduction. In: Burley J, Evans E, Younqist JA (eds) Encyclopaedia of forest sciences. Academic Press, London, p 1655–1668

Leakey RRB, Newton AC (1994) Tropical Trees: the Potential for Domestication and the Rebuilding of Forest Resources. HMSO, London, p 284

Leakey RRB, Simons AJ (1998) The domestication and commercialization of indigenous trees in agroforestry for the alleviation of poverty. Agrofor Syst 38:165–176

Leakey RRB, Tchoundjeu Z (2001) Diversification of tree crops: domestication of companion crops for poverty reduction and environmental services. Exp Agr 37:279–296

Leakey RRB, Page T (2006) The 'ideotype concept' and its application to the selection of cultivars of trees providing agroforestry tree products. Forests Trees Livelihoods 16(1):5–16

Leakey RRB, Last FT, Longman KA (1982) Domestication of tropical trees: an approach securing future productivity and diversity in managed ecosystems. Common Forestry Rev 61:33–42

Leakey RRB, Mesén JF, Tchoundjeu Z, Longman KA, Dick JMcP, Newton A, Matin A, Grace J, Munro RC, Muthoka PN (1990) Low-technology techniques for the vegetative propagation of tropical trees. Common Forestry Rev 69:247–257

Leakey RRB, Fondoun J-M, Atangana A, Tchoundjeu Z (2000) Quantitative descriptors of variation in the fruits and seeds of *Irvingia gabonensis*. Agrofor Syst 50: 47–58

Leakey RRB, Atangana AR, Kengni E, Waruhiu AN, Usoro C, Anegbeh PO, Tchoundjeu Z (2002) Domestication of *Dacryodes edulis* in Western and Central Africa: chatacterisation of genetic variation. Forest Trees Livelihoods 12:57–71

Leakey RRB, Tchoundjeu Z, Smith RI, Munro RC, Fondoun J-M, Kenhue J, Anegbeh PO, Atangana AR, Waruhiu AN, Asahh E, Usoro C, Ukafor V (2004) Evidence that subsistence farmers have domesticated indigenous fruits (*Dacryodes edulis* and *Irvingia gabonensis*). Agrofor Syst 60(2):101–111

Leakey RRB, Greenwell P, Hall MN, Atangana AR, Usoro C, Anegbeh PO, Fondoun J-M, Tchoundjeu Z (2005a) Domestication of Irvingia gabonensis: 4. Tree-to-tree variation in food-thickening properties and in fat and protein contents of dika nut. Food Chem 90(3):365–378

Leakey R, Shackleton S, Duplessis P (2005b) Domestication potential of Marula (*Sclerocarya birrea* subsp. *Caffra*) in South Africa and Namibia: 1. Phenotypic variation in fruit traits. Agrofor Syst 64:25–35

Leakey RRB, Pate K. Lombard C (2005c) Domestication potential of Marula (Sclerocarya birrea subsp caffra) in South Africa and Namibia: 2. Phenotypic variation in nut and kernel traits. Agrofor Syst 64:37–49

Leakey RRB, Fuller S, Treloar T, Stevenson L, Hunter D, Tio Nevenimo, Binifa J, and Moxon J (2008) Characterization of tree-to-tree variation in morphological, nutritional and medicinal properties of *Canarium indicum* nuts. Agrofor Syst 73:77–87

Le Thomas A (1969) Les Annonacées. In: Aubreville A (ed) Flore du Gabon, Muséum national d'histoire naturelle-Phanérogamie, vol 16. Paris, p 371

Lebot V, Walter A, Sam C (2008) The domestication of fruit and nut tree species in Vanuatu, Oceania. In: Akinnifesi FK, Leakey RRB, Ajayi O (eds) Indigenous fruit tree in the tropics: domestication, utilization and commercialization. CABI, pp 120–136

Lovett PN, and Haq N (2000) Diversity of the Sheanut tree (Vitellaria paradoxa C.F. Gaertn.) in Ghana. Genet Res Crop Evol 47:293–304

Lowe AJ, Gillies ACM, Wilson J, Dawson IK (2000) Conservation genetics of bush mango from central/west Africa: implications from random amplified polymorphic DNA analysis. Mol Ecol 9(7):831–841

Lowe J (1984) Gnetum in West Africa. Nigerian Field 49:1–4

Lynch M, Milligan BG (1994) Analysis of population genetic structure with RAPD markers. Mol Ecol 3:91–99

Maghembe JA (1995) Achievement in the establishment of indigenous fruit trees of the miombo woodlands of southern Africa. In: Maghembe JA, Ntupanyama Y, Chirwa PW (eds) Improvement of Indigenous Fruit Trees of the Miombo Woodlands of Southern Africa. ICRAF, Nairobi, Kenya, pp 39–49

Maghembe J, Simons AJ, Kwesiga F, Rarieya M (1998) Selecting indigenous trees for domestication in southern Africa. ICRAF, Nairobi, Kenya, p 94

Mapongmatsem P-M, Duguma B, Nkongmeneck BA, Puig H (1999) The effect of various seed pretreatments to improve germination in eight indigenous tree species in the forests of Cameroon. Annal Forest Sci 56: 679–684

Mateke S (2000) Tree improvement in the domestication programme of indigenous fruit trees native to southern Africa: the Botswana experience. In: Shumba EM, Lusepani E, Hangula R (eds) The domestication and commercialization of Indigenous fruit trees in the SADC region. SADC Tree Seed Centre Network, Gaborone, Botswana, pp 29–35

Mateke S (2003) Cultivation of native fruit trees of Kalahari Sandveld: studies on the commercial potential, interactions between soil and biota in Kalahari sands of southern Africa. Veld Products, Botswana, p 49

Maundu PM, Ngugi GW, Kabuye CHS (1999) Traditional food plants of Kenya. National Museums of Kenya, Nairobi, p 270

Mesén F, Newton AC, Leakey RRB (1997) The effects of propagation environment and foliar area on the rooting physiology of *Cordia alliodora* (Ruiz & Pavon) Oken cuttings. Trees 11:404–411

Mialoundama F, Paulet P (1985) Développement et rythme d'émergence foliaire chez le *Gnetum africanum* Welw. en relation avec la photopériode et les propriétés de rayonnement. Acta Oecol/Oecolog Plantar 6(20):309–318

Mkonda A, Lungu S, Maghembe JA, Mafongoya PL (2003) Fruit- and seed germination characteristics of *Strychnos cocculoides* an indigenous fruit tree from natural populations in Zambia. Agrofor Syst 58:25–31

Moran GF, Butcher PA, Glaubitz JC (2000) Application of genetic markers in the domestication, conservation and utilisation of genetic resources of Australian tree species. Australian J Botany 48:313–320

Muchugi A, Lengkeek AG, Kadu CAC, Muluvi GM, Njagi ENM, IK Dawson (2006) Genetic variation in the threatened medicinal tree *Prunus Africana* in Cameroon and Kenya: Implications for current management and evolutionary history. South African J Botany 72:498–506

Muthulashkmi P, George ST, Mathew KL (1999) Morphological and biochemical variations in different sex forms of kodampuli (*Garcinia gummi-gutta* L.). J Trop Agricul 37:28–31

Nair KKN, Mohanan C, Mathew G (2005) Preliminary observations on seeds and plantation technology for Garcinia gummi-gutta (Guttiferae): an indigenous and economic tree crop of the Indian peninsula. Forests Trees Livehoods 15:75–87

Nanthachai S (1994) Durian: fruit development, postharvest physiology, handling and marketing in ASEAN. ASEAN Food Handling Bureau, Kuala Lumpur, Malaysia

Nazeem PA, Gopikumar K, Kumaran K (1984) Vegetative propagation in jack (*Artocarpus heterophyllus* Lam.). Agricul Res J Kerala 22:149–154

Ndam N, Nkefor JP, and Blackmore P (2001) Domestication of *Gnetum africanum* and *G. buccholzianum* (Gnetaceae), over-exploited wild forest vegetables of the Central African Region. Syst Geography Plants 71(2):739–745

Ndoye O, Pérez MR, Eyebe A (1997) The markets of non-timber forest products in the humid forest zone of Cameroon. Rural Development Forestry Network Paper 22c, ODI, London

Newton AC, Jones AC (1993) The water status of leafy stem cuttings of four tropical tree species in mist and non-mist propagation systems. J Horticul Sci 68(5):653–663

Ngo Mpeck ML, Asaah E, Tchoundjeu Z, Atangana AR (2003) Strategies for the domestication of *Ricinodendron heudelotii*: Evaluation of variability in natural populations from Cameroon. Food Agricul Environ 1: 257–262

Ngo Mpeck ML, Atangana AR, Omgba AM (2009) Temporal variation in the rooting of *Annickia chlorantha*. New Forests 38:273–283

Ngo Mpeck ML, Atangana A (2007) Rooting of leafy stem cuttings of *Baillonella toxisperma*. Forest Sci 53(5):571–579

Ngono LD, Ndoye O (2004) Njansang and bush mango: Cameroonian seeds in national and international markets. In: Lopez C, Shanley P (eds) Riches of the forests: For health life and spirit in Africa, pp 21–24

Nkefor JP (2000) From the forest to the kitchen: following the market chain of Eru, an important non-timber forest product in Cameroon. Limbe Botanic Garden Internal Report, Limbe, Cameroon

Nkongmeneck B (1985) Le Genre Cola au Cameroun. Revues Sciences et Techniques (Serie de la Science Agronomique, Yaoundé, Cameroon) 1(3):57–70

Noble IR, Dirzo R (1997) Forests as human-dominated ecosystems. Science 227(5325):522–525

Nordström AC, Jacobs FA, Eliasson L (1991) Effect of exogenous indole-3-acetic acid and indole-3-butyric acid on internal levels of the respective auxins and their conjugation with aspartic acid during adventitious root formation in pea cuttings. Plant Physiol 96:856–861. doi:10.1104/pp.96.3.856

Nwadinigwe CA (1982) Nutritional value and mineral contents of *Chrysophyllum albidum* fruit. J Sci Food Agricul 33:283–286

Ofori DA, Peprah T, Cobbinah JR, Atchwerebour HA, Osabutey F, Tchoundjeu Z, Simons AJ, Jamnadass R (2011) Germination requirements of *Allanblackia parviflora* seeds and early growth of seedlings. New Forests 41:337–348

Ogu E, Agu RC (1995) A comparison of some chemical properties of *Garcinia kola* and hops for assessment of *Garcinia kola* brewing value. Biores Technol 54(1):1–4

Okafor JC (1974) Varietal delimitation in *Irvingia gabonensis* (Irvingiaceae). Nigerian J Forestry 4:80–87

Okafor JC (1980) Edible indigenous woody plants in the rural economy of the Nigerian forest zone. Forest Ecol Manag 3:45–55

Omoti U, Okiy A (1987) Characteristics and composition of the pulp oil and cake of the African pear, *Dacryodes edulis* (G.Don) H.J. Lam (Burseraceae). J Sci Food Agricul 38:67–72

O'Neill GA, Dawson I, Sotelo-Montes C, Guarino L, Guariguata M, Current D, Weber JC (2001) Strategies for genetic conservation of trees in the Peruvian Amazon. Biodiver Conserv 10:837–850

Potters J (1997) 'Farmers' knowledge and perceptions about tree use and management: the case of Trancayacu, a Peruvian community in the Amazon', MSc thesis dissertation, Wageningen Agricultural University, Wageningen, The Netherlands

Puri S, Swamy SL (1999) Geographical variation in rooting ability of stem cuttings of *Azadirachta indica* and *Dalbergia sissoo*. Genet Res Crop Evol 46: 29–36

Pye-Smith C (2010) The fruits of Success: A programme to domesticate West and Central Africa's wild fruit trees is raising incomes, improving health and stimulating the rural economy. ICRAF Trees for Change no.4. Nairobi: World Agroforestry Centre

Raebild A, Søndergård Larsen A, Svejgaard Jensen J, Ouedraogo M, Sitske De Groote, Van Damme P, Bayala J, Ousmane Diallo B, Sanou H, Kalinganire A, Dahl Kjaer E (2010) Advances in domestication of indigenous fruit trees in the West African Sahel. New Forests. doi:10.1007/s11056-010-9237-5

Roshetko J, Verbist B (2000) Tree domestication. ICRAF-SE Asia Lecture No 6, p 12

Russell JR, Weber JC, Booth A, Powell W, Montes C, Dawson IK (1999) Genetic variation of *Calycophyllums pruceanum* in the Peruvian Amazon Basin, revealed by amplified fragment length polymorphism (AFLP) analysis. Mol Ecol 8:199–204

Russell JR, Kadu CAC, Jamnadass R, Booth A, Cordeiro NJ, Woodhead M, Dawson IK (2009) AFLP and ssr diversity in the African fruit *Allanblackia*: implications for management of a genus newly subject to domestication for the edible oil industry. Tree Genet Genom 5:517–527

Sanou H, Lovett PN, Bouvet JM (2000) Comparison of quantitative and molecular variation in agroforestry populations of the shea tree (*Vitellaria paradoxa* CF Gaertn) in Mali. Mol Ecol 14:2601–2610

Sanou H, Kambou S, Teklehaimanot Z, Dembele M, Yossi H, Sina S, Djingdia L, Bouvet J-M (2004) Vegetative propagation of *Vitellaria paradoxa* by grafting. Agrofor Syst 60:93–99

Sanou H, Picard N, Lovett PN, Denbélé M, Korbo A, Diarisso D, Bouvet J-M (2006) Phenotypic variation of agromorphological traits of the shea tree, *Vitellaria paradoxa* C.F. Gaertn., in Mali. Genet Res Crop Evol 53:145–161

Sanchez PA (2002) Soil fertility and hunger in Africa. Science 295:2019–2020

Shiembo PN, Newton AC, Leakey RRB (1996) Vegetative propagation of Gnetum africanum Welw., a leafy vegetable from West Africa. J Horticul Sci 7:149–155

Shiembo PN, Newton AC, Leakey RRB (1997) Vegetative propagation of *Ricinodendron heudelotii*, a West African fruit tree. J Trop Forest Sci 9: 514–525

Sigaud P, Hald S, Dawson I, Ouedraogo A (1998) FAO/IPGRI/ICRAF workshop on the conservation, management, sustainable utilization and enhancement of forest genetic resources in dry-zone sub-Saharan Africa. Forest Genetic Resources No. 26, FAO, Rome, pp 9–12

Simons AJ (2003) Concepts and principles of tree domestication. In: Simons AJ, Beniest J (eds) Tree domestication in agroforestry. World Agroforestry Centre, Nairobi, Kenya, pp 244

Simons AJ, Leakey RRB (2004) Tree domestication in tropical agroforestry. Agrofor Syst 61:-167–181

Slovin JP, Bandurski RS, Cohen JD (1999) Auxin. In Hooykaas PJJ, Hall MA, Libbenga KR (eds) Biochemistry and molecular biology of plant hormones. Elsevier, Amsterdam, pp 115–140

Sotelo Montes C, Weber JC (1997) Priorizacio'n de especies arboreas para sistemas agroforestales en la selva baja del Peru. Agroforesterı'a en las Americas 4(14):12–17

Sotelo Montes C, Vidaurre H, Weber JC, Simons AJ, Dawson IK (2000) Produccion de semillas a partir de la domesticacion participativ de 6 arboles agroforestales en la amazonia peruana', In: Salazar R (coord.) Memorias del Segundo Symposio sobre Avances en la Produccion de Semillas Forestales en America Latina, Turrialba, Costa Rica: Centro de Agricultura Tropical y de Ensenianza

Sunderland T, Njiamnshi A, Koufani A, Ngo Mpeck M-L, Obama C (1997) The ethnobotany, ecology and natural distribution of Yohimbe (Pausinystalia johimbe (K. Schum) Pierre ex Beille), an evaluation of the sustainability of current bark harvesting practices, and recommendations for domestication and management. A report prepared for the International Council for Research in Agroforestry (ICRAF), p 31

Tabuna H (1999) The market for Central African non-wood forest products in Europe. In: Sunderland TCH, Clark LE, Vantomme P (eds) Current research issues and prospects for conservation and development. FAO, Rome, pp 251–263

Thiong'o MK, Kingori S, Jaenicke H (2002) The taste of the wild: Variation in the nutritional quality of marula fruits and opportunities for domestication. Acta Horticul 575:237–244

Tchoundjeu Z, Leakey RRB (1996) Vegetative propagation of African mahogany: Effects of auxin, node position, leaf area and cutting length. New Forests 11:125–136

Tchoundjeu Z, Atangana AR (2006) Ndjanssang *Ricinodendron heudelotii*. Fruits from the Future. Southampton Centre for Underutilised Crops, ICUC, University of Southampton, Southampton, UK

Tchoundjeu Z, Kengue J, Leakey RRB (2002) Domestication of *Dacryodes edulis*:State-of-the-art. Forest Trees Livelihoods 12:3–13

Tchoundjeu Z, Ngo Mpeck M-L, Asaah E, Amougou A (2004) The role of vegetative propagation in the domestication of *Pausinystalia johumbe* (K. Schum), a highly threatened medicinal species of West and Central Africa. Forest Ecol Manag 188(1–3):175–183

Tchoundjeu Z, Atangana AR, Degrande A (2005) Indigenous methods in preserving Bush mango kernels in Cameroon. American J Appl Sci 2(9):1337–1342

Tchoundjeu Z, Asaah EK, Anegbeh PO, Degrande A, Mbile P, Facheux C, Tsobeng A, Atangana AR, Ngo Mpeck ML, Simons AJ (2006) Putting participatory domestication into practice in West and Central Africa. Forests Trees Livelihoods 16:53–69

Teklehaimanot Z (1997) Germplasm conservation and improvement of Parkia biglobosa (Jacq.) Benth. for multipurpose use. Final report of EU-INCO research contract TS3*-CT92–0072, 1993–1997. University of Wales, Bangor

Teklehaimanot Z, Tomlinson H, Lemma T, Reeves K (1996) Vegetative propagation of Parkia biglobosa (Jacq) Benth, an undomesticated fruit tree from West Africa. J Horticul Sci 71:205–215

Tewari RK, Bajpai CK (2005) Bench grafting in aonla (Emblica officinalis Gaertn.). Indian J Agrofor 7:65–67

Tiki Manga T, Fondoun JM, Kengue J, Tchiegang C (2000) Chemical composition of *Ricinodendron heudelotii*: an indigenous fruit tree in southern Cameroon. African Crop Sci J 8: 195–201

Tio Nevenimo, Moxon J, Wemin J, Johnston M, Bunt C, Leakey RRB (2007) Domestication potential and marketing of *Canarium indicum* nuts in the Pacific: 1. A literature review. Agrofor Syst 69:117–134

Troupin G (1950) Les Burseracées du Congo Belge et du Ruanda-Urundi. Bulletin Soc. Bot. Royale de Belgique 83:111–126

Versteegh CPC, Sosef MSM (2007) Revision of the African genus *Annickia* (Annonaceae). Syst Geogr Plants 77(1):91–118

Villachica H (1996) Frutales y Hortalizas Promisorios de la Amazonia. Tratado de Cooperacion Amazonica, Lima, Peru, p 367

Virtanen P, Lassila V, Njimi T, Ekotto Mengata D (1989) Effect of splenectomy on Hepasor treatment in allyl-alcohol-traumatized rat liver. Acta Anat (Basel) 134:301–304. doi:10.1159/000146706

Vivien J, Faure JJ (1985) Arbres des forêts denses d'Afrique Centrale. Agence de Coope'ration Culturelle et Technique, p 565

Vivien J, Faure JJ (1996) Fruitiers sauvages d'Afrique (espèces du Cameroun). CTA, and Nguilou-Kirou Paris, France, p 416

Voon YY, Hamid NSA, Rusul G, Osman A, Quek SY (2007) Characterisation of Malaysian Durian (*Durio zibethinus Murr.*) cultivars: Relationship of physico-chemical and flavour properties with sensory properties. Food Chem 103(4):1217–1227

Wahuriu AN (1999) Characterization of fruit traits towards domestication of an indigenous fruit tree of West and Central Africa: A case study of *Dacryodes edulis* in Cameroon. MSc Thesis dissertation, University of Edinburg, UK, p 58

WCFSD (1999) Our forests—our future. Summary Report: World Commission on Forests and Sustainable Development, WCFSD, Winnipeg, p 38

Weber JC, Sotelo Montes C, Vidaurre H, Dawson IK, Simons AJ (2001) Participatory domestication of agroforestry trees: an example from the Peruvian Amazon. Develop Prac 11 (4):425–433

Yazzie D, Jagt DJ van der, Pastuszyn A, Okolo A, Glew RH (1994) The amino acid and mineral content of baobab (*Adansonia digitata L.*) leaves. J Food Comp Anal 7:189–193

Young A (1997) Agroforestry for soil management. CAB International, Wallingford, UK, p 320

Zobel B, Talbert J (1991) Applied Forest Tree Improvement. Waveland Press, Illinois

Part II
The Benefits and Services of Agroforestry Systems

Chapter 7
Ecological Interactions and Productivity in Agroforestry Systems

Abstract Ecological Interactions (hereafter, interactions) between the components of an agroforestry system occur when one component influences the performance of other components, and that of the whole system. Interactions in agroforestry can be positive, neutral or negative. A positive interaction is the complementarity between woody and herbaceous components in resource acquisition. Competition between components for water, nutrients and light is one example of negative interactions. Allelopathy, damage caused by animals or pests, and disease transmission are other negative interactions. Neutral interactions occur in agroforestry systems when the different components of the system exploit the same pool of resources, and increases in capture by one species result in a proportional decrease in capture by the associated species. Trees contribute to productivity in agroforestry through fertilization, and soil conservation. Tree contribution to fertilization is through the symbiotic fixation of nitrogen, root turnover, nutrient cycling, and their involvement in the formation of the humus layer. Maintenance of the physical properties of the soil is accomplished through erosion control and the stabilization of the soil by the roots.

7.1 Introduction

Agroforestry systems are complex mixtures of woody perennials and agricultural crops or animals, where special emphasis is placed on the associated interactions of the different components. The success of an agroforestry system is highly dependent upon the efficient interactions of the system's components.

7.2 Interactions of Components in Agroforestry Systems

Few studies have been conducted on the theoretical and experimental aspects of interactions in agroforestry, in part due to the complexity and lengthy time-frame which such research would be conducted (Nair 1993). The interaction of components can be defined as the influence of one component within the system on the performance, both of the other components and of the whole system. Interactions

A. Atangana et al., *Tropical Agroforestry*, DOI 10.1007/978-94-007-7723-1_7,

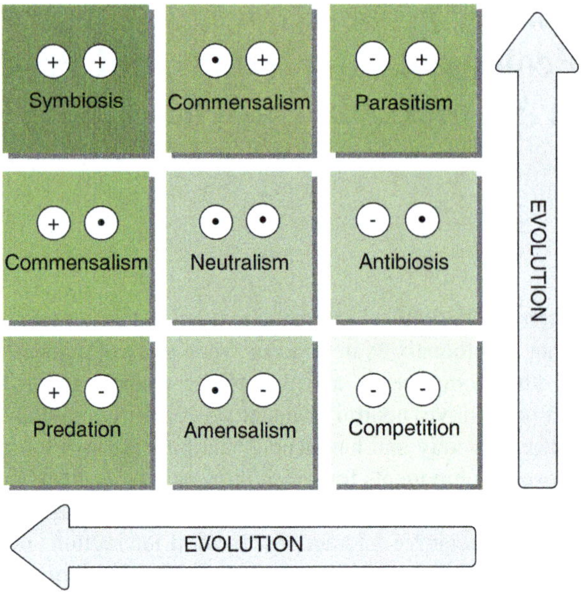

Fig. 7.1 Whittaker's graph illustrating the different possible interactions between two organisms that live in the same environment (Fortin et al. 2008)

between components can be classified into five types: (1) amensalism, which is an interaction where one individual negatively affects another without gaining any advantage; (2) allelopathy, which occurs when one species secretes chemicals that reduce the growth, survival and fitness of another; Black walnut (Juglans nigra L.) is native to eastern North America, and produces juglone (5-hydroxy−1,4-naphthalenedione), which inhibits the growth of other plants growing under its canopy; 3) (3) parasitism, which involves species feeding off or gaining benefit from the host, frequently killing the host; (4) commensal relationships, where the host species provides food or shelter for the guest species, but gains no benefit itself; and 5) symbiosis and its variant, mutualism. In a mutualistic relationship, both organisms are physiologically independent, but together assume roles akin to an organic function, and their respective survival is interrelated. The aforementioned interactions are characteristics of those that exist between two organisms living in the same environment (Fig. 7.1).

Interspecific interactions can be complementary, competitive, supplementary or neutral, and take place belowground, in the form of roots competing for water and nutrients, or aboveground, for light. These interactions can be classified as tree-herbaceous or tree-animal, and an interaction can be positive or negative. Positive interactions include the reduction of stress due to shading, biomass inputs to the system, water and soil conservation, shade, and manure inputs. Negative interactions typically result from the competition for light, allelopathy, animal damage, pests, and disease interactions (Nair 1993).

7.2.1 Positive Interactions

Interactions between woody and herbaceous plants in agroforestry systems usually improve the microclimate and nutrient availability in the soil. The belowground presence of trees affects moisture availability and soil temperature and these, in turn, affect transpiration and energy conversion of nearby plants (Rosenberg et al. 1983). Agroforestry is peculiar in that species are selected to interact effectively in the conservation of soil resources and amelioration of the environment. Agroforestry systems are designed in a way that optimizes resource use spatially, physically, and temporally by maximizing positive and minimizing negative interactions (Jose et al. 2000). Agroforestry is based on the assumption that 'the trees must acquire resources that the crop would not otherwise acquire' (Cannell et al. 1996). Such an interaction is apparent in the complementarity of woody and herbaceous components in resource acquisition. Complementary resource use in agroforestry is spatial or temporal (e.g., reverse phenology of *Faidherbia albida*; Ong and Leakey 1999). This same complementarity is also observed in terms of light usage. Some species that are used in agroforestry systems, such as cocoa plants (especially young ones), perform well under shade. For this reason, some trees are spared during forest clearing for the establishment of cocoa farms, to provide shade for the cocoa seedlings. The canopy intercepts radiant energy and rainwater, and affects the amount of light that reaches the ground. Soil temperature is reduced, and the water content of the soil is affected. For example, *Leucaena*-millet (*Pennisetum glaucum*) alley cropping trials allowed greater light interception throughout the year and, hence, greater biomass production at Hyderabad in southern India (Monteith et al. 1991). When the roots of woody and herbaceous plants occupy different soil layers, completion between those species is minimized, allowing for greater biomass production (Huxley 1983). While this holds true for nitrogen, it is not necessarily true for phosphorus (Kho 2000). The choice of species for inclusion in agroforestry systems should reflect the ecological status of species, strata that are occupied by these species, type of root system (taproot or lateral root system), and the soil horizons that are exploited by the roots of these species. As more of the soil surface is covered by plants, thermal fluctuations are reduced temperature becomes more constant.

Woody plant-crop interactions in agroforestry systems are also manifested in changes in the uptake of soil nutrients by legumes, either through nitrogen fixation or by the addition of organic matter through mulching, and the improvement of soil physical properties. The soil is improved through stabilization by the plant roots and the legume's involvement in the formation of the humus layer by depositing twigs and leaves on the ground. Legumes in alley cropping systems supply nutrients to the soil, thereby allowing a reduction in the fallow period. Alley cropping is also a good practice for weed control. Interactions can also be sequential, like in an improved fallows system with *Cajanus cajan*, or the *Taungya system*. In alley cropping systems, legumes also take up nutrients from soils. For this reason, a balance should be established between the input of nutrients into the soil, and their uptake by legume trees and shrubs. Pruning woody legume trees that are grown in alley cropping has likely reduced the amounts of nitrogen that are taken up from soils by these trees. In the humid zone of southwestern Nigeria, decreasing pruning frequency (from six- to tri-, bi- and monthly) and increasing pruning height (from 25 cm to 50 and 100 cm) increased plant biomass, dry wood production, and nitrogen yield from *Leucaena leucocephala* (Lam.) de Wit that were grown in hedgerows of alley cropping systems on an alfisol (Duguma et al. 1988). The same study revealed similar trends for *Gliricidia sepium* (Jacq.) Kunth ex Walp. and *Sesbania grandiflora* (L.) Poiret (Duguma et al. 1988), indicating the effectiveness of pruning in minimizing tree-crop competition for nutrients in agroforestry. For that reason, the efficient contribution of legumes to supply nitrogen to soil in agroforestry systems requires that the trees and shrubs are pruned regularly.

Plants-animal interactions also can be positive. Some agroforestry species, such as *Calliandra calothyrsus,* can be used as livestock fodder (Patterson et al. 1998). A diet based on forage that is obtained from legume trees has been shown to increase digestibility and milk production in cattle (Camero et al. 2001). Leguminous trees provide further advantages in that they can produce fodder even during the dry season. Their rooting systems are deeper than those of grasses and they will continue to grow, even as the grasses die. The trees provide shade for grazing animals. Otherwise, excess exposure to the sun creates heat stress in animals, leading to decreased milk production (Roman-Ponce et al. 1977; Mitlohner et al. 2001). Animal waste is used to fertilize the plants, and animals help to control weed infestation by grazing, which optimizes nutrient uptake by the crops.

7.2.2 *Negative Interactions*

Interactions can be positive aboveground and negative belowground. For example, in the semi-arid tropics, the interactions between components of alley cropping were negative belowground and positive aboveground (Ong et al. 1991). An analysis of the root systems showed an abundance of *L. leucocephala* roots in the top 30 cm of soil, where the roots of annual plants are also found, which lead to competition for nutrients (Ong et al. 1991). Similarly, observations made on the root systems of safou or *Dacryodes edulis* H.J. Lam showed that lateral roots were densest in the top 20 cm of soil, and extended as much as 5 m from the seedling (Asaah et al. 2010). Root densities of *D. edulis* were high in the top 20 cm of soil for seedlings and 5-year-old plants that had been obtained from marcotting (air layering) of shoots (Asaah et al. 2010), indicating that any association with this species for cultivation should allow spacing greater than 5 m. A preliminary study of the root system of the species or germplasm line that is selected to be included in a tree-crop association with *D. edulis* is necessary to limit competition for soil nutrients and water.

Plants that emit chemical substances that are harmful to other plants (allelopathy) are not suitable for use in agroforestry. Some cases of allelopathy have been observed in some agroforestry species, most notably *Alnus nepalensis*, *Casuarina equisetifolia*, *Eucalyptus tereticornis*, *G. sepium*, *Grevillea robusta*, and

L. leucocephala (Nair 1993, p. 252). An allelopathic compound (3-Hydroxyuridine) was also isolated from Moabi or African pearwood (*Baillonella toxisperma* Pierre; Ohigashi et al. 1989), which is an endemic tree species from the lowland rainforests of Central Africa, and that has high agroforestry potential. While the substance negatively affected the growth of numerous weeds, it did not have any appreciable effect on the growth of some important food crops such as maize (Ohigashi et al. 1989).

Negative aboveground interactions may be more important than negative belowground interactions. For instance, competition for light in a maize-*Tectona grandis* agroforestry system in Nigeria was found to have a greater effect upon the growth of the maize than between-root competition (Verinumbe and Okali 1985). Similarly, shading was of greater importance than root competition in a *Pennisetum glaucum*-peanut agroforestry system in India (Willey and Reddy 1981). It would be wise to learn from the experiences of intercropping that is practiced locally to develop or refine an agroforestry system that minimizes any negative interactions that may occur.

Competition for water is intense, especially in agroforestry systems in semi-arid regions. In semi-arid zones, alley cropping induces competition for moisture between trees and herbaceous plants (Singh et al. 1989). In *Leucaena*-alley cropping trials in the semi-arid tropics, tree competition for water with castor beans (*Ricinus communis* L.), cowpeas (*Vigna unguiculata* (L.) Walp.), and sorghums (*Sorghum spp.*) appeared to be more important than the effects of shading (Singh et al. 1989). Competition for water has also been observed in the temperate zone in a trial associating maize and silver maple in an alley cropping system (*Acer saccharinum* L.; Miller and Pallardy 2001).

Microclimate modification may also have an effect on pests and diseases. Shading promotes the development of microorganisms that attack crops, and the increased humidity found in shady areas can result in an increase in bacterial and fungal growth (Huxley and Greenland 1989). In shady conditions, cocoa (*Theobroma cacao* L.) plants are more likely to be infected by *Phytophthora palmivora* Butler, an oomycete fungus that causes bud rot in palms (Arecaceae or Palmae family), and fruit rot and root rot in papaya (*Carica papaya* L.) and coconut (*Cocos nucifera* L.), and cocoa black pod rot (Alvim 1977; Erwin and Ribeiro 1996; Drenth and Guest 2004).

Cattle can cause mechanical damage to crops, directly by trampling the plants and indirectly by compacting the soil. In addition, the high levels of urea contained in animal liquid waste can further damage plants. Plants, in turn, can negatively affect their herbivores through constitutive and inducible chemical defenses that are mounted in their tissues (Rhodes and Cates 1976). Tannins and other polyphenolics have dispersed biological properties, which are related to their molecular structure and mass (Ayres et al. 1997). Their ingestion may benefit the herbivores feeding them (Aerts et al. 1999; Fernández-Salas et al. 2011), but tannins also have pronounced anti-nutritional effects. Tannins bind problems, thereby reducing both food intake and the digestibility of plant tissues once they have entered the gut (Fahey and Jung 1989; Ayres et al. 1997; Aharoni et al. 1998). For example, the levels of condensed tannins that are found in some plants, such as *Senna siamea* and

Table 7.1 Management techniques to reduce negative interactions among agroforestry species

Interaction	Technique reducing negative effects	Source
Belowground competition	Root pruning	Cannell and Grace (1993)
Belowground competition for *Erythrina*, *Inga*, *G. sepium*, cowpea, castor bean, *Senna siamea*	Root pruning/trenching or pollarding	Muschler (1993); Singh et al. (1989); Schroth (1999)

G. sepium, reduce palatability and, therefore, the acceptability of their leaves as fodder for animals (Copper and Owen-Smith 1985). In addition to their role in herbivore deterrence by directly influencing the host's animal behavior and physiology (Copper and Owen-Smith 1985; Austin et al. 1989), tannins have antimicrobial properties that adversely affect the gut microflora of ruminant mammals (Scalbert 1991).

The toxicity of mimosine or leucanol, a non-protein amino-acid that inhibits the initiation of DNA replication in mitosis and that is found in *Leucaena* forage, has been discussed by Nair (1993). Similarly, some substances found in fodder from *Acacia* or *Robinia* species may be toxic for cattle (Ivory 1990). Toxic substances from agroforestry trees and shrubs have been reported by Nair (1993, adapting data from Devendra 1990 and Lowry (1980)). Toxic substances include cyanoglucosides, fluoracetate and tannins in *Acacia* plants, tannins from *C. calothyrsus, G. sepium,* and *Prosopis* spp., and mimosine found in *Leucaena* spp.

Negative agroforestry interactions such as belowground competition can be minimized through root pruning or trenching (Singh et al. 1989; Schroth 1999; Muschler 1993). Negative interactions in agroforestry can also be minimized through the adoption of cultural practices and designs that minimize specific interactions while maximizing environmental benefits (Jose et al. 2004). Optimal management of agroforestry system components can increase the benefits of their interactions (Table 7.1); however, an understanding of the root system and the needs of the associated species are of prime importance.

7.2.3 Neutral Interactions

Neutral interactions in agroforestry occur when trees and crops exploit the same pool of resources so that increased capture by one species results in proportionally decreased capture by the associated species (Fig. 7.2; Ong and Leakey 1999). This usually occurs in dry areas. For example, savanna grasses exploit water from topmost soil layers, whereas tree roots have exclusive access to deeper water, creating a clear niche separation (Weltzin and Coughenour 1990; Deans et al. 1995). Because agroforestry systems are built on the assumption that trees must acquire resources that crops would not otherwise acquire, neutral interactions allow better utilization of soil nutrients and water.

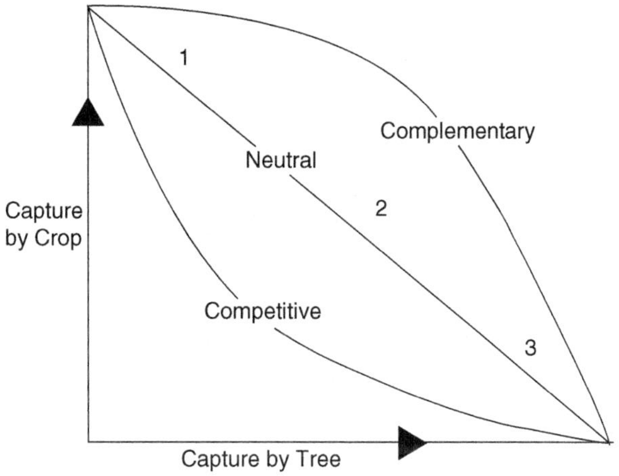

Fig. 7.2 Resource capture by trees and crops showing competitive, complementary and neutral interactions. *1* Parklands and savannas, *2* Boundary planting, *3* Alley cropping (Ong and Leakey 1999)

7.3 Soil Productivity in Agroforestry

Rationales for agroforestry include the sustainable management of soil, and increased and diversified agricultural production. The assumption underlying agroforestry is that *appropriate agroforestry systems improve soil physical properties, maintain soil organic matter, and promote nutrient cycling* (Sanchez 1987). The beneficial effects of trees on soils are summarized in Fig. 7.3.

Trees play two key roles in agroforestry: fertilization and soil conservation. Fertilization occurs via symbiotic N_2-fixation, nutrient cycling, and involvement in the formation of the humus layer. Soil conservation is accomplished through erosion control and soil stabilization by roots which maintain soil physical properties. In the humid tropics, soils are dominated by Oxisols and Ultisols (40 %), which are acidic and infertile, and highly leached. Soils of moderate fertility, such as Alfisols, Vertisols, Mollisols and Andisols, support about 23 % of tropical forests (Nair 1993; Bekunda et al. 2010). In the humid lowlands of tropical Africa, soils are not only poor, but exhibit toxicity due to their high aluminum contents. Inherent low fertility, coupled with shifting agricultural practices and the reduced lengths of fallow periods that are attributable to increasing demographic pressure, has led to the development and popularization of improved fallows and alley cropping techniques that utilize legumes (Table 7.2 and 7.3).

Tropical soils are diverse, and many do not contain rhizobia, which perform symbiotic nitrogen fixation (Lal and Sanchez 1992; Nwaga et al. 2010). The role of trees in productivity of soils and soil conservation is widely documented (Kang and Wilsom 1987; Sanchez 1987; Juo 1989; Kang et al. 1990; Avery et al. 1990; Szott et al. 1991; Rhoades 1996), as well as the role of agroforestry in soil conservation in general (Wiersum 1986; Lundgren and Nair 1985). Leaves falling from trees, together with twigs and leaves that accumulate following pruning, are

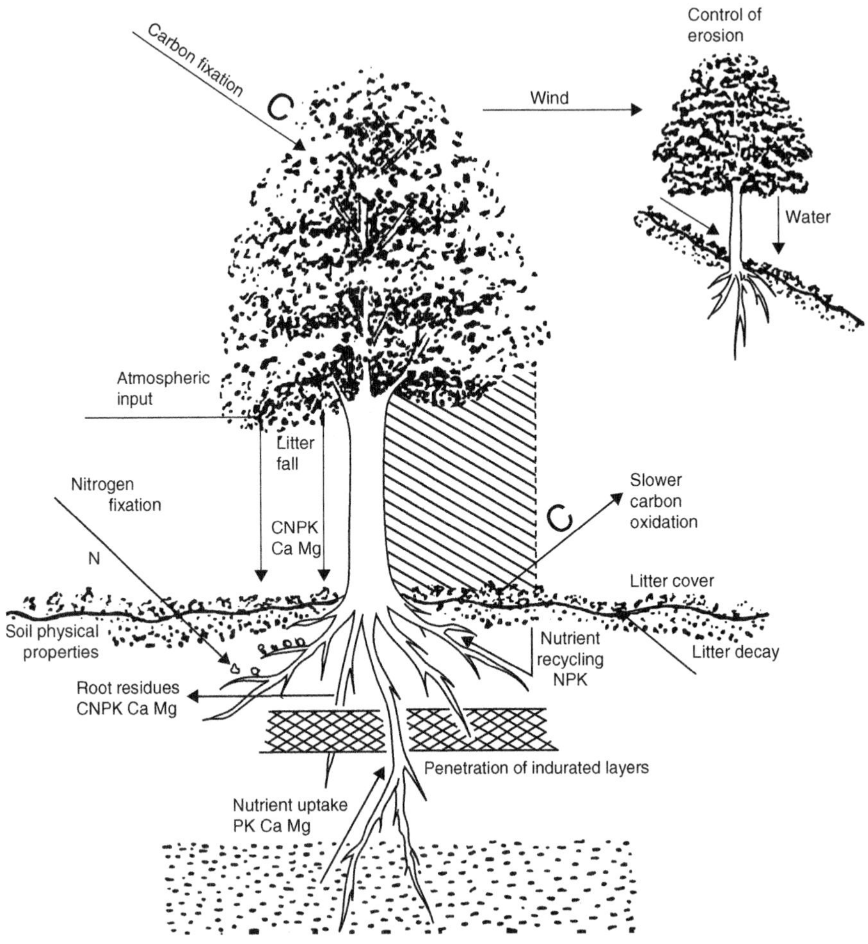

Fig 7.3 Processes by which trees improve soils (Young 1989b)

Table 7.2 Geographic distribution of acid soils in the humid tropics (Szott et al. 1991)

	Humid tropical America		Humid tropical Africa		Humid tropical Asia and Pacific		Total humid tropics	
Soil order	(10^6) ha	%	(10^6) ha	%	(10^6) ha	%	(10^6) ha	%
Oxisols	332	50	179	40	14	4	525	35
Ultisols	213	32	69	16	131	35	413	28
Psamments	6	1	67	15	17	4	90	6
Spodosols	10	2	3	1	6	2	19	1

important to humus formation and to soil carbon budgets (Table 7.4). However, trees that produce small quantities of litter seem to have a favorable effect on soil C compared to monocultures (Schroth et al. 2002). Agroforestry practices may be responsible for the accumulation of up to 9 tons of C ha^{-1} of soil in semi-arid

Table 7.3 Main constraints of tropical soils (Sanchez and Logan 1992)

Soil constraint	Humid tropics		Arid savannas		Semiarid tropics		Tropical steeplands		Tropical wetlands		Total	
	million ha and (%)											
Low nutrient reserves	939	(64)	287	(55)	166	(16)	279	(26)	193	(16)	803	(6)
Al toxicity	808	(56)	261	(50)	132	(13)	269	(29)	23	(4)	1493	(32)
Acidity without Al toxicity	257	(18)	264	(50)	298	(29)	177	(16)	164	(29)	1160	(25)
High P fixation by Fe oxides	537	(37)	166	(32)	94	(9)	221	(20)	0	(0)	1018	(22)
Low CEC	165	(11)	19	(4)	63	(6)	2	(–)	2	(–)	251	(5)
Calcareous reaction	6	(0)	0	(0)	80	(8)	60	(6)	6	(1)	152	(5)
High soil organic matter	29	(2)	0	(0)	0	(0)	–	(0)	40	(7)	69	(1)
Salinity	8	(1)	0	(0)	20	(2)	–	(0)	38	(7)	66	(1)
High P fixation by allophane	13	(1)	2	(0)	5	(0)	26	(2)	0	(0)	50	(1)
Alkalinity	5	(0)	0	(0)	12	(1)	–	(0)	33	(0)	50	(1)
Total area	1444	(100)	525	(100)	1012	(100)	1086	(100)	571	(100)	4637	(100)

[a] Percentages are in brackets

regions, 21 tons of C ha^{-1} in sub-humid areas, and 50 tons C ha^{-1} in humid zones (Schroeder 1994).

The presence of trees in alley cropping systems helps to recycle nutrients, reduce nutrient leaching losses, stimulate the activities of soil fauna, improve soil fertility, maintain high levels of crop production, and control soil erosion (Kang 1997). Maintaining the productivity of tropical soils is possible by applying the following techniques (Kang 1997):

- Maintenance of organic matter and adequate coverage on the soil surface
- Minimal disruption of the soil surface
- Appropriate use of fertilizers
- Alley cropping
- Fallow and crop rotation

The maintenance of soil organic matter content is possible if burning before crop establishment is eliminated, but that involves extra work. Burning is seen as a means of clearing land with little effort in tropical rainforests, and suppressing weeds. Alley cropping contributes efficiently to the deposition of organic matter such as leaves, twigs and fruit onto the soil surface. Long-term fallows also contribute to restoring soil fertility.

The nutrient cycle involves continual transfers among the different compartments of the ecosystem. The cycle encompasses the disintegration of minerals, activities

Table 7.4 Dry matter and nutrient input via litterfall or pruning in production systems in the humid tropics (Szott et al. 1991)

Systems	Dry matter (tonnes ha^{-1} yr^{-1})	Nutrient inputs (kg ha^{-1} yr^{-1})					Source
		N	P	K	Ca	Mg	
Fertile soils							
Rainforests	10.5	162	9	41	171	37	1
High input cultivation							
Alley cropping							
L. leucocephala[1]	22.0	200–280					3[a]
G. sepium[1]	11.0	171–205					3[a]
Sesbania[1]	7.5	25–110					3[a]
L. leucocephala	5–6.5	160	15	150	40	15	4[b]
Erythrina poeppigiana[1]	9.6	278	24	216	120	52	5[c]
G. sepium[1]	12.3	358	28	232	144	60	5[c]
L. leucocephala	8.1	276	23	122	126	31	6[b]
Erythrina spp.	8.1	198	25	147	111	26	6[b]
Shade systems							
Coffee/*Erythrina*	172.2 (13.5)	366 (182)	30 (21)	264 (156)	243 (131)	48 (27)	7
Coffee/*Erythrina*/*Cordia*	15.8 (9.1)	331 (75)	22 (8)	162 (45)	328 (46)	69 (12)	7
Coffee/*Erythrina* pruned[1]	20.0 (12.2)	461 (286)	35 (24)	259 (158)	243 (121)	76 (43)	8[d]
Coffee/*Erythrina* non-pruned[1]	7.6 (2.0)	175 (55)	11 (4)	75 (14)	122 (40)	33 (9)	8[d]
Cacao/ *Erythrina*[1]	6.5 (2.5)	116 (62)	6 (4)	40 (13)	116 (47)	41 (12)	8[d]
Cacao/*Cordia*[1]	5.8 (2.9)	95 (60)	11 (8)	57 (33)	108 (58)	43 (23)	8[d]
Cacao/mixed shade	8.4	52	4	38	89	26	9
Cacao/ *Erythrina*[1]	6.0	81	14	17	142	42	10
Infertile soils	8.8	108	3	22	53	17	1
Rainforest/Oxisols-Ultisols	7.4	48	2	22	63	10	1
Spodosol							
Savanna-Oxisol	3.5	25	5	31	10	11	2
Low input cultivation-Ultisol	6.0	77	12	188	27	12	2
Alley cropping-Ultisol							
Inga edulis	5.6	136	10	52	31	8	11[b]
Erythrina spp.	1.9	34	4	19	8	4	11[b]
Inga edulis	12.5						12[b]
Cassia reticulata	6.5						13[b]
G. sepium	1.4						13[b]
Shade systems							
Erythrina spp. Inceptisol	11.8–18.4	170–238	14–24	119–138	84–222	27–56	14[c]

[1] Fertilized and limed; originally an acid, infertile soil

+The numbers in parentheses represent litter production by *Erythrina*; the number to the left of the parentheses is total litter production

[a] Based on 2 m hedge spacing

[b] Based on 4 m hedge spacing

[c] Based on 6 m hedge spacing in 1st year, 3 m in other years. *Erythrina* spacing was 3 m x 6 m

[d] Plant densities: coffee (5000 ha^{-1}), *Erythrina* (555 ha^{-1}), *Cordia* (278 ha^{-1})

[e] Plant densities: coffee (4300 ha^{-1}), *Erythrina* (280 ha^{-1})

1 Source: (1) Vitousek and Sanford (1987); (2) Sanchez et al. (1989); (3) Duguma et al. (1988); (4) Kang et al. (1984); (5) Kass et al. (1989); (6) A. Salazar (unpublished); (7) Glover and Beer (1986); (8) Alpizar et al. (1983); (9) Boyer (1973); (10) FAO (1985); (11) Szott (1987); (12) Palm (1988); (13) A. Salzar (unpublished); (14) Russo and Budowski (1986)

of soil organisms, and changes that occur in the rest of the biosphere, and in the atmosphere, lithosphere, and hydrosphere (Golley et al. 1975; Jordan 1985). Plants require nutrients for growth. These nutrients come from the weathering of rocks and the mineralization of organic matter. Released nutrients are accumulated in plant tissues, and when the plants shed their leaves (or other parts), the litter that is produced decomposes, and the resulted processed or organic matter mineralization provide nutrients to the roots. This is the key process of nutrient cycling (Fig. 7.4). The cycles of nitrogen, phosphorus, and potassium vary considerably. Nutrients enter the cycle through rainfall, dust, weathering of rocks and nitrogen fixation. Losses occur through erosion, leaching, harvesting of crops, burning, denitrification and volatilization, particularly that of nitrogen (Nair 1993). A comparison of nutrient cycles in agricultural and forest systems is shown in Fig. 7.5.

Some agroforestry species are more efficient at absorbing nutrients from the soil. Sharma et al. (1995) observed greater production in a mandarin-*Albizia* agroforestry system due to the increased efficiency of nutrient utilization under *Albizia*. In any agroforestry system, the contribution of nutrients by trees must be synchronized with the needs of the crop. Field trials assessing the contributions of agroforestry trees to the nutrient requirements of intercropped plants showed that as much as 80 % of the nutrients are released during annual crop growth, but less than 20 % is absorbed by the crop (Palm 1995). The nutrients that are not used by the crops are most frequently lost through leaching, and efforts should be made to limit the system's loss of nutrients. Nutrient inputs affect productivity. A homegarden can produce more than double of what a rice farm in the same region can produce (Jensen 1993). A comparison of production from a forest and two agroforestry systems in India showed that an *Alnus*-cardamom system had the highest levels of nitrogen and phosphorus in the soil (Table 7.5).

Another study showed that in an agroforestry system (*Acacia*, *Eucalyptus*, *Populus*, or rice-based), inorganic soil nitrogen pools were higher by 8 to 74 % higher, and mineralization was 12–37 % higher than in a rice monoculture system (Kaur et al. 2000). Similarly, the presence of trees in the system increased soil carbon content by 11 to 52 % (Kaur et al. 2000). Litter that is produced in an agroforestry system consisting of polyculture system is roughly equal to that produced in the Amazon primary rainforest (Martius et al. 2004), but the rate of litter decomposition was higher in the agroforestry system. This indicates a qualitative and quantitative change of the active elements (N, C and P) in the system's decomposition in Amazonia. For a system to be sustainable, production must be higher than losses. If mulching is done in alley cropping, some species are able to introduce as much as 100–200 kg of nitrogen ha^{-1}, which is roughly the amount of nitrogen that is exported in a mixed cereal/legume system (Young 1989).

The quantity of certain nutrients that are stored in the soil is also important in agroforestry. For example, palms accumulate potassium in the soil (Folster et al. 1976), *Gmelina arborea* Roxb. trees accumulate calcium (Sanchez et al. 1985), and *Cecropia* spp. that are grown in acidic soils accumulate calcium and phosphorus (Odum and Pigeon 1970). An increase in the number of trees in an agroforestry system leads to an increase in the amount of litter, and carbon and nitrogen in the soil

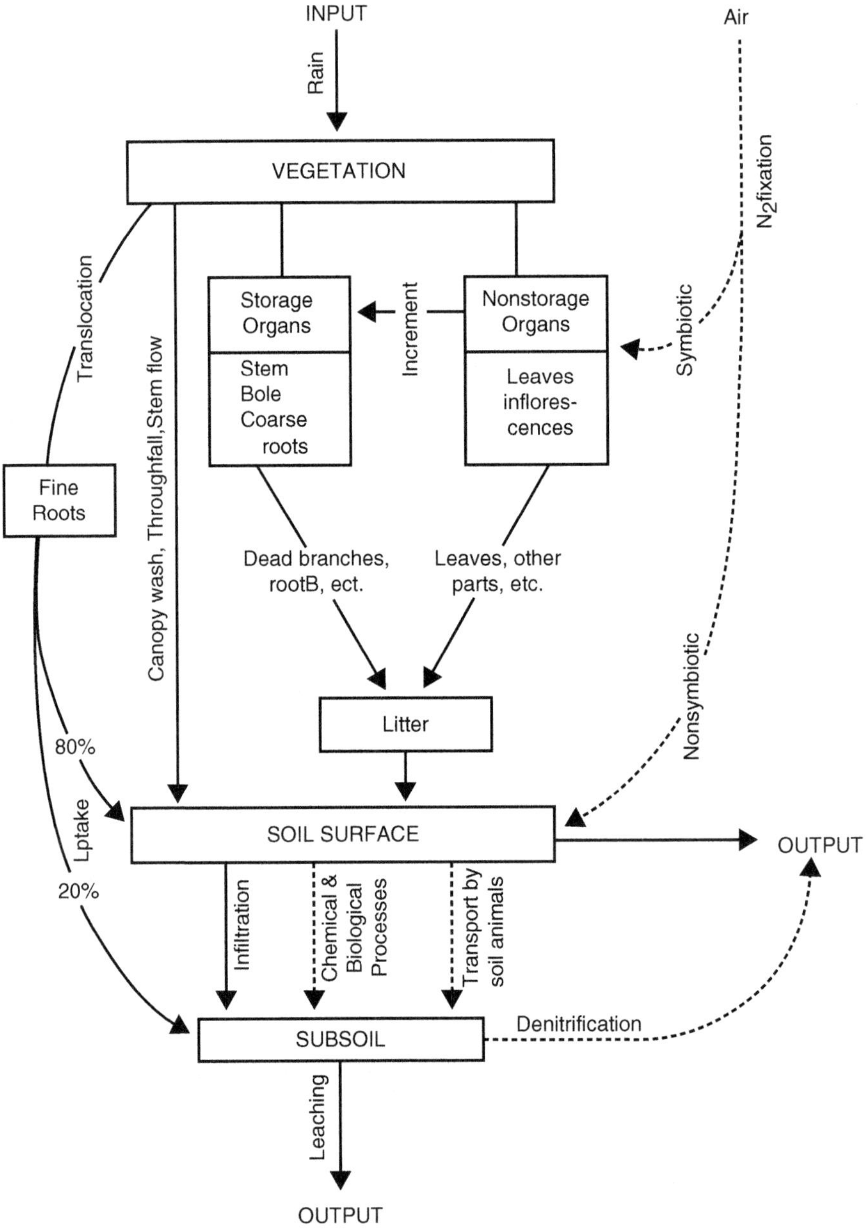

Fig. 7.4 A simplified model of nutrient cycling in a forest ecosystem (Nair 1984)

(Table 7.6). This increase, however, results in a decrease in agricultural production and less carbon being absorbed by the roots (Table 7.7). The decomposition of this organic matter varies with the specific agroforestry species being used (Constantinides and Fownes 1994).

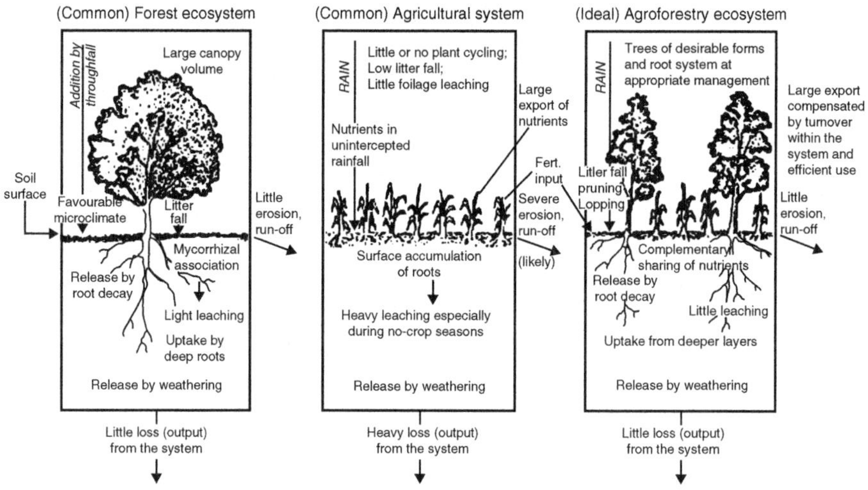

Fig. 7.5 Schematic representation of nutrient relations and advantages of "ideal" agroforestry systems in comparison with common agricultural and forestry systems (Nair 1984)

Table 7.5 Annual litter production, floor litter and nutrient contribution in different agroforestry systems in the Mamlay Watershed, Sikkim, India (Sharma et al. 1997)

Litter	Agroforestry systems			
	Forest-cardamom	*Alnus*-cardamom	*Albizia*-mandarin	Mandarin
Production (kg ha^{-1} $year^{-1}$)				
Tree leaf and twig[a]				
Biomass	2905	2851	1511	1657
Nitrogen	38.43	75.89	25.34	22.87
Phosphorus	3.02	3.62	2.10	2.12
Cardamom/crop residue[b]				
Biomass	1669	4419	2238	2172
Nitrogen	14.71	31.98	40.51	39.31
Phosphorus	1.54	4.99	5.19	5.04
Stand total				
Biomass	4574	7270	3749	3829
Nitrogen	53.14	107.87	65.85	62.18
Phosphorus	4.56	8.61	7.29	7.16
Floor standing state (kg ha^{-1})[c]				
Biomass	5280	6881	3942	4074
Nitrogen	66.72	137.87	45.75	37.62
Phosphorus	5.36	7.93	6.65	6.77
Production: floor ratio (biomass)	0.866	1.057	0.951	0.939

Each value given above is the mean of at least three replicates, with the standard error of the estimate being ach v in all the cases

[a] Monthly values over a 2-year period were used for calculating leaf and twig litter production.

[b] Cardamom/crop residue at the time of crop harvest.

[c] Includes aboveground tree litterfall mass and cardamom/crop residues.

Table 7.6 Leaf-litter fall, litter fall-derived organic carbon and total nitrogen to soils in a Dalbergia sissoo Roxb. plantation in 1995 and cumulative additions (Chander et al. 1998)

Spacing of D. sisso planting (m)	Number of trees planted ha^{-1}	Cumulative additions[1] ($kg\ ha^{-1}$)			Cumulative additions[1] ($kg\ ha^{-1}$)		
		Litterfall	Organic C	Total N	Litterfall	Organic C	Total N
Number of trees (control)	0	0	0	0	0	0	0
10 × 10	100	521	225	14	3390	1465	88
10 × 5	200	813	350	21	5276	2274	136
5 × 5	400	1351	582	35	10044	4329	242

Table 7.7 Average yield of intercrops and cumulative root-derived organic carbon and total nitrogen additions to soils (Chander et al. 1998)

Spacing of D. sissco plantings (m)	Average above ground dry matter yields[a] ($kg\ ha^{-1}$)		Average root biomass ($kg\ ha^{-1}$)		Estimated root-derived organic C[b] and total N[c] inputs ($kg\ ha^{-1}$)		Total estimated organic C and total N inputs to soils by the crop tree cultivation system ($kg\ ha^{-1}$)	
	Wheat[d]	Cowpea[e]	Wheat	Cowpea	Organic C	Total N	Organic C	Total N
No trees (control)	11290	3860	3387	1544	6135	120	6135	120
10x10	9450	3220	2835	1288	5565	101	7026	189
10x5	8720	2800	2616	1120	5043	90	7317	226
5x5	6600	1310	1980	393	3204	47	7533	289

[a] Values are those by averaging the harvesting above-grounded dry matter for three conservative year (1993–1995)
[b] Values were calculated on the basis that both wheat and cowpea roots contained an average of 45% organic C
[c] Values were calculated on the basis that wheat and cowpea roots contained an average of 0.5 and 1.5% total N, respectively
[d] Values include yields of both wheat grain and straw
[e] Values represent the dry-fodder yield of 70-day-old crops

The carbon content of the system after one year is given by the following formula (Young 1989):

$$C_1 = C_{0-} kC_0,$$

where k is the decay rate, C_1 is the carbon content after one year, and C_0 is the initial carbon content. The overall reduction in soil humus follows a simple exponential decay function, which is described by the equation below:

$$C_t = C_0 e^{-rt}$$

where C_t is the carbon content after t years and r is a constant (equal to k). Based on this relationship, the half-life of the soil humus (during which half of the carbon is oxidized) is calculated by the formula:

$$\text{Half-life} = 0.693/r$$

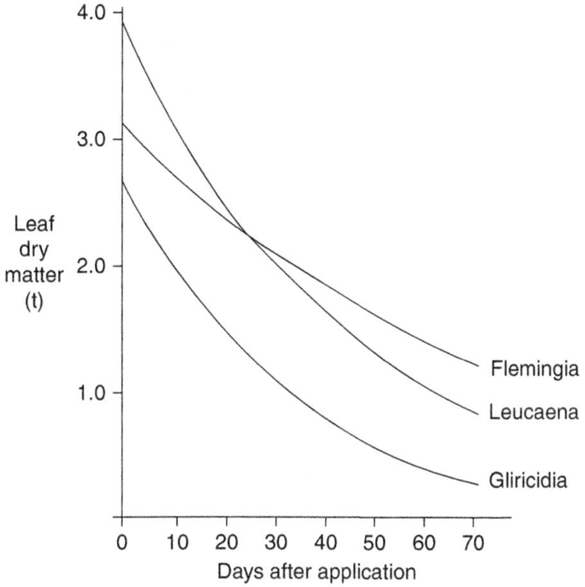

Fig. 7.6 Mulch decomposition patterns of three multipurpose trees of Côte d'Ivoire (Budelman 1988)

Improved methods for estimating carbon storage are recommended (Nair et al. 2009). About 5 to 10 kg C ha^{-1} can be stored after 25 years in an extensive tree-annual plants system in arid and semi-arid zones, and 100 to 200 kg of C ha^{-1} after 10 years in a multi strata perennial system under shade and in homegardens in the tropics (Nair et al. 2009).

Generally, litter obtained from plants that are rich in nitrogen breaks down quickly, and is considered to be of good quality. Litter from lignified material that decomposes slowly is considered to be of poor quality. Material that is rich in nitrogen with a low

C: N ratio decomposes quickly and releases large amounts of nitrogen. In contrast, carbon-rich material with a high C:N ratio decomposes slowly and allows a rapid increase of microbial growth due to excess carbon, an excellent energy source. Microbes will eventually reduce reserves of soil nutrients such as nitrogen. When the source of carbon is used up, the microbial population declines and the nitrogen that was incorporated into microbial tissues is released and made available to plants (Nair 1993). Branches and leaves obtained from pruning of trees used in alley cropping are rich in nitrogen, and make excellent litter, which are of benefit to the crops. The rate of litter decomposition varies, depending on the species that are used (Fig. 7.5). Concentrations of soluble polyphenols in the leaves of legumes vary according to species, and influence the rate of decomposition of these leaves and, thus, soil nitrogen content (Palm and Sanchez 1991). This trend is also observed for potassium. The release of potassium into the soil is fastest with L. leucocephala leaves, followed by leaves of G. sepium and Flemingia macrophylla (Willd.) Merr. (Budelman 1988; Fig. 7.6 and 7.7).

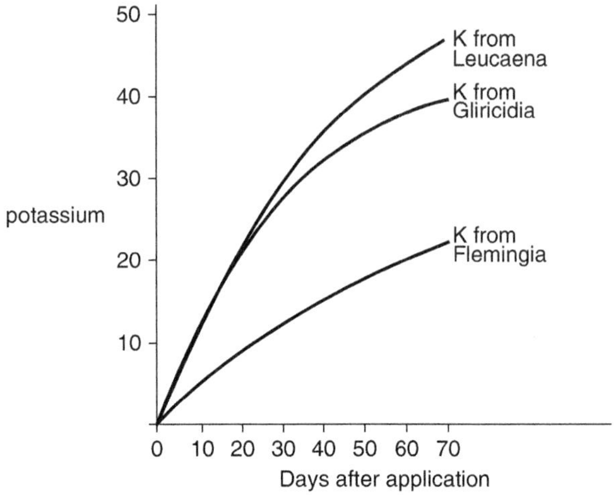

Fig. 7.7 The release of potassium from the mulches of three multipurpose trees in the humid lowlands of Côte d'Ivoire (Budelman 1988)

Generally, the release of P, K, Ca and Mg is faster from decomposing leaves of Erythrina leaves compared to those of Inga edulis or Cajanus cajan (Palm and Sanchez 1990; Szott et al. 1991). The decomposition of litter can also vary depending on whether it is derived from the aerial parts of a plant, or from its roots. Vanlauwe et al. (1996) observed higher lignin content in the roots of Leucaena (8 %) and Dactyladenia (26 %) compared to nitrogen (2 and 1 %, respectively). The leaves of these two species contain 2 and 13 % lignin, and 2.9 and 4 % nitrogen, respectively. The difference in lignin content in roots and leaves could explain why leaves decompose faster than roots, and release nutrients more rapidly into the soil (Lehman et al. 1998). Because of their rapid decomposition, leaves can be used for nitrogen inputs in the short-term, whereas woody materials such as roots would supply soil nutrients more slowly over time. Incorporation of the leaves of two agroforestry species into the soil showed quantitative differences in nutrients released into the soil after decomposition (Teklay and Malmer 2004). Species with leaves that decompose rapidly, such as Leucaena, Gliricidia, and Erythrina, can be used in combination with fast-growing crops like maize. Species with leaves that decompose slowly (Senna siamea, Flemingia macrophylla, Dactyladenia barteri, for example) can be used to control weeds and reduce soil moisture loss. Moreover, they and can be grown in association with slow-growing plants to provide greater synchrony between rates of nutrient release and nutrient uptake by plants. The choice of an agroforestry species should be based on the decomposition rate of mulch that it provides. Pruning period adjustments are also necessary to regulate the timing of mulch additions, as well as the method of mulch application (ground cover, or incorporation of leaves and twigs into the soil).

One promising mulching agrotechnology for regenerating soils is Ramial Chipped Wood (RCW) that was developed in the Department of Wood and Forest Sciences, Université Laval, Quebec city (Quebec, Canada). RCW consists of young wood from woody plant species, preferentially that belonging to nitrogen-fixing trees (such as *Leucaena leucocephala*) as a source of organic fertilizer, which are chipped to provide a mulch. Depending on the growing conditions, tree or shrub stems reach appropriate dimensions for coppicing with diameters of 5 cm (1–1.5 m in height). Small-diameter branches or stems less than 7 cm in diameter (more than 75 % of nutrients are stored in twigs) are chipped using mobile mechanical equipment. Carbon-nitrogen ratios (C/N) of the resulting chipped materials should be <20:1, as higher ratios incur the risk of temporary nitrogen shortages for the crops.

RCW maintains and stimulate soil biodiversity and water balance, and contributes to erosion control. Legume and actinorhizal trees or shrubs that have been planted in alley cropping can be inoculated with strains of Rhizobium or Frankia, respectively, to promote nitrogen fixation these species can be further inoculated with arbuscular mycorrhiza fungi (AMF) to promote absorption of other nutrients, including phosphorus. RCW can be transformed into a "soil food" for feeding soil microfauna and microflora, thereby bringing mid- and long-term benefits to both agricultural and forest ecosystems at the lowest cost (Lemieux 1993; Caron et al. 1998). RCW is a good tool that is available to all societies, even the poorest ones, to reverse soil degradation and desertification. Further, the use of RCW is the key to understanding the biological basis of our terrestrial ecosystems and of pedogenetic processes.

One of the important benefits of agroforestry is the production of tree biomass, especially in terms of sequestering atmospheric carbon. Tree biomass consists of stem, fruit, leaves and roots. Productivity varies from biome to biome as a function of climate. In Sahelian Africa, the biomass of carbon stock in trees in an improved agroforestry system can reach 54 Mg ha^{-1} (Takimoto et al. 2008). In humid tropical zones, the biomass that is produced by trees in an agroforestry system is similar to that of natural ecosystems (Nair 1993, p. 297), reaching values as high as 20 to 40 Mg ha^{-1} year^{-1}. In the Cameroon rainforest, Terminalia ivorensis A. Chev. agrisilviculture can accumulate 71 Mg ha^{-1} year^{-1} in the ninth year after establishment, and 84 Mg ha^{-1} year^{-1} from the twentieth year onward (Norgrove and Hauser 2002). In terms of soil fertility management, trees branches and leaves are usually pruned to provide mulch. Thus, the growth rate of trees, together with the frequency of their pruning, are important factors to be considered in the choice of woody species to be introduced. Thus, the growth rate of trees, together with the frequency of their pruning, are an important factors to be considered in the choice of woody species to be introduced. For example, Acacia has a faster rate of biomass production than does Leucaena, which is a much slowly growing plant (Table 7.8).

Root growth constitutes a vital component of tree production. Roots add fixed carbon to the soil through exudate production and tissue death, while providing structural stability to the plant. Root turnover plays an important role in the release of litter into the soil, because renewal of roots depends on litter quantity. For example, production of 1.66 tons of fine roots per ha per year was observed in Brazilian Eucalyptus plantations (Jourdan et al. 2008). Root biomass also varies with plant density (Puri et al. 1994). Populus deltoides plants produce 71.5 tons of roots per ha at a density of 6 × 6 m, and 251.5 tons per ha at a density 2 × 2 m (Puri et al. 1994). Alley cropping tends to reduce root length density in crops such as cassava (Lose

Table 7.8 Mean biomass accumulation (Mg ha^{-1}) in multipurpose trees of three age sequences in Kerala, India (Kumar et al. 1998)

Species/age	Bole	Branch	Foliage	Roots	Total aboveground		
					Biomass	MAI	n
Woodlot (8.8 years of age)							
[1]Acacia auriculiformis	274.93a	42.55a	8.95a	17.73a	362.43a	37.09	31
Ailanthus triphysa	27.78bc	7.68b	4.08b	7.40a	40.54bc	4.61	30
Artocarpus heterophyllus	54.38bc	19.85a	7.78ab	10.13a	82.01b	9.32	32
Artocarpus hirsutus	32.15bc	14.83b	11.95d	11.15a	58.93bc	6.70	28
[1]Casuarina equisetifolia	73.25b	16.63b	5.70b	5.60a	95.58b	10.86	26
Emblica officinalis	46.20a	18.23b	4.43b	12.63a	68.86bc	7.83	17
[1]Leucaena leucocephala	15.28e	6.25b	1.28c	3.23a	22.81e	2.59	18
[1]Paraserianthes falcatharia	141.18d	37.25a	5.05b	13.78a	183.48d	20.85	19
Pterocarpus marsupium	52.60bc	10.08b	3.43b	7.3a	66.11bc	7.51	30
Silvopasture (7 years of age)							
Acacia auriculiformis	13.119a	31.17a	19.18a	nd	183.54a	26.22	113
Ailanthus triphysa	15.79b	1.45b	2.05b	nd	19.38b	2.77	120
Casuarina equisetifolia	27.21b	4.72b	1.75b	nd	33.68b	4.81	86
Leucaena leucocephala	51.91b	10.49b	1.11b	nd	63.51b	9.07	109
Silvopasture (5 years of age)							
Acacia auriculiformis	111.8a	17.5a	11.1a	16.3a	140.5a	28.1	21
Ailanthus triphysa	16.8c	1.2b	1.8b	4.2a	19.8bc	3.96	20
Casuarina equisetifolia	28.2b	4.3b	3.4b	3.4a	35.9b	7.18	15
Leucaena leucocephala	51.7b	10.0c	4.1b	12.0a	65.8b	13.16	20

MAI: Mean annual increment (Mg ha^{-1} $year^{-1}$); values followed by the same superscript do not differ significantly (age classes were analyzed separately); [1] =exotics; nd no determined; n=number of trees sampled

et al. 2003). However, the average root biomass of trees is generally lower than that of grasses, although exact root density is dependent on site conditions.

The highest densities in root length ([1]) in an alley cropping system of Acacia saligna and Sorghum bicolor in an arid zone were observed between 0 and 15 cm of depth (Lehman et al. 1998). Branch pruning had reduced the root length density of the trees in this system by 47% (Lehman et al. 1998). However, root length density during the dry season was higher than that observed in the crop alley during the wet season (Lehman et al. 1998). The combination of trees and herbaceous plants uses water more efficiently than a monoculture system. Root density is increased and the distance between plant roots is reduced, increasing the likelihood of inter-plant competition in agroforestry systems (Young 1989).

Bibliography

Aerts RJ, Barry TN, McNabb WC (1999) Polyphenols and agriculture: beneficial effects of proanthocya- nidins in forages. Agricult Ecosys Environ 75:1–12

Aharoni Y, Gilboa N, Silanikove N (1998) Models of suppressive effect of tannins. Analysis of the suppressive effect of tannins on ruminal degradation by compartmental models. Animal Feed Sci Technol 71:251–267

Alvim P de T (1977) Cacao. In: Alvim P de T, Kozlowski TT (eds) Ecophysiology of Tropical Crops. Academic Press, New York, pp 279–313

Asaah EK, Tchoundjeu Z, Wanduku TN, Van Damme P (2010) Understanding structural roots system of 5-year-old African plum tree (D. edulis (G. Don) H. J. Lam) of seed and vegetative origins. Trees 24:789–796

Austin PJ, Suchar LA, Robbins CT, Hagerman AE (1989) Tannin-binding proteins in saliva of deer and their absence in saliva of sheep and cattle. J Chem Ecol 15:1335–1339

Avery ME, Cannell MGR, Ong CK (eds) (1990) Biophysical research for Asian agroforestry. Oxford Press, India

Ayres MP, Clausen TP, MacLean Jr SF, Redman AM, Reichardt PB (1997) Diversity of structure and antiherbivore activity in condensed tannins. Ecology 78:1696–1712

Bekunda M, Sanginga N, Woomer PL (2010) Restoring soil fertility in sub-Sahara Africa. Advances Agronomy 108. ISSN 0065-2113, DOI: 10.1016/S0065-2113(10)08004-1

Budelman A (1988) The decomposition of the leaf mulches of *Leucaena leucocephala, Gliricidia sepium* and *Flemingia macrophylla* under humid tropical conditions. Agrofor Syst 7:33–45

Camero A, Ibrahim M, Kass M (2001) Improving rumen fermentation and milk production with legume-tree fodder in the tropics. Agrofor Syst 51:157–166

Cannell MGR, Grace J (1993) Competition for light: detection, measurement and quantification. Can J For Res 23: 1969–1979

Cannell MGR, Noordwijk M van, Ong CK (1996) The central agroforestry hypothesis: the trees must acquire resources that the crop would not otherwise acquire. Agrofor Syst 34:27–31

Caron C, Lemieux G, Lachance L (1998) Regenerating soils with ramial chipped wood. Publication N 83, Coordination Group on Ramial Wood, Department of Wood and Forestry Science, Université Laval. Dépôt légal: Bibliothèque Nationale du Québec, ISBN 2-921728-32–X

Chander K, Goyal S, Nandal DP, Kapoor KK (1998) Soil organic matter, microbial biomass, and enzyme activities in a tropical agroforestry system. Biol Fert Soils 27:168–172

Constantinides M, Fownes JH (1994) Nitrogen mineralization from leaves and litter of tropical plants: Relationship to nitrogen, lignin and soluble polyphenol concentrations. Soil Biol Chem 26(1):49–55

Cooper SM, Owen-Smith N (1985) Condensed tannins deter feeding by browsing ruminants in South African savanna. Oecologia 67:142–146

Deans JD, Lindley DK, Munro RC (1995) Deep beneath trees in senegal, annual report 1993–1994, 12–14. Institute of Terrestrial Ecology, Penicuik Scotland pp 12–14

Drenth A, Guest DI (2004) Diversity and management of phytophthora in Southeast Asia. ACIAR Monograph No 114, pp 238

Duguma B, Kang BT, Okali D (1988) Effects of pruning intensities of tree woody leguminous species grown in alley cropping with maize and cowpea on an alfisol. Agrofor Syst 6:19–35

Erwin DC, Ribeiro OK (1996) Phytophthora diseases worldwide. American Phytopathological Society Press, St Paul

Fahey GC, Jung HG (1989) Phenolic compounds in forages and fibrous feedstuffs. In: Cheeke PR (ed). Toxicants of plant origin. Vol. IV Phenolics. CRC Press Inc., Florida pp 123–190

Fernández-Salas A, Alonso-Díaz MA, Acosta-Rodríguez R, Torres-Acosta JFJ, Sandoval-Castro CA, Rodríguez-Vivas RI (2011) In vitro acaricidal effect of tannin-rich plants against the cattle tick Rhipicephalus (Boophilus) microplus (Acari: Ixodidae). Veterinary Parasitol 175(1–2):113–118

Folster H, de las Salas G, Khanna P (1976) A tropical evergreen forest site with perched water table. Oecologia Plantarum 11: 297–320

Golley FB, McGinnis JT, Clements RG, Child GI, Duever MJ (1975) Mineral cycling in a tropical forest moist ecosystem. University of Georgia Press, Athens

Huxley PA (1983) Some characteristics of trees to be considered in agroforestry. In: Huxley PA (ed), Plant research and agroforestry. ICRAF, Nairobi, pp 3–12

Ivory DA (1990) Major characteristics, agronomic features, and nutritional value of shrubs and tree fodders. In: Devendra C (ed) Shrub and tree fodders for farm animals: Proceedings of a workshop in Denpasar, Indonesia, July 1989. IDRC, Ottawa

Jensen M (1993) Productivity and nutrient cycling of a Javanese homegarden. Agrofor Syst 24:187–201
Jordan CF (1985) Nutrient cycling in tropical forest ecosystems. John Wiley, New York
Jose S, Gillespie AR, Seifert JR, Biehle DJ (2000) Defining competition vectors in a temperate alley cropping system in the mid-western USA. 2. Competition for water. Agrofor Syst 48:41–59
Jose S, Gillespie AR, Pallardy SG (2004) Interspecific interactions in temperate agroforestry. Agrofor Syst 61:237–255
Jourdan C, Silva EV, Gonçalves JLM, Ranger J, Moreira RM, Laclau JP (2008) Fine root production and turnover in Brazilian *Eucalyptus* plantations under contrasting nitrogen fertilization regimes. For Ecol Manag 256:396–404
Juo ASR (1989) New farming systems development in the wetter tropics. Experi Agric 25:145–163
Kang BT (1997) Alley cropping-soil productivity and nutrient recycling. For Ecol Manag 91:75–82
Kang BT, Wilson GF (1987) The development of alley cropping as a promising agroforestry technology. In: Steppler HA, Nair PKR (eds) Agroforestry: A decade of development. ICRAF, Nairobi pp 227–243
Kang BT, Reynolds L, Atta-Krah AN (1990) Alley farming. Advances in Agronomy 43:315–359
Kang BT, Wilson GF, Sipkens L (1981) Alley cropping maize (*Zea mays* L.) and Leucaena (*Leucaena leucocephala* Lam.) in southern Nigeria. Plant Soil 63:165–179
Kho RM (2000) A general tree-environment-crop interaction equation for predictive understanding of agroforestry systems. Agric. Ecosys Environ 80(1–2):87–100
Kaur B, Gupta SR, Singh G (2000) Soil carbon, microbial activity and nitrogen availability in agroforestry systems on moderately alkaline soils in northern India. Applied Soil Ecol 15(3):283–294
Kumar MB, Suman JG, Jamaludheen V, Suresh TK (1998) Comparison of biomass production, tree allometry, and nutrient use efficiency of multipurpose trees grown in woodlot and silvopastoral experiments in Kerala, India. For Ecol Manag 112:145–163
Lal R, Sanchez PA (1992) Myths and science of soils of the tropics. SSSA Special Publication 29, Soil Science Society of America, Madison
Lehmann J, Peter I, Steglich C, Gebauer G, Huwe B, Zech W (1998) Below-ground interactions in dryland agroforestry. For Ecol Manag 111:157–169
Lemieux G (1993) A universal pedogenesis upgrading process: RCWs to enhance biodiversity and productivity. Food and Agriculture Organization (FAO) Rome, ISBN 2-921728-05-2, pp 6
Lose SJ, Hilger TH, Leihner DE, Kroschel J (2003) Cassava, maize and tree root development as affected by various agroforestry and cropping systems in Benin, West Africa. Agric. Ecosys Environ 100:137–151
Lundgreen B, Nair PKR (1985) Agroforestry for soil conservation. In: El-Swaify SA, Moldenhauer WC, Lo A (eds) Soil erosion and conservation. Soil Conservation society of North America, Ankeny, IA. pp 703–717
Martius C, Hofer H, Garcia MVB, Rombke J, Hanagarth W (2004) Litter fall, litter stocks and decomposition rates in rainforest and agroforestry sites in central Amazonia. Nutr Cycl Agro Ecosyst 68:137–154
Miller AW, Pallardy SC (2001) Resource competition across the tree-crop interface in a maize-silver maple temperate alley cropping stand in Missouri. Agrofor Syst 53:247–259
Mitlohner FM, Morrow JL, Dailey JW, Wilson SC, Galyean ML, Miller MF, McGlone JJ (2001) Shade and water misting effects on behavior, physiology, performance and carcass traits of heat-stressed feedlot cattle. J Animal Sci 79(9):2327–2335
Monteith JL, Ong CK, Corlett JE (1991) Microclimatic interactions in agroforestry systems. For Ecol Manag 45(1–4):31–44
Muschler RG (1993) Biomass production, light transmission and management of *Erythrina berteroana, Erythrina Fusca* and *Gliricidia sepium* used as living supports in Costa Rica, Talamanca. In Westley SB, Powell MH (Eds.), *Erythrina* in the New and old Worlds. Nitrogen Fixing Tree Research Reports Special Issue, Paia, Hawaii, pp 192–199
Nair PKR (1984) Soil productivity aspects of agroforestry. ICRAF, Nairobi, Kenya
Nair PKR (1993) An introduction to agroforestry. Kluwer Academic Publishers, Dordrecht, pp 520

Nair PKR, Nair VD, Mohan Kumar B, Haile SG (2009) Soil carbon sequestration in tropical agroforestry systems: a feasibility appraisal. Environ Sci Policy 2(8):1099–1111

Nordgrove L, Hauser S (2002) Measured growth and tree biomass estimates of *Terminalia ivorensis* in the three years after thinning to different stand densities in an agrisilvicultural system in southern Cameroon. For Ecol Manag 166:261–270

Nwaga D, Jansa J, Abossolo Angue M, Frossard E (2010) The potential of soil beneficial microorganisms for slash-and-burn agriculture in the humid forest zone of Sub Saharan Africa. Chap. 5. In: Dion P (ed) Soil Biology and Agriculture in the Tropics, Soil Biology 21, Springer-Verlag, Heidelberg, pp 81–107

Odum HT, Pigeon RF (1970) A tropical rainforest. Vol. III. Office of Information Services, U.S. Atomic Energy Commission, Washington DC

Ohigashi H, Kaji M, Sakai M, Koshimizu K (1989) 3-Hydroxyuridine, an allelopathic factor of an African tree, *Baillonella toxisperma*. Phytochemistry 28(5):1365–1368

Ong CK, Corlett JE, Singh RP, Black CR (1991) Above and below-ground interactions in agroforestry systems. For Ecol Manag 45(1–4):45–57

Ong CK, Leakey RRB (1999) Why tree-crop interactions in agroforestry appear at odds with tree-grass interactions in tropical savannahs? Agrofor Syst 45:109–129

Palm CA (1995) Contribution of agroforestry trees to nutrient requirements of intercropped plants. Agrofor Syst 30:105–124

Palm CA, Sanchez PA (1990) Decomposition and nutrient release patterns of three tropical legumes. Biotropica 22:330–338

Palm CA, Sanchez PA (1991) Nitrogen release from leaves of some tropical legumes as affected by their lignin and polyphenol contents. Soil Biol Biochem 23:83–88

Patterson RT, Karanja GM, Roothaert RL, Nyaata OZ, Kariuki IW (1998) A review of tree fodder production and utilization within smallholder agroforestry systems in Kenya. Agrofor Syst 41:181–199

Puri S, Singh V, Bhushan B, Singh S (1994) Biomass production and distribution of roots in three stands of *Populus deltoides*. For Ecol Manag 65:135–147

Rhoades CC (1996) Single-tree influences on soil properties in agroforestry: lessons from natural forest and savanna ecosystems. Agrofor Syst 35:71–94

Rhodes DF, Cates RG (1976) Toward a general theory of plant antiherbivore chemistry. Recent Advances in. Phytochemistry 10:168–172

Roman-Ponce H, Thatcher WW, Buffington DE, Wilcox CJ, Van Horn HH (1977) Physiological and production responses of dairy cattle to a shade structure in a subtropical environment. J Dairy Sci 60(3):424–430

Rosenberg NJ, Blad BL, Verma DB (1983) Microclimate: The biological environment, 3rd edition. Wiley and Sons, New York

Sanchez PA (1987) Soil productivity and sustainability in agroforestry systems. In: Steppler HA, Nair PKR (eds) Agroforestry: A decade of development. ICRAF, Nairobi pp 205–223

Sanchez PA, Logan TJ (1992) Myths and science about the chemistry and fertility of soils in the tropics. Special Publication No 29, SSSA, Madison

Sanchez PA, Palm CA, Davey CB, Szott LT, Russell CE (1985). Trees as soil improvers in the humid tropics? In: Cannell MGR, Jackson JE (eds) Attributes of Trees and Crop Plants. Institute of Terrestrial Ecology, Huntingdon

Scalbert A (1991) Antimicrobial properties of tannins. Phytochemistry 12:3875–3883

Schroeder P (1994) Carbon storage benefits of agroforestry systems. Agrofor Syst 27:89–97

Schroth, G (1999) A review of below-ground interactions in agroforestry: focusing on mechanism and management options. Agrofor Syst 42: 5–24

Schroth G, Agra d'Angelo S, Teixeira WG, Haag D, Lieberei R (2002) Conversion of secondary forest into agroforestry and monoculture plantations in Amazonia: consequences for biomass, litter and soil carbon stocks after 7 years. For Ecol Manag 163(1–3):131–150

Sharma R, Sharma E, Purohit AN (1995) Dry matter production and nutrient cycling in agroforestry systems of mandarin grown in association with *Albizia* and mixed tree species. Agrofor Syst 29:165–179

Sharma R, Sharma E, Purohit AN (1997) Cardamom, mandarin and nitrogen-fixing trees in agroforestry systems in India's Himalayan region. I. Litterfall and decomposition. Agrofor Syst 35:239–253
Singh RP, Ong CK, Saharan N (1989) Above and below-ground interactions in alley cropping in semi-arid India. Agrofor Syst 9:259–274
Szott LT, Fernandes ECM, Sanchez PA (1991) Soil-plant interactions in agroforestry systems. In: Jarvis PG (ed) Agroforestry: Principles and practice. Elsevier, Amsterdam, pp 127–152
Takimoto A, Nair PKR, Nair VD (2008) Carbon stock and sequestration potential of traditional and improved agroforestry systems in the West African Sahel. Agric. Ecosys Environ 125:159–166
Teklay T, Malmer A (2004) Decomposition of leaves from two indigenous trees of contrasting qualities under shaded-coffee and agricultural land-uses during the dry season at Wondo Genet, Ethiopia. Soil Biol Biochem 36:777–786
Vanlauwe B, Nwoke OC, Sanginga N, Merckx R (1996) Impacts of residue quality on the C and N mineralization of leaf and root residues of three agroforestry species. Plant Soil 183:221–231
Verinumbe I, Okali DUU (1985) The influence of coppiced teak (*Tectona grandis* L.f.) regrowth and roots on intercropped maize (*Zea mays* L.). Agrofor Syst 3:381–386
Warren M, Zou X (2002) Soil macrofauna and litter nutrients in three tropical tree plantations on a disturbed site in Puerto Rico. For Ecol Manag 170(1–3):161–171
Weltzin JF, Coughenour MB (1990) Savanna tree influence on understorey vegetation and soil nutrients in north western Kenya. J Vegetation Sci 1:325–334
Wiersum KF (1986) Ecological aspects of agroforestry with special emphasis on tree-soil interactions: lecture notes. FONC Project Communication 1986–16. Fakultas Kerhuanan Universiti Gadja Mada, Yogyakarta
Willey RW, Reddy MS (1981) A field technique for separating above and below-ground interactions in intercropping: an experiment with pearl millet/groundnut. Experimen Agric 17:257–264
Young A (1989) Agroforestry for soil conservation. CAB International, Wallingford
Young A (1989b) The environmental basis of agroforestry. In: Darnhofer T, Reifsnyder WE (eds) Meteorology and agroforestry: proceedings of an international workshop on the application of meteorology to agroforestry systems planning and management. ICRAF, Nairobi, pp 29–48
Young A (1997) Agroforestry for soil management, 2nd edn. CABI Publishing, CAB International, Wallingford, pp 320

Chapter 8
Biological Nitrogen Fixation and Mycorrhizal Associations in Agroforestry

Abstract Biological nitrogen fixation is performed through both symbiotic and non-symbiotic means. Symbiotically, the most common association is that of roots and *Rhizobium* bacteria or filamentous actinomycete *Frankia.* Nearly all legumes used in alley cropping fix significant amounts of nitrogen in association with *Rhizobium*. Non-legume shrubs or trees such as Casuarinas and alders also fix significant amounts of nitrogen in association with *Frankia.* The amount of nitrogen fixed by legumes is variable. *Leucaena leucocephala,* which forms abundant nodulation, fixes between 100 and 550 kg of nitrogen per ha per year. The potential for nitrogen fixation by *Acacia* is also high, with up to 200 kg per ha per year. The contributions of fixed nitrogen to native as well as managed ecosystems by the actinorhizal symbioses (*Frankia*-non legume symbioses) are comparable to those of the more extensively studied *Rhizobium*-legume interactions. For instance, the roots of *Casuarina equisetifolia* and *C. junghuhnina* produce nodules where the bacteria fix atmospheric nitrogen (362 kg ha^{-1} $year^{-1}$). The main selection criteria for provenances or species for introduction into an agroforestry system include the rate of nitrogen fixation. The species selected should have the highest possible rate for a range of climatic conditions, and must also be able to tolerate environmental constraints such as pests and low nitrogen levels in the soil. Agroforestry species form symbiotic associations with mycorrhiza, typically arbuscular mycorrhizas, to enhance nutrient and water uptake and plant growth. Possible topics of interest to researchers in agroforestry are the efficient use of ecologically adapted biofertilizers (nitrogen fixing and mycorrhizal inoculants) in relation to plant species and soil fertility (N and P availability), the quality assurance of commercial inoculants, as well as the response to inoculation (improvement of methods to estimate nitrogen fixation and P uptake).

8.1 Introduction

Biological nitrogen fixation is one of the more important benefits of agroforestry, and almost all tree species used in alley cropping and improved fallows are nitrogen fixing. Nitrogen fixation has long been used in traditional agriculture. For example, peanuts were grown together with maize and cassava in the humid lowlands of West and Central Africa. The amount of nitrogen fixed by legumes is variable, and can

A. Atangana et al., *Tropical Agroforestry*, DOI 10.1007/978-94-007-7723-1_8,

range from 30 – 500 kg N ha^{-1} $year^{-1}$ (Brady and Weir 2010, p. 408). Those rates are higher to rates of nitrogen fixation in evergreen rainforests, which is fixed by herbaceous plants and also highly variable, that average between 0.26 kg ha^{-1} $year^{-1}$ in dry season and 2.71 kg ha^{-1} $year^{-1}$ in wet season (Reed et al. 2007). Biological nitrogen fixation is performed through symbiotic and non-symbiotic means. *Rhizobium* bacteria form association with numerous leguminous species, and *Frankia* species (actinomycetes) are symbionts of non-leguminous species. Non-symbiotic fixation is performed by free-living soil organisms and is generally of low importance. The rates are highly variable depending on the organic matter content and the microbial activity in the soil.

Agroforestry species, which are mainly Rhizorhizal plants (*Rhizobium*-legume symbioses) and Actinorhizal plants (*Frankia*-non legume symbioses), form, in addition, symbiotic associations with mycorrhizal fungi, typically arbuscular mycorrhizas. These hypersymbiotic associations most often enhance nutrient uptake, and subsequently improve plant growth. These associations also affect root nodulation, increasing nitrogen fixation by the host plant. Symbiotic associations of plants with *Rhizobacterium*, *Frankia* and mycorrhiza are often interrelated. Agroforestry species can be classified in three groups, with regard to mycorrhizal dependency. Mycorrhizal infections increase the absorption of phosphate and other non-mobile ions and water, and resistance to abiotic and non-biotic stresses. Species with high mycorrhizal dependency include: *Azadirachta indica, Leucaena leucocephala, Gliricidia maculata* (Kunth) Steud., *Sesbania grandiflora, Cassia siamea* Lam. [also known as *Senna siamea* (Lam.) Irwin et Barneby] and *Acacia melanoxylon* R.Br. Non mycorrhizal dependent species, including *Diospyros melanoxylon* Roxb., *Mangifera indica, Murraya koenigii* (L.) Sprengel, *Polyalthia longifolia* Sonn., *Psidium guajava* L., *Saraca indica* L. syn. *Saraca asoca* (Roxb.) Wilde and *Ziziphus mauritiana.* Species with moderate mycorrhizal dependency include the vast majority of agroforestry species (Pindi et al. 2000). The arbuscular mycorrizal fungi (AMF) are the most important group of mycorrhizae in tropical agroforestry systems. Other groups include the ectomycorrhiza, orchid mycorrhiza, and ericoid mycorrhiza (de Carvalho et al. 2010). Farmers in the tropical rainforests of Brazil use several mycorrhizal tree and shrub species in their agroforestry systems (Table 8.1). AMF increases the soil volume exploited by the plant, enhancing nutrient uptake beyond the root depletion zone. AMF species also protect the root system against pathogens, phytotoxic elements and heavy metals, aid in the formation and maintenance of soil structure, increase C input to soils, and help maintain plant biodiversity (de Carvalho et al. 2010). Mycorrhizal incidence in deep soil layers is greater in agroforestry systems than in monoculture crop systems (de Carvalho et al. 2010).

8.2 Plant Species Forming Associations with *Rhizobium* or Mycorrhizal Fungi

Rhizobia are soil bacteria that fix nitrogen by associating with plants in root nodules. A distinction is made between α-proteobacteria (Rhizobiales) and β-proteobacteria (Bukholderiales). The roots of tropical legumes form nitrogen fixing nodules with

Table 8.1 Mycorrhizal tree and shrub species used by smallholder farmers in agroforestry coffee systems, Zona da Mata of Minas Gerais, Atlantic coastal rainforest, Brazil (modified from de Carvalho et al. 2010)

Family	Species (mycorrhizal status)	Reference
Anacardiaceae	*Mangifera indica* L. (AMF)	Silveira and Gomez (2007)
	Myracrodruon urundeuva M.Allemão (AMF)	Siqueira et al. (2007)
	Schinus terebinthifolius Raddi (AMF)	Zangaro et al. (2003)
Annonaceae	*Annona muricata*[a] L. (AMF)	Silveira and Gomes (2007)
Apocynaceae	*Aspidosperma polyneuron* Müll.Arg. (AMF)	Zangaro et al. (2002)
Araucariaceae	*Araucaria angustifolia* (Bertol.) Kuntze (AMF)	Siqueira et al. (2007)
Bignoniaceae	*Tabebuia impetiginosa* (Mart. Ex DC.) Standl. (AMF)	Carneiro et al. (1998)
	T. chysotricha (Mart. Ex DC.) Standl. (AMF)	Zangaro et al. (2003)
	T. serratifolia (Vahl) Nichols. (AMF)	Siqueira and Saggin-Júnior (2001)
Bixaceae	*Bixa orellana*[a] L. (AMF)	Carneiro et al. (1998)
Bombacaceae	*Ceiba speciosa* (A.St.-Hil.) Ravenna syn. *Chorisia speciosa* A.St.-Hil. (AMF)	Zangaro et al. (2003)
Cannabaceae	*Trema micrantha* (L.) Blume (AMF)	Carneiro et al. (1996)
Caricaceae	*Carica papaya*[a] L. (AMF)	Silveira and Gomez (2007)
Casuarinaceae	*Casuarina equisetifolia* L. (AMF/ECM)	Carneiro et al. (1998)
Cecropiaceae	*Cecropia glaziovii* Snethl. (AMF)	Zangaro et al. (2002)
	C. pachystachya Trécul (AMF)	Zangaro et al. (2003)
Ebenaceae	*Diospyros kaki* Thunb. (AMF)	Silveira and Gomes (2007)
Euphorbiaceae	*Manihot esculenta*[a] Crantz (AMF)	Silveira and Gomes (2007)
Lauraceae	*Persea americana* Mill. (AMF)	Silveira and Gomes (2007)
Leguminosae-Caesalpinioideae	*Caesalpinea ferrea* Mart. Ex Tul. (AMF)	Siqueira et al. (2007)
	C. peltophoroides Benth. (ECM)	Siqueira et al. (2007)
	Copaifera langsdorffii Desf. (AMF)	Carneiro et al. (1998)
	Hymenaea courbaril L. (AMF)	Zangaro et al. (2003)
	Pterogyne nitens Tul. (AMF)	Zangaro et al. (2002)
	Schizolobium parahyba (Vell.) S.F.Blake (AMF)	Carneiro et al. (1998)
	Senna macranthera (Collad.) H.S.Irwin & Barneby (AMF)	Carneiro et al. (1998)
	S. multijuga (Rich.) Irwin & Barneby (AMF)	Carneiro et al. (1998)
Leguminosae-Mimosaideae	*Anadenanthera peregrina* Speg. (AMF)	Siqueira et al. (2007)
	Enterolobium contortisiliquum (Vell.) Morong (AMF)	Zangaro et al. (2003)
	Inga sessilis (Vell.) Mart.(AMF)	Zangaro et al. (2003)
	Leucaena leucocephala (Lam.) de Wit (AMF)	Siqueira and Saggin-Júnior (2001)
	Piptadenia gonoacantha (Mart.) J. F. Macbr. (AMF)	Carneiro et al. (1998)
	Mimosa caesalpiniaefolia Benth. (AMF)	Siqueira et al. (2007)
Leguminosae-Papilionoideae	*Cajanus cajan*[a] (L.) Millsp. (AMF)	Siqueira et al. (2007)
	Machaerium nictitans (Vell.) Benth. (AMF)	Carneiro et al. (1998)
	M. stipitatum (DC.) Vogel. (AMF)	Zangaro et al. (2003)
Malpighiaceae	*Malpighia emarginata*[a] DC. (AMF)	Silveira and Gomes (2007)
Malvaceae	*Luehea divaricata* Martius et Zuccarini (AMF)	Zangaro et al. (2003)

Table 8.1 (continued)

Family	Species (mycorrhizal status)	Reference
	L. grandiflora Martius et Zuccarini (AMF)	Siqueira and Saggin-Júnior (2001)
Melastomataceae	*Tibouchina granulosa* (Desr.) Cogn. (AMF)	Siqueira and Saggin-Júnior (2001)
Meliaceae	*Azadirachta indica* A.Juss. (AMF)	Siqueira et al. (2007)
	Cedrela fissilis Vell. (AMF)	Carneiro et al. (1998)
	Melia azedarach L. (AMF)	Carneiro et al. (1998)
Musaceae	*Musa* sp.[a] L. (ARM)	Silveira and Gomez (2007)
Myrsinaceae	*Rapanea ferruginea* (Ruiz & Pav.) Mez (AMF)	Siqueira et al. (2007)
Myrtaceae	*Campomanesia xanthocarpa* O.Berg (AMF)	Zangaro et al. (2002)
	Eugenia uniflora L. (AMF)	Zangaro et al. (2003)
	Psidium guajava L. (AMF)	Zangaro et al. (2002)
Palmae	*Euterpe edulis* Mart. (AMF)	Zangaro et al. (2003)
	Syagrus romanzoffiana (Cham.) Glassman (AMF)	Zangaro et al. (2003)
Rhamnaceae	*Colubrina glandulosa* Perkins (AMF)	Zangaro et al. (2003)
	Hovenia dulcis Thunb. (AMF)	Carneiro et al. (1998)
Rutaceae	*Citrus* sp. L. (AMF)	Silveira and Gomez (2007)
Solanaceae	*Solanum argenteum* Dunal ex Poir. (AMF)	Zangaro et al. (2002)
	S. granulosum-leprosum Dun. (AMF)	Siqueira and Saggin-Júnior (2001)
Verbenaceae	*Aegiphila sellowiana* Cham. (AMF)	Zangaro et al. (2003)
	Cytharexilum myrianthum Cham. (AMF)	Zangaro et al. (2002)
	Vitex montevidensis Cham. (AMF)	Zangaro et al. (2003)

AMF Arbuscular mycorrhizal; *ECM* Ectomycorrhiza. [a] Shrub species

rapidly multiplying *Rhizobium sensu stricto*, slow multiplying *Bradyrhizobium* (Elkan 1984), *Azorhizobium* (Dreyfus et al. 1988; Moreira et al. 2006), and *Sinorhizobium* (Chen et al. 1988). *Rhizobium* species form symbiotic relationships with most *Leucanea leucocephala*, and *Sesbania grandiflora*. *Azorhizobium* also associate with species in *Sesbania*, and *Sinorhizobium* and form symbioses with *Medicago*, *Melilotus* and *Trigonella*. *Acacia mearnsii* De Wild. and *Faidherbia albida* (Delile) A.Chev. form nodules with *Bradyrhizobium*, and *Acacia seyal* Del. associates with *Rhizobium* and *Bradyrhizobium*.

Phosphorus (P) is the second most important macronutrient for plants after nitrogen. However, P is mostly found in immobile forms in soils. P is moved by diffusion in the soil, and roots may explore large a volume of soil to increase P uptake. Mycorrhizas are important for P uptake as fungal hyphae increase the volume that roots explore. Therefore, mycorrhizas provide roots with available P.

Table 8.2 Reported mycorrhiza status in the genus *Acacia* (Haselwandter and Bowen 1996)

Mycorrhiza type for *Acacia* species	
Ectomycorrhiza (ECM)	*A. delbeata, decurrens, melanoxylon, mitchellii, pycnantha, retinoides, rubida, salicina, sophorae, sparciflora, verticillata*
Arbuscular mycorrhiza (AMF)	*A. alata, albida, aulospora, arabica, bancrofii, caregna, concurrens, confusa, constricta, cyanophylla, dunii, extensa, farnesiana, fimbriata, floribunda, goetzei, greggii, horrida, inaequilatera, lateriticola, latescens, leiocalyx, mearnsii, mellifera, mucronata, nigrescens, nilotica, nubica, oxycedrus, pendula, pyrifolia, raddiana, richi, salicina, saligna, senegal, seyal, spectabilis, stenophylla, suaveolens, torulosa, tumida, urophylla, yirrkalensis*
ECM+AMF	*A. anastrocarpa, aneura, auriculiformis, bivenosa, christolmii, coriacea, cowleana, eriopoda, harbophylla, hilliana, hippuroides, holosericea, lysiphlora, macrodenia, mangium, monticola, myrtifolia, pellita, platycarpa, plectocarpa, potalyniifolia, retoxylon, retivenea, rothii, simsii, trachycarpa, translucens*

8.2.1 *Acacia Sensu Lato*

Faidherbia albida (formerly *Acacia albida* Delile) and *Acacia longifolia* are known to form nodules with *Bradyrhizobium* strains (Dreyfus and Dommergues 1981; Rodriguez-Echeverria et al. 2007). *Acacia nilotica* and *A. raddiana* form nodules with *Rhizobium,* and *A. seyal* forms nodules with *Rhizobium* and *Bradyrhizobium* (Dreyfus and Dommergues 1981). *Bradyrhizobium* strains are present in many soils, but inoculation of *Faidherbia albida* seedlings has produced a poor response. In addition, the sequestration potential of this species is low. The capitalization of genetic variation (Rodriguez-Echeverria et al. 2007-genetic variation observed in *A. longifolia*) can be used to enhance this potential. *Acacia senegal* only forms nodules with *Rhizobium* strains that are less ubiquitous than *Bradyrhizobium. A. senegal* often requires more inoculations than *F. albida.* The potential for nitrogen fixation in both *A. senegal* and *F. albida* is low. These two species fix less than half of the nitrogen fixed by either *Acacia seyal* or *Acacia raddiana* (Ndoye et al. 1995). *Acacia auriculiformis* has a high potential for nitrogen fixation (Domergues 1983a), as well as *A. mangium*, which naturally hybridizes with *A. auriculiformis. Acacia mearnsii* forms abundant nodules with *Bradyrhizobium* strains when the pH is above 4.5 (Halliday and Somasegaran 1983). The sequestration potential of *A. mearnsii* is high, estimated to be about 200 kg of nitrogen per ha per year (Orchard and Darby 1956).

The mycorrhizal status of the genus *Acacia* was reviewed by Haselwandter and Bowen (1996). Most of the species in the genus *Acacia* are of arbuscular mycorrhizal type (Table 8.2). *Acacia auriculiformis*, *Acacia leptocarpa* and *Acacia mangium* are colonized by arbuscular mycorrhizae (Bakarr and Janos 1996, Khasa et al. 1994), and *Acacia auriculiformis* and *A. mangium* were found to be highly dependent on mycorrhizal colonization. (De La Cruz et al. 1992; Khasa et al. 1992; Haselwandter and Bowen 1996). In agroforestry systems in southwestern Ethiopia, *Glomus* and *Aucalospora* spores were found to be the most common spores under *Acacia abyssinica* canopies; other common genera being *Entrophospora, Gigas-*

Table 8.3 Root concentration (cm 100 cm^{-3}), mycorrhizal colonization (%) and arbuscular mycorrhiza spore number (per 100 g of dry weight soil) associated with four different tree species growing at Bandia, Senegal (Ingleby et al. 1997)

	Prosopis julilora	*Acacia nilotica*	*Acacia tortilis*	*Acacia aneura*	*P value*
Root concentration	32 bc[a]	58 a	47 ab	26 c	0.003
Mycorrhizal colonization	64 a[b]	55 b	31 c	47 b	0.013
Total spore number	8.4 d[c]	32 b	19 c	51 a	<0.001
"Live" spore number	0.5 c[c]	2.8 b	2.1 b	9.5 a	<0.001

[a] Letters indicate significant differences within each row at $P<0.05$ as determined by ANOVA and Fischer's LSD test

[b] Arcsine transformations were performed on mycorrhizal percentages for statistical analysis; significance is given against untransformed data

[c] Log (n+1) transformations were performed on spore numbers for statistical analysis; significance is given against untransformed data

Table 8.4 Effect of tree species on (a) root concentration, mycorrhizal infection and spore concentration found in alley cropping soils at Thiénaba, Senegal, and (b) growth and mycorrhizal infection of millet seedlings grown in similar soils (Diagne et al. 2001)

	Tree species			
	A. nilotica	*A. tortilis*	*P. juliflora*	*P* value
(a) Field samples				
Root concentration (cm/100 cm^3)	181 a[a]	184 a	142 b	<0.018
Root infection (%)	42.2 b	45.9 b	142 b	<0.001
Spore concentration (No./100 g soil)	164	167	187	0.786
(b) Bioassay plants				
Shoot dry weight (mg)	36.1	39.0	34.9	0.719
Root fresh weight (mg)	442	478	388	0.334
Root infection (%)	12.6 b	17.9 a	11.0 b	<0.001

[a] Letters indicate significant differences within each row at $p<0.05$ as determined by ANOVA and Fisher's LSD test

pora and *Scutellospora* (Muleta et al. 2008). *Acacia nilotica* has better potential for mycorrhizal colonization than *A. tortilis* and *A. aneura* (Ingleby et al. 1997; Table 8.3). However, mycorrhizal inoculum potential is greater in *A. tortilis* than in *A. nilotica* (Table 8.4), and seedlings grown in top layer soils, between 0 and 25 cm of depth, have higher levels of infection than seedlings grown in deep soils, between 25 and 50 cm (Diagne et al. 2001; Table 8.5). Inoculation of *Acacia nilotica* with *Glomus clarum* could increase nodulation under unstressed conditions (Osonubi et al. 1992). In a screening with 13 different AM fungi, *Acacia nilotica* seedlings were found to respond best to inoculation with *Glomus mosseae* (Reena and Bagyaraj 1990). Inoculated seedlings were found to grow better, and nitrogen and phosphorus uptake was enhanced. Some *Acacia* species are dependent on mycorrhizal fungi for their growth. In a study carried out in lateritic soils in India by Ghosh and Verma (2006), it was determined that the growth of *Acacia mangium* is 57 % dependent on the AM fungus *Glomus occultum*. Greenhouse inoculations combining the arbuscular mycorrhizal fungus strain *Rhizophagus irregularis* DAOM 181602 and

Table 8.5 Effect of soil depth on (a) root concentration, mycorrhizal infection and spore concentration found in alley cropping soils at Thiénaba, Senegal, and (b) growth and mycorrhizal infection of millet seedlings grown in similar soils (Diagne et al. 2001)

	Soil depth		
	0–25 cm	25–50 cm	*P* value
(a) Field samples			
Root concentration (cm 100 cm^{-3})	236 a[a]	102 b	<0.001
Root infection (%)	50.0	52.4	0.598
Spore concentration (Number 100 g^{-1} soil)	279 a	66 b	<0.001
(b) Bioassay plants			
Shoot dry weight (mg)	48.2 a	25.1 b	<0.001
Root fresh weight (mg)	561 a	311 b	<0.001
Root infection (%)	17.6 a	10.1 b	<0.001

[a] Letters indicate significant differences within each row at $p < 0.05$ as determined by ANOVA and Fischer's LSD test

Bradyrhizobium strains improved the growth of *Acacia auriculiformis* and *Acacia mangium* (Diouf et al. 2005), a potentially useful technique for soils with salt constraints, because high concentrations of Na, Cl, Mg and SO_4 ions in soils inhibit the growth of numerous plant species (Lambers 2003). However, field trials should be undertaken to confirm this greenhouse finding. The ectomycorrhizal fungus *Pisolithus alba* was found to form a symbiotic association with *Acacia holosericea* in Senegal, facilitated by bacteria known as mycorrhiza helper bacteria (Founoune et al. 2002).

8.2.2 *Albizia*

Two species of *Albizia* (*A. lebbeck* (L.) Benth. and *Paraserianthes falcataria* (L.) Nielsen, formerly *Albizia falcataria* (L.) Folsberg), among the hundreds of species found in the tropics, are reputed to be beneficial in soil improvement due to their abundant nodulation. However, *P. falcataria* only nodulates abundantly when the soil has a profile that reflects the plant's preferred growing conditions. *Albizia ferruginea* (Gull. & Perr.) Benth. and *P. falcataria* are colonized by arbuscular mycorrhizal fungi (AMF; Bakarr and Janos 1996), and *A. ferruginea* is included among highly mycorrhizal dependent species (Habte and Musoko 1994). Inoculation of *P. falcataria* with AM fungi and *Rhizobium* was found to enhance phosphorus uptake and nitrogen fixation (Pindi 2011a). Also, *Glomus fasciculatum* was found to act as a very good colonizer of *P. falcataria*, and to induce better growth in the species (Pindi 2011). *Albizia lebbeckoides* (DC.) Benth. was also found to be highly mycorrhizal dependent (De La Cruz et al. 1992; Haselwandter and Bowen 1996). In agroforestry systems of southwestern Ethiopia, *Albizia gummifera* (Gmel) C.A.Sm is mostly colonized by *Glomus* and *Aucalospora* spores, other colonizers being from the *Entrophospora*, *Gigaspora* and *Scutellospora* genera (Muleta et al. 2008). Wubet et al. (2003) also reported arbuscular mycorrhizal colonization in the roots of *A. gummifera* and

Albizia schimperiana Oliv. in Ethiopia's dry Afromontane forests. The study did not find any evidence of ectomycorrhizal colonization in the roots of *A. gummifera* or *A. schimperiana*. In Brazil, arbuscular mycorrhizal fungi were found to colonize the roots of *A. hassleri* (Chodat) Burkart (Zangaro et al. 2005). Inoculation of AMF was also found to reduce the wilting of *A. procera* (caused by *Fusarium* spp.) in India (Chakavraty and Mishra 1986). AMF inoculation stimulated phosphorus and nitrogen uptake in *P. falcataria* in a low phosphorus soil. Ectomycorrhizal inoculation did not have the same effect, but this may have been due to possible parasitic associations between *Albizia* and ectomycorrhiza (Osonubi et al. 1991).

8.2.3 *Calliandra calothyrsus*

Calliandra calothyrsus is slower than *Sesbania sesban* in forming nodules (Purwantari et al. 1995), suggesting a low potential for nitrogen fixation. However, it responds positively to infection by *Rhizobium* (Lesueur et al. 2001) and *Bradyrhizobium* (Purwantari et al. 1995).

Calliandra calothyrsus roots were colonized by AMF in agroforestry systems in Brazil (Cardoso et al. 2003). Inoculation of *C. calothyrsus* with a strain of *Glomus etunicatum* (Becker and Gerdemann) BEG 176 isolate and *Rhizobium* strain KWN35 enhanced tree growth, but the effect did not last long (Lesueur and Sarr 2008). Inoculation of *C. calothyrsus* with the same strain of *G. etunicatum* enhanced nodulation (Lesueur and Sarr 2008). Inoculation of *C. calothyrsus* with the AMF *G. etunicatum* (Becker and Gerdemann) BEG 176 isolate and *Gigaspora albida* (Schenck and Smith) was found to enhance growth of *C. calothyrsus* intercropped with maize or beans (Fig. 8.1; Lesueur and Sarr 2008). *Calliandra calothyrsus* and maize share the same fungus, and *G. etunicatum* (Becker and Gerdemann) BEG 176 isolate was found to be more mobile and spread more rapidly. The increased rate of spread formed higher levels of colonization at increasing distances from the tree, and was responsible for most of the mycorrhizal cross-contamination (Ingleby et al. 2007). Cross-contamination of trees and crops by mycorrhiza in agroforestry systems could be used to increase crop production. In a screening of 13 AMF, seedlings of *C. calothyrsus* were found to respond best to *Glomus velum* and *Glomus merredum* (Reena and Bagyaraj 1990). In Uganda, inoculation of *Glomus* and *Acaulospora* to roots was found to enhance *C. calothyrsus* growth, and inoculation of both mycorrhiza increased tree biomass (Sebuliba et al. 2010).

8.2.4 *Erythrina*

Several species of *Erythrina* form nodules with a strain of *Bradyrhizobium* (Hallyday and Somasegaran 1983; Milnytski et al. 1997). The nodules of *E. poeppigiana* tend to be large, spherical and clumped around the central root system (Allen and Allen 1981). The biomass of root nodules varies from 80 mg dm^{-3} of soil to 205 mg dm^{-3}

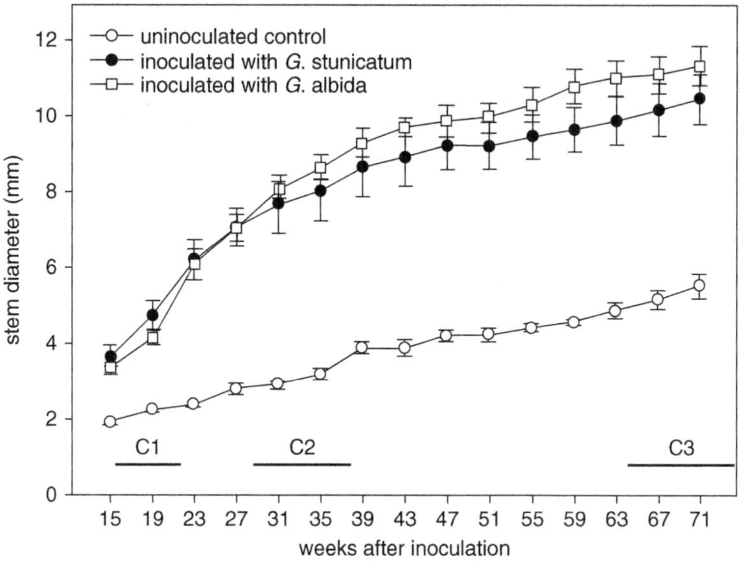

Fig. 8.1 Stem diameter of inoculated and uninoculated *Calliandra calothyrsus* trees in a greenhouse experiment during 2004–2005 [*error bars* indicate SE; *horizontal bars* indicate cropping periods C1–3] (Ingleby et al. 2007)

of soil (dry matter) (Lindbald and Russo 1986). Pruning of *E. poeppigiana*, usually done for mulching, has led to nodule mortality (Nygren and Ramirez 1995).

Erythrina poeppigiana (Walp.) O.F.Cook can be inoculated with the arbuscular mycorrhizal (AM) species *G. etunicatum*, *G. mosseae*, and *Rhizophagus irregularis* (formerly *G. intraradices*). Inoculated plants respond well to nitrogen fertilization (Cuenca and Azcon 1994), displaying growth enhancement N source is NO_3^-. The yield increase in AM inoculated plants fertilized with 3 and 6 mM NO_3^--N can go up by 255 % and 268 % respectively for *G. etunicatum-colonized* plants, 201 % and 164 % respectively for *G. mosseae-colonized* plants, and 286 % and 218 % respectively for *G. intraradices-colonized* plants (Cuenca and Azcon 1994). The authors suggested that AM mycelium has a capacity for NO_3 absorption, and that AM symbiosis affects nitrogen metabolism in *E. poeppigiana*. Inoculation of *Erythrina berteroana* Urb. seedlings with AM fungi increased shoot biomass by 10.6 % versus non-inoculated seedlings, whereas AM-inoculated cuttings of the same species exhibit a 16 % decrease in shoot biomass (Cooperband et al. 1994). This difference in responsiveness was attributed to the cost-benefit relationship between *E. berteroana* and the fungal symbiont with respect to energy and nutrient reserves (Cooperband et al. 1994).

8.2.5 Gliricidia sepium

Gliricidia sepium (Jacq.) Kunth ex Walp. forms nodules with *Rhizobium* (Hallyday and Somasegaran 1983; Bala and Giller 2001). Based on nodule biomass and the rates of nitrogenase activity, the estimated level of nitrogen fixation is approximately

Table 8.6 Vegetative growth and biomass yield of *Glomus deserticola*-inoculated and non-inoculated *Gliricidia sepium* after 16 weeks of growth in non-disinfected soils (Fagbola et al. 2001)

Soil	Mycorrhizal inoculation	Watering regime	Height (cm)[a]	Stem girth (cm)[a]	Leaf dry weight (g)[b]	Stem dry weight (g)[b]	Root length (m)[b]
Topsoil	Without	Watered	61.17a	1.11b	12.99b	16.46a	113a
		Drought	42.27bc	0.83d	7.10c	8.13b	22d
	With	Watered	65.27a	1.46a	15.48a	19.84a	101a
		Drought	46.00b	0.81de	6.91c	7.92bc	60b
Subsoil	Without	Watered	30.87e	0.70e	4.41d	3.99cd	38c
		Drought	25.50e	0.57f	2,74d	2.63d	8e
	With	Watered	38.83cd	0.97c	6.32c	8.98b	55b
		Drought	36.17d	0.74d	3.77d	5.57bc	25d
ANOVA							
Watering (W)			***	***	***	***	***
Mycorrhizal inoculation (I)			***	***	*	*	***
Soil type (S)			***	***	***	***	***
Interactions							
W x I			NS	**	NS	NS	***
W x S			***	***	***	**	***
I x S			NS	NS	NS	NS	NS
W x I x S			NS	**	NS	NS	***

[a] For each variate, values followed by the same letters are not significantly different at $P \leq 0.05$ according to Duncan's Multiple Range Test
[b] For each variate, values followed by the same letters are not significantly different at $P \leq 0.01$ according to Duncan's Multiple Range Test
* $P \leq 0.05$;
** $P \leq 0.01$;
*** $P \leq 0.001$
NS non-significant

13 kg ha^{-1} year^{-1}, in conditions prevalent in Mexico (Roskoski et al. 1982). *Gliricidia* may fix up to 50% of the nitrogen derived from the atmosphere (Ndfa) into the soil (Ladha et al. 1993). Nitrogen fixation by this species may become limited if roots are attacked by root nematodes. The exploitation of intraspecific genetic variation in *G. sepium* (Sanginga et al. 1992) has been recommended to address this issue.

Gliricidia sepium is a species colonized by AMF (Bakarr and Janos 1996), and the species was found to be mycorrhizal dependent (Osonubi et al. 1991). Inoculation of *G. sepium* with ectomycorrhiza or AMF enhances plant growth and nutrient uptake (Osonubi et al. 1991). *Glomus deserticola*, an AM fungus, was found to increase the dry matter accumulation and nutrient uptake of lone *G. sepium* hedgerow trees in a fallowed alley cropping, done on degraded Alfisols in Nigeria (Okon et al. 1996). Inoculation of *G. sepium* by *Glomus deserticola* enhanced plant growth (Table 8.6) and increased N-fixing activity (Fagbola et al. 2001). Rhizobial strains were found to influence the frequency of mycorrhization of *Glomus aggregatum* in *G. sepium* (Diouf et al. 2008). High spore density in soil under alley cropping involving cassava and *Gliricidia sepium* (or *Leucaena leucocephala* or *Cassia siamea*) increases the uptake of nutrients by trees, and improves cassava tuber yield

(Atayese et al. 1993). Nutrient uptake by roots in alley cropping featuring cassava, *G. sepium*, *Leucaena*, and *Cassia siamea* was influenced by AM inoculation, but the leaves were not affected, except by the increased P uptake (Osonubi et al. 1995). Greater uptake of nitrogen, potassium and phosphorus was observed in roots of inoculated, alley-cropped cassava grown in association with *Gliricidia* and *Leucaena* than that grown with *Cassia*. The productivity of cassava was speculated to be regulated by the amount of nutrients the roots could absorb (Osonubi et al. 1995). Cassava tuber yield following plant inoculation with *Glomus deserticola* in a *Gliricidia sepium*-based alley cropping system, in a degraded Alfisol, can be as high as 101 % (Okon 2011). Nutrient transfer from *G. sepium* to an associated crop may occur several ways. Nitrogen transfer from *Gliricidia sepium* to an associated grass (*Dichanthium aristatum* (Poir.) C.E. Hubbard) was found to vary between 3.7–14.0 % and 0.7–2.5 % of the total nitrogen in the grass. This transfer was speculated to occur via root exudation or to be driven by a source/sink relationship between the plants (Jalonen et al. 2009).

8.2.6 *Inga edulis*

A study revealed that integration of *Inga edulis* Mart. into a plot of *Terminalia amazonia* (J.F.Gmel.) Exell resulted in a significant increase in soil nitrogen (Nichols and Carpentier 2006), indicating that *I. edulis* may fix nitrogen to restore soil fertility. A closely related species, *Inga jinicuil* Schltdl., fixes about 35–40 kg nitrogen ha^{-1} $year^{-1}$, which corresponds to a nodule biomass of 71 ± 14 kg of dry matter ha^{-1}. For a density of 205 trees per ha, the approximate nodule biomass per tree is 346 g of dry matter, a value similar to that reported by Akkermans and Houvers (1983) for *Alnus* (Roskoski 1981; 1982). Nodule formation in *Inga* is slow, as according to a study on *Inga oerstediana* Benth. Ex Seem. (Grossmann et al. 2006).

Inoculation of *Inga edulis* seedlings with AMF significantly increases plant growth (Iglesias et al. 2011). AM fungi from the *Glomus mosseae/intraradices* group, the *Glomus etunicatum/claroideum* group, and *Acaulosporaceae sensu stricto* were found to colonize the roots of *Inga edulis* (Shepherd et al. 2007). Seedlings of *Inga leiocalycina* Willd. were well colonized by AM fungi (Whitbeck 2001). Light was found to influence AM fungal colonization and development of symbiotic structures (Table 8.7).

8.2.7 *Leucaena leucocephala*

Leucaena leucocephala fixes between 100 and 500 kg of nitrogen per ha per year (Sanginga et al. 1985; 1986; 1988; 1989). Sanginga et al. (1989) reported that *L. leucocephala* fixed 98–134 kg of nitrogen per ha in 6 months when grown on Alfisols (pH = 6.1). The abundant nodulation of this species under certain soil conditions explains the high fixing potential. The dry weight of nodules in

Table 8.7 AM colonization. Root length, AM length, and percent AM values are back transformed means. Values for AM structures are presence/absence data, where the first number is the number of samples possessing the structure and the second number is the total number of samples assayed. For each study, values within a column followed by different letters are significantly different at $P<0.05$ level, as determined by the Scheffé post hoc tests of significant ANOVA factors for the length data and by Chi-square test for the anatomical data (Whitbeck 2001)

Treatment (% full sum)	Root length (cm)	AM length (cm)	% AM				
(a) Shadehouse							
0.6	91.3 a	21.5 a	19.2 a				
12.4	281.8 b	82.6 b	30.0 b				
28.3	497.3 c	233.1 c	46.7 c				
(b) Field study							
Habitat	Root length (cm)	AM length (cm)	% AM	Hyphae	Vesicles	Coils	Spores
Understory	58.7 x	13.8 x	30.2 x	30/30 x	11/30 x	20/30 x	1/30 x
Gap	105.9 y	17.5 x	25.3 y	30/30 x	16/30 x	28/30 y	5/30 x

L. leucocephala can reach up to 51 kg on a plot of 830 trees per ha (Högberg and Kvarnström 1982) and 63 kg on a plot of 2,500 trees per ha (Lulandala and Hall 1986). *Leucaena leucocephala* usually forms nodules in the presence of *Rhizobium sensu stricto* (Halliday and Somasegaran 1983), and occasionally in the presence of *Bradyrhizobium* (Dreyfus and Domergues 1981). The *Rhizobium* strain specific to *L. leucocephala* is rare, explaining the positive response to inoculation obtained in soils with nutrient levels high enough to meet the species' needs, with the exception of nitrogen (Brewbaker 1987). *Sinorhizobium morelense sp. Nov.* seems to be associated with *L. leucocephala* (Wang et al. 2002). A study of the genetic diversity of isolates from nodules of *L. leucocephala* identified strains belonging to *Rhizobium*, *Mesorhizobium*, and *Sinorhizobium* (Wang et al. 1999). *Leucaena leucocephala* can also form nodules with *Rhizobium tropici* (Martinez-Romero et al. 1991). When the soil is infertile, *L. leucocephala* does not form nodules.

Leucaena leucocephala was found to be highly mycorrhizal dependent (Habte and Manjunath 1987; Osonubi et al. 1992; Haselwandter and Bowen 1996). *L. leucocephala* is colonized by AM (Bakarr and Janos 1996). However, no significant contribution was evident in the growth of *L. leucocephala* due to arbuscular mycorrhiza in simulated eroded soil conditions in Nigerian Alfisols (Fagbola et al. 2001). In depleted soils in Kenya, *Leucaena leucocephala* roots were found to be infected by *Scutellospora* spp. and *Acaulospora* spp. Nodulation was absent or poor in all soils studied, indicating the need for rhizobia inoculation of the species belonging to the Leucaena cross inoculation group (Shepherd et al. 1996). *Glomus fasciculatum* (Thaxter sensu Gerd.) Gerd and Trappe, an AM fungus, was found to enhance growth of *L. leucocephala* by increasing shoot and root dry weight, leaf area and root length (Huang et al. 1985). The growth promoting effect of AM fungi was found to equal the effect of 12 weeks of phosphorus fertilization on *L. leucocephala* seedlings (Michelsen and Rosendahl 1990). On a typical Eutrustox, colonization of *L. leucocephala* with *Glomus fasciculatum* was found to significantly increase top growth, nodule fresh weight, nitrogenase activity, and soil K, as well as to linearly increase low-soluble P fertilization to 300 mg P per kg of soil (Purcino et al. 1986). The study also reported that interactions between K and mycorrhiza increased nodulation and nitrogenase activity. Mycorrhizal and P and K interactions were also found to affect nodule fresh weight (Purcino et al. 1986). In sandy soils, *L. leucocephala* seedlings inoculated with *G. fasciculatum* and *Rhizobium* develop abundant AM and root nodules without NaCl amendments (Dixon et al. 1993). An isolate of *Glomus mosseae* was found to be the best mycorrhizal inoculant of Leucaena, allowing healthy, vigorous plants to grow on red sandy loam (Bagyaraj et al. 1989).

8.2.8 *Mimosa*

Mimosa species respond positively to inoculation (Dobereiner 1984), and form nodules with a fast-growing strain of *Rhizobium* (Halliday and Somasegaran 1983). Five strains of *Rhizobium* infect *Mimosa scabrella* Benth., and can fix 285 mg of N per plant in 81 days (Franco and De Faria 1997). Other species in the genus *Mimosa*, such as *M. acutistipula* (Mart.) Benth., *M. bimucronata* (DC.) Kuntze, *M. caesalpiniifolia* Benth., *M. flocculosa* Burkart, and *M. tenuiflora* (Will.) Poir., can also form nodules in the presence of *Rhizobium* strains (Franco and De Faria 1997). *Burkholderia* strains also form nodules with species in the genus *Mimosa* (Chen et al. 2005). Strains of *Burkholderia phymatum* and *Cupriavidus taiwanensis* form symbioses with *M. pudica*. However, the *Burkholderia* strain has a higher nitrogenase activity than the *Cupriavidus* strain (Elliott et al. 2007). The sequestration potential of *Mimosa* species needs to be properly evaluated.

Mimosa scabrella trees in the Brazilian Araucaria forest are colonized by AMF (Andrade et al. 2000). Soil taken from below the canopies of six Mimosa species (*M. adenantheroides* (M. Martens & Galeotii) Benth., *M. calcicola* B.L.Rob., *M. luisana* Brandegee, *M. polyantha* Benth., *M. lacerata* Rose and *M. texana* var. *filipes* (Britton? Rose) Barneby) was found to contain more AM fungal spores than soil from non-vegetated areas in the semiarid valley of Tehuacán-Cuicatlán, Mexico, indicating that endemic *Mimosa* species can serve as mycorrhizal 'resource islands' in semiarid areas (Camargo-Ricalde and Dhillion 2003). Inoculation of *M. adenantheroides* (M.Martens & Galeotti) Benth., *M. lacerata*, *M. luisana*, *M. polyantha* and *M. texana* var. *filipes* seedlings resulted in a higher shoot and total dry weight than non-mycorrhizal seedlings (Camargo-Ricalde et al. 2010). *Mimosa pudica* L. in southern India (Muthukumar et al. 2006) and Thailand (Higo et al. 2011) are colonized by AM fungi. Nutrient application on *Mimosa* does not seem to affect interaction between AM and rhizobia inoculations. In fact, application of phosphorus was found to increase mycorrhizal infection by *Glomus* spp. and *Gigaspora* spp., as long as the *Mimosa caesalpiniaefolia* plants were not also inoculated with rhizobia strains (*Bradyrhizobium* spp.) and grown in an acid soil (Stamford et al. 1997).

8.2.9 *Sesbania*

Sesbania species form nodules with fast-growing strains of *Rhizobium*. *Sesbania sesban* (Jacq.) W. Wight plants form nodules in abundance, and may prove to be a good nitrogen fixing species (Domingo 1983; Evans and Rotar 1987). In 60 days, *Sesbania rostrata* can fix 0.68–0.78 g nitrogen per plant in flooded soils and 0.59–0.62 g of nitrogen per plant on well-drained soil. Comparatively, *S. sesban* can only fix between 0.05 and 0.06 g of nitrogen per plant in flooded soils, and between 0.12 and 0.13 g of nitrogen in well-drained soils over the same time-period, demonstrating the much higher nitrogen fixing potential of *Sesbania rostrata* Bremek. & Oberm. compared to *S. sesban* (Ndoye and Dreyfus 1998). In some soils, the roots of *S. grandiflora* (L.) Poiret are susceptible to nematode attacks. In response, *S. sesban* seeds produce a nematicidic substance (Khurma and Mangotra 2004).

Sesbania spp. in Thailand were found to be colonized by AM fungi (Higo et al. 2011). Growth and nutrient uptake in *S. grandiflora* was enhanced by inoculation with *Glomus mosseae* and *Glomus fasciculatum*. In sterile soils, the effect of *G. fasciculatum* was more pronounced than the effect of *G. mosseae* (Habte and Aziz 1985). In southern Malawi, the diversity of glomalean mycorrhizal fungi in *Sesbania* (*S. sesban* and *S. macrantha* Welw. Ex E. Phillips & Hutch.)-maize based agroforestry systems was found to be lower than in monocrop maize systems, indicating that mycorrhizal diversity is influenced by agroforestry combinations (Jefwa et al. 2006). The same study reported that mycorrhizal diversity is increased by the addition of an inorganic N fertilizer. The mycorrhizal species recorded were members of the *Acaulospora*, *Glomus*, *Gigaspora* and *Scutellospora* genera (Jefwa et al. 2006). *Sesbania grandiflora* was also found to be highly mycorrhizal dependent (Pindi et al. 2000). A simultaneous application of both arbuscular mycorrhiza and a natural rock phosphate enhances the growth of *Sesbania sesban* seedlings, and can result in more than 200 % increase in weight (Ndiaye et al. 2009). The inoculation of *Sesbania grandiflora* plantlets *in-vitro* with *Glomus* spp. and *Rhizobium* enhanced the establishment and growth of the plantlets, alleviating transplantation shock (Pindi 2011b). Similar results were found with *Sesbania sesban*. Subhan et al. (1998) reported a 100 % survival rate in micropropagated plantlets inoculated with *Glomus fasciculatum*. Only 30 % of the non-inoculated plantlets survived transplantation. The growth of *Sesbania rostrata* was also found to be enhanced by an inoculation of *Glomus mosseae* and *Azorhizobium caulinodans* along with the application of a rock phosphate (Rahman and Parsons 1997).

8.3 Actinomycorrhizal Plants

Actinomycorrhizal plants are characterized by their ability to form nitrogen-fixing nodules in their roots by partnering with filamentous Gram-positive bacteria of the genus *Frankia* (Frankiaceae). Actinomycetes association with root nodules of woody plants is called actinomycorrhiza. About 200 non-legume plants, in 24 genera belonging to the Betulaceae, Casuarinaceae, Coriaceae, Cycadaceae, Eleagnaceae, Myriaceae, Rhamnaceae, Rosaceae, and Ulmaceae families form symbiotic associations with *Frankia*. The main actinomycorrhizal agroforestry species in the tropics are members of the *Alnus*, *Casuarina*, *Allocasuarina* and *Crotalaria* genera (Akkermans and Houvers 1983; Bond 1983; Gauthier et al. 1984; Roy et al. 2007). *Frankia* species can be classified into four cluster groups (Chaia et al. 2010). Cluster 1 strains include *Frankia alni*, and colonize *Alnus*, *Casuarina* and *Myrica*. In Cluster 2 Frankia are found in the nodules of *Dryas*, *Coriaria*, *Dastica* and *Ceanothus*. Cluster 3 is composed of *Elaeagnaceae* and *Rhamnaceae* strains, excluding *Ceanothus* strains. Cluster 4 contains atypical non-N_2 fixing strains, or strains that are not able to re-colonize the original host but have been isolated from actinorhizal nodules, thus failing to fulfill Koch's postulates. The nitrogen-fixing potential of actinomycorrhizal plants is high, but the amount of nitrogen fixed in the soil is often low, mainly due to unfavorable environmental conditions or inappropriate management practices (Dommergues 1997).

8.3.1 *Alnus acuminata* (syn. *Alnus jorullensis*)

Alnus acuminata Kunth has a great capacity for nitrogen fixation (Budowski 1983), and is popular in agroforestry pastures in Costa Rica. In one study, two *Frankia* strains were isolated from *Alnus acuminata*. One of the strains, AacIII, produced a large number of nodules (greater than 15) per plant (Caru et al. 2000), confirming the high nodulation potential of *Alnus*.

Frankia and *Glomus intraradices* have positive interactions with *A. acuminata* seedlings, affecting nodule weight at moderate (50 ppm) phosphorus levels (Russo 1989).

8.3.2 *Casuarinaceae*

The Casuarinaceae family consists of a group of 82 native species, mostly indigenous to Australia, but includes members originating in Southeast Asia and the Pacific Islands (NAS 1984). Four genera are documented, namely *Casuarina*, *Allocasuarina*, *Gymnostoma* and *Ceuthostoma* (Sogo et al. 2001). *Casuarina* species typically show good nodulation, whereas nodulation in *Allocasuarina* species is variable and sometimes non-existent. Strains of *Frankia* isolated from *Casuarina* usually do not infect *Allocasuarina*, and vice-versa.

8.3.3 *Coriaria*

All 15 species of the genus *Coriaria* form nodules, indicating that nodulation is a generic character of the genus. Two species, namely *C. sinica* Maxim and *C. arborea* Linds., are effective components of agroforestry systems. *Coriaria sinica* is deciduous, fast-growing, and is a source of green manure and food for silkworms. *Coriaria arborea* used in *Pinus radiata* D.Don plantations in New Zealand can fix up to 192 kg of nitrogen per ha per year, although its effects on *P. radiata* growth need to be studied. Endophytes from *Coriaria* are a discrete *Frankia* lineage (Nick et al. 1992).

8.4 Quantification of Nitrogen Fixation

The methods and basic principles used in the estimation of nitrogen fixation are subject to debate (LaRue and Patterson 1981; Herridge 1982; Bergersen 1988; Peoples et al. 1989; Peoples and Herridge 1990). Some of the most widely used methods were reviewed by Danso et al. (1992), and are presented here.

8.4.1 Total Nitrogen Difference

This method estimates the difference between the total nitrogen harvested from a nitrogen fixing plant with nodules, and that harvested from a non-N-fixing plant, which has no nodules, preferably of the same species. The type of root system of both study plants strongly influences the accuracy of the estimate. This method is less commonly used because it has many weak points, and serious errors might affect the estimates of nitrogen fixation (Danso 1985).

8.4.2 Acetylene Reduction Assay

An acetylene reduction assay is the most sensitive method for estimating nitrogen fixation in trees, and is simple and inexpensive (Danso et al. 1992). The nitrogen-fixing system is placed in an atmosphere enriched with 10% acetylene. After an incubation period of one to two hours, a sample is removed from the atmosphere and the ethylene that results from the nitrogenase activity is analyzed. The reduction in the acetylene percentage is converted into estimates of nitrogen fixation using a conversion ratio (C_2H_2: N_2) that was originally evaluated at 3:1. It is now recognized that this ratio is variable, and should be tested for each system. Several techniques of this method have been described in detail by Bergersen (1980). This method, previously used to test *Inga jinicuil* (Roskoski 1981), and *Leucaena leucocephala* (Högberg et Kvarnström 1982) fixing, has been questioned by Witty and Minchin (1988), and is increasingly less frequently used for attempting to measure nitrogen fixation in the field. However, the method continues to be useful for comparative and short-term studies. There are two major drawbacks to the method: (i) it is an instantaneous assay, and may not truly reflect nitrogen fixation over long durations (Fried et al. 1983), and (ii) the conversion ratio varies.

A method of measuring acetylene reduction has been described by Staal et al. (2001), and is regularly used for sediment and water samples. It consists of a gas flow cell connected to a gas mixing system and an automatic loop control in the gas chromatograph. Alternatively, ethylene can be estimated by using a laser to detect trace gasses. The laser detection technique used to analyze trace gas has a detection limit, which is three orders of magnitude higher than gas chromatography, and is commonly used in the natural environment, especially water or sediments.

8.4.3 ^{15}N Enrichment

Also known as the direct method of isotope dilution, ^{15}N enrichment method refers to the comparison of non-fixing and fixing plants in a soil containing ^{15}N, usually by adding as labeled urea, nitrate or ammonium. Nitrogen-fixing plants obtain nitrogen from both the soil and the air, and consequently contain a lower content of the ^{15}N isotope than non-fixing plants, that only absorb nitrogen in the soil. The fixed nitrogen amount is obtained from the excess ^{15}N atom percentage in non-fixing compared to fixing plants. This method can yield results similar to the difference method (Gauthier et al. 1985; Sanginga et al. 1989).

8.4.4 Natural Abundance in ^{15}N

This method studies the differences in the natural abundance of ^{15}N between non-N-fixing and N-fixing plants. Soil nitrogen is typically slightly richer in ^{15}N than atmospheric nitrogen. Because most biological responses favor the lighter nitrogen isotope, fixed nitrogen has a ^{15}N content that is slightly lower than soil nitrogen. Consequently, the natural abundance of ^{15}N is lower in nitrogen-fixing plants than in non-fixing plants (Knowles 1983). This method was successfully used to quantify nitrogen fixed by deep roots in *Prosopis* (Virginia et al. 1981), and demonstrated successful results when tested in improved fallows systems (Gathumbi et al. 2002).

8.4.5 Sap Nitrogen Solute Analysis

The sap that rises in the xylem of nitrogen-fixing plants transports compounds created by the fixation of inorganic soil nitrogen (most often NO_3^-) and absorbed by the roots and nodules. Sap nitrogen solute analysis measures the ureide content in xylem sap (Herridge et al. 1990). Leguminous plants can be ranked in two categories: ureide exporters including *Vigna unguiculata* (L.) Walp. and *Glycine max* (L.) Merr., that export fixed nitrogen as allantoin or allantonic acid, and amide exporters such as *Lupinus albus* L. and *Trifolium* spp., that export the fixed nitrogen as

Table 8.8 Amounts of N_2 fixation by some woody species used in agroforestry (Nair 1993)

Species	N_2 fixation (kg ha^{-1} year^{-1})	Source
Acacia mearnsii	200	Dommergues (1987)
Casuarina equisetifolia	60–110	Dommergues (1987)
Erythrina poeppigiana	60	Escalante et al. (1984)
Faidherbia (Acacia) albida	20	Nair (1984)
Gliricidia sepium	13	Roskoski et al. (1982) Szott et al. (1991)
Inga jinicuil	35–40	Roskoski (1982)
Leucaena leucocephala	100–500	Högberg and Kvarnström (1982), Sanginga et al. (1985)
Sesbania rostrata	83–109	Peoples and Herridge (1990)

asparagine, glutamine, or amid substitutes. The sap also carries nitrates or other organic products produced in the roots from nitrate reduction.

Xylem nitrogen is found in the form of free, non-reduced nitrate or ureide nitrogen in ureide exporters, and as nitrate and amino acids in non-fixing plants. The relative abundance of ureide in sap can be taken as an indication of nitrogen-fixing activity in plants. This method measures nitrogen fixation over short periods of time.

Other methods used for nitrogen fixation estimation include the selection of reference plants (Danso et al. 1992) and the ^{15}N isotope dilution method (e.g., N derived from atmosphere as a percentage of the total plant N is a function of the ratio of ^{15}N atom percentage excess in a nitrogen fixing plant over the ^{15}N atom percentage excess in a non-fixing plant (Warembourg 1993)). These methods are suitable for estimating nitrogen fixation, though a few drawbacks have been reported (Busse 2000).

In summary, nitrogen fixation by plants can be reliably estimated using the various existing methods under carefully controlled conditions. If possible, two methods should be used simultaneously to crosscheck the estimated rates. Table 8.8 summarizes some rates of nitrogen fixation by trees.

8.5 Technologies for the Exploitation of Nitrogen-Fixing Mycorrhizal Plants in Agroforestry

Mechanisms by which fixed nitrogen is transferred from nitrogen-fixing plants to non-fixing crops, as well as nitrogen turnover rates are not yet fully understood. The hypothesis that the degeneration of nodules and roots releases nitrogen, making it available to adjacent plants, needs to be tested under various soil types and climatic conditions. The processes that allow the transmission of fixed nitrogen to non-fixing plants also needs to be identified, and the percentage of the nitrogen made available by nitrogen-fixing plants that is reabsorbed by non-fixing plants needs to be calculated. A discussion of the principles regulating the choice of species and the provenances of nitrogen-fixing trees, and recommended practices to alleviate some environmental problems follow.

8.5.1 *Selecting Species and Provenances of Nitrogen-Fixing Trees*

The main selection criteria for provenances or species for introduction into an agroforestry system is the rate of nitrogen fixation. The rate should be the highest possible over a range of climatic conditions. The selected provenances or species must also be able to tolerate environmental constraints such as low nitrogen levels and the presence of pests. Therefore, *Calliandra calothyrsus*, which fixes less nitrogen than *Leucaena leucocephala* in fertile soil, has been screened and selected from among ten nitrogen-fixing species for introduction into agroforestry systems in the lowland humid tropics of Cameroon (Duguma and Tonye 1994). The other nine candidates were *Acacia auriculiformis, Acacia mangium, Cassia javanica, Cassia siamea, Gliricidia sepium, Leucaena leucocephala, Paraserianthes falcataria, Sesbania grandiflora* and *Sesbania sesban*. Results on nitrogen fixation from the available data rank nitrogen-fixing trees into two categories (Nair 1993):

- Species with a high potential fixation rate, between 100 and 300 kg nitrogen per ha per year. Species in this category include: *Acacia mangium*, *Casuarina equisetifolia*, *Leucaena leucocephala*, and *Coraria arborea.*
- Species with a low potential for nitrogen fixation, less than 20 kg nitrogen per ha per year. Examples include: *Faidherbia albida*, *Acacia raddiana* and *A. seyal.*

Species with high nitrogen fixing potential were further divided in two groups by Nair (1993):

- intolerant or demanding species, such as *Leucaena leucocephala* and *Calliandra calothyrsus,* which require large amounts of P, K and Ca; and
- tolerant or non-demanding species, such as *Acacia mangium*, which can flourish in poor or acidic soils with low nutrient levels.

8.5.1.1 Inoculation with *Rhizobium* or Frankia

The technique of inoculating a host plant using infected soil or crushed nodules, though strongly recommended, is risky. Seed contamination with pathogens such as *Rhizoctonia solani* or *Pseudomonas solanacearum* (in the case of *Casuarina equisetifolia*) (Liang Zichao 1986) or nematode infection (as seen in Australian Acacias introduced in West Africa) (Dommergues 1987) can occur. The best techniques of *Rhizobium* inoculation are either mixing the seeds with clean inoculum before planting, or spraying the inoculum into the container when planting.

The inoculation of actinorhizal plants with pure cultures of *Frankia* strains is becoming more common due to recent progress in understanding the physiology of the *Frankia* species (Oliveira et al. 2005, Roy at al. 2007). Inoculation of *Frankia* may be best achieved by mixing the soil or substrate being used with the *Frankia* inoculum.

8.5.1.2 Inoculation with Mycorrhizal Fungi

Mycorrhizal inoculations positively affect the absorption of phosphate and other non-mobile ions in the soil, such as Zn^{2+}, $Cu^{2+}$and Mo^{2+}. Mycorrhizal fungi form symbiotic associations with the roots of nitrogen-fixing trees. The most common fungi are arbuscular mycorrhizae, which penetrate host roots. Ectomycorrhizae, which remain external, are less common. AM are the most common type of mycorrhizas, and exhibit the best performance of plant nutrition facilitation. Nodulation and nitrogen fixation require a high level of phosphorus in the host plant, which can be facilitated by the mycorrhizal symbiont by increasing the available P in the soil. Mycorrhizas amplify the effects of small amounts of phosphate fertilizer added to highly P deficient soil (Ganry et al. 1985). Mycorrhizas also allow the host plant to increase its water absorption, improve hormonal balance, and break cutting dormancy (Hayman 1986).

8.5.1.3 Fertilizers

Nitrogen-fixing trees and shrubs need to be fertilized just like other crops (Yadav 1983; de Souza Moreira et al. 2010). Some species such as *Leucaena leucocephala* have very high nutrient demands, but can produce a tremendous amount of biomass if the required nutrients are available (Waring 1985). A three year old *L. leucocephala* plantation can absorb 11–27 kg of P, 174–331 kg of K, 138–305 kg of Ca, and 31–62 kg of Mg per ha (Hu and Kiang 1983). Waring (1985) reported that *Casuarina* might also have high Ca demands. Low P supply inhibits nodulation by limiting plant growth, and therefore nitrogen demand. *Frankia* is affected by low P supply in the rhizosphere in the early stages of nodule formation, but the reason for the low P effect was not explained (Reddel et al. 1986).

Phosphorus is the most limiting factor for *L. leucocephala* and *Sesbania virgata* (de Souza Moreira et al. 2010). Application of mineral nitrogen at high levels inhibits nodulation and nitrogen fixation. Work is needed to quantify the exact nitrogen requirements of nitrogen-fixing trees.

8.5.1.4 Acidity Control

Tropical soils are mostly acidic, and exhibit aluminum and magnesium toxicity. For example, the soils in the Congo Basin are often aluminum toxic. Acidity influences nitrogen fixation by its effects on plants and symbiotic micro-organisms. *Acacia mearnsii* do not form nodules in the highlands of Burundi, where soils are acidic and have a high exchangeable aluminum content (Nair 1993). To circumvent adverse effects, it is recommended to choose species and provenances of plants and micro-organisms that are tolerant to acidity (Hutton 1984; Halliday et Somasegaran 1983; Franco 1984, Kernaghan et al. 2002; Campagnac et al. 2013). The application of appropriate amendments such as lime or organic matter, either applied directly on the soil or pelleted with the seeds, can also help control soil acidity effects.

It should be noted that, in the long term, nitrogen-fixing plants can generate acidity, reducing the soil pH value in slightly amended soil. Periodic lime application would help maintain high productivity (Franco 1984). High levels of organic matter content in soils under nitrogen-fixing trees can also stimulate a satisfactory production, even when the pH is lower than recommended values.

An alternative to lime application on pastures is seed coating with calcium carbonate or phosphate. This technique has proven to be effective in maintaining pH levels and maintaining productivity when establishing fodder legumes in pastures (Williams 1984).

8.6 Areas of Research Concerning Nitrogen Fixation and Mycorrhizae in Agroforestry

More and more work is being done on the quantification of biological nitrogen fixation by legumes in agroforestry systems. A review of research that has been done on nitrogen-fixing trees reported that these trees can add more than 60 kg of nitrogen to the soil per ha per year through biological nitrogen fixation, and that their biomass contribution can reduce non-organic nitrogen requirements by 75 % (Akinnifesi et al. 2010). Improving the rate of nitrogen fixation and understanding mechanisms for the redistribution of N from nitrogen-fixing to non-fixing plants are areas that require further research, along with the influence of available N in the soil and nodulations. Nitrogen fixation can be improved through choice of both the host tree (West et al. 2005), and the symbiotic micro-organisms. To date, few *Rhizobium* strains have been isolated that are effective in nodule formation with nitrogen-fixing trees. Strains that have been isolated include *L. leucocephala* (Roskoski 1986; Sanginga et al. 1986, 1989) and *Acacia nilotica* (Woldemeskel and Sinclair 1998). Other effective *Rhizobium* strains have been isolated from *Inga vera* (Maia and Scotti 2010). More work needs to be done to collect *Rhizobium* strains from nitrogen-fixing trees. Screening tests also need to be performed, especially for genetic compatibility, efficiency in nitrogen fixation, and tolerance to environmental stresses, especially soil acidity under natural conditions.

Frankia strains that form symbiotic associations with Casuarinaceae have high variability in genetic compatibility and effectiveness (Zhang et al. 1984; Puppo et al. 1985; Roy et al. 2007). Even if only associated with a single *Casuarina* species, *Frankia* strains exhibit large differences in effectiveness. *Frankia* strain effectiveness can further vary in different Casuarinaceae species (Reddell 1986). Mansour and Baker (1994) postulated that changes in performance and nitrogen fixation are related to the interaction between the *Casuarina* cultivar and the *Frankia* strain. This symbiotic compatibility should be tested to improve production (Mansour and Baker 1994). Using molecular biology techniques, the genes in new strains of *Rhizobium* and *Frankia* can be studied for their involvement in symbiosis, specifically, those regulating nitrogen fixation and nodulation, and competition between strains (Ligon and Nakas 1987; Rouvier et al. 1996).

Nitrogen fixing potential (NFP) describes the ability of legumes to fix nitrogen in the absence of limiting factors. Basically, it is the amount of nitrogen that can be fixed by any nitrogen-fixing plant (Halliday 1984). NFP is genetically controlled by both the host plant and the associated symbiont. Consequently, a nitrogen-fixing tree with a great ability to fix nitrogen, and which is highly tolerant to environmental stresses like temperature, drought, and nitrogen supply is more likely to be useful in an agroforestry system. However, under field conditions, the amount of nitrogen that is fixed may be lower than the NFP. High levels of soil mineral N availability have an inhibitory effect on nitrogen fixation. More research needs to be done to understand how to overcome such negative effects.

Methods used for estimating nitrogen fixation also need to be improved. For example, the natural abundance of ^{15}N did not reveal the presence of N derived from biological fixation by *Ceratonia siliqua* trees on a farm (La Malfa et al. 2010). The ^{15}N method also failed to provide a reliable estimate of the nitrogen biologically fixed by legumes in mixed forests of the Amazon (Gehring and Vlek 2004).

Bibliography

Akinnifesi FK, Ajayi OC, Sileshi G, Chirwa PW, Chianu J (2010) Fertiliser trees for sustainable food security in the maize-based production systems of East and Southern Africa. A review. Agron. Sustain. Dev 30:615–629

Akkermans ADL, Houvers H (1983) Morphology of nitrogen fixers in forest ecosystems. In: Gordon JC, Wheeler CT (eds) Biological Nitrogen Fixation in Forest Ecosystems: Foundations and Applications. Njihoff/Junk, The Hague, The Netherlands, pp 7–53

Allen ON, Allen EK (1981) The Leguminosae. A source Book of Characteristics, Uses, and Nodulation. University of Wisconsin Press, Madison, WI, USA.

Andrade ACS, Queiroz MH, Hermes RAL, Oliveira VL (2000) Mycorrhizal status of some plants of the Araucaria forests and the Atlantic rainforest in santa Catarina, Brazil. Mycorrhiza 10:131–136

Atayese MO, Awotoye OO, Osonubi O, Mulongoy K (1993) Comparisons of the influence of vesicular-arbuscular mycorrhiza on the productivity of hedgerow woody legumes and cassava at the top and the base of a hillslope in alley cropping system. Biol Fert Soils 16:198–204

Bagyaraj DJ, Byra Reddy MS, Nalini PA (1989) Selection of an efficient inoculant VA mycorrhizal fungus for *Leucaena*. For Ecol Manag 27:81–85

Bakarr MI, Janos DP (1996) Mycorrhizal associations of tropical legume trees in Sierra Leone, West Africa. For Ecol Manag 89:89–92

Bala A, Giller KE (2001) Symbiotic specificity of tropical tree rhizobia for host legumes. New Phytol 149(3):495–507

Bergersen FJ (1980) Methods for evaluating nitrogen fixation. John Wiley, New York, USA

Bergersen FJ (1988) Measurements of dinitrogen fixation. In: Shamsuddin ZH, Othman WMW, Marziah M, Sundram J (eds) Biotechnology of Nitrogen Fixation in the Tropics. Universiti pertanian Malaysia, Serdang, Malaysia, pp 105–115

Bond G (1983) Taxonomy and distribution of non-legume nitrogen-fixing systems. In: Gordon JC, Wheeler TC (eds), Biological Nitrogen Fixation in Forest Ecosystems: Foundations and Applications. Njihoff/Junk, The Hague, The Netherlands, pp 55–87

Brady NC, Weir RR (2010) Elements of the nature and properties of soils. 3rd edn. Prentice Hall, Upper Saddle River, New Jersey, USA,p 614 with index

Brewbaker JL (1987) Significant nitrogen fixing trees in agroforestry systems. In: Gholz HL (ed) Agroforestry: realities, possibilities and potentials. Martinus Njihoff, Dordrecht, The Netherlands, pp 31–45

Budowski G (1983) An attempt to quantify some current agroforestry practices in Costa Rica. In: Huxley PA (ed) Plant Research and Agroforestry. ICRAF, Nairobi, Kenya, pp 43–62

Busse MD (2000) Suitability and use of the 15N-isotope dilution method to estimate nitrogen fixation by actinorhizal shrubs. For Ecol Manag 136(1–3):85–95

Camargo-Ricalde SL, Dhillion SS (2003) Endemic *Mimosa* species can serve as mycorrhizal 'resource islands' within semiarid communities of The Tehuacán-Cuicatlán Valley, Mexico. Mycorrhiza 13:129–136

Camargo-Ricalde SL, Montaño NM, Reyes-Jaramillo I, Jimenéz-González C, Dhillion SS (2010) Effects of mycorrhizae on seedlings on six endemic *Mimosa* L. species (Leguminosae-Mimosoideae) from the semi-arid Tehuacán-Cuicatlán Valley, Mexico. Trees 24:67–78

Campagnac E, Piché Y, Khasa DP (2013) Relationship between genetic variability in *Rhizophagus irregularis* and tolerance to saline conditions. Mycorrhiza (In révision)

Cardoso IM, Boddington C, Janssen BH, Oenema O, Kuyper TW (2003) Distribution of mycorrhizal fungal spores in soils under agroforestry and monocultural cropping systems in Brazil. Agrofor Syst 58:33–43

Caru M, Becerra A, Sepulveda D, Cabello A (2000) Isolation of infective and effective *Frankia* strains from root nodules of *Alnus acuminata* (Betulaceae). World J Microb Biot 16:647–651

Chaia EE, Wall LG, Huss-Danell K (2010) Life in soil by the actinorhizal root nodule endophyte *Frankia*. A review. Symbiosis 51:201–226

Chakravarty P, Mishra RR (1986) The influence of VA mycorrhizae on the wilting of Albizia procera and Dalbergia sissoo. Eur J Forest Pathol 16:91–97

Chen WX, Yan GH, Li JL (1988) Numerical taxonomic study of fast-growing soybean rhizobia and a proposal that *Rhizobium fredii* be assigned to *Sinorhizobium* gen. nov. Int J Syst Bacteriol 38:392–397

Cooperband LR, Boerner REJ, Logan TJ (1994) Humid tropical leguminous tree and pasture grass responsiveness to vesicular-arbuscular mycorrhizal infection. Mycorrhiza 4:233–239

Cornet F, Diem HG, Dommergues YR (1982) Effet de l'inoculation avec Glomus mosseae sur la croissance d'Acacia holosericea en pepiniere et apres transplantation sur le terrain. In: Les Mycorhizes : Biologie et Utilisation. INRA, Paris, France

Chen W-M, Faria SM de, Straliotto R, Pitard RM, Simões-Araùjo JL, Chou J-H, Yi-Ju C, Barrios E, Prescott AR, Elliott GN, Sprent JI, Young JPW, James EK (2005) Proof that *Burkholderia* strains form effective symbioses with legumes: a study of novel *Mimosa*-nodulating strains from South America. Appl Environ Microbiol 71(11):7461–7471

Cuenca G, Azcón R (1994) Effects of ammonium and nitrate on the growth of vesicular-arbuscular mycorrhizal *Erythrina poeppigiana* O.I. Cook seedlings. Biol Fert Soils 18:249–254

Danso SKA (1985) Methods for estimating biological nitrogen fixation. In: Sali H, Keya SO (eds) Biological Nitrogen Fixation in Africa. MIRCENS, Nairobi, Kenya pp 213–234

Danso SKA, Bowen GD, Sanginga N (1992) Biological nitrogen fixation in trees in agro-ecosystems. PlantSoil 141:177–196

de Carvalho AMX, de CTavarerR, Cardoso IM, Kuyper TW (2010) Mycorrhizal associations in agroforestry systems. In: Dion P (ed) Soil Biology and Agriculture in the Tropics, Soil Biology 21. doi:10.1007/978–3-642–05076-3_9. Springer-Verlag, Berlin Heidelberg

De LCruzRE, Zarate JT, Aggangan NS, Lorilla EB (1992) Differential mycorrhizai dependency of some agricultural, horticultural and forestry crops to inoculation with arbuscular mycorrhizal fungi. In: Jasper D, Thomson B (eds) International Symposium on Management of Mycorrhizas in Agriculture, Forestry and Horticulture, 28 September–2 October 1992, Perth, W. Australia, Abstracts, p 54

De SMoreiraFM, Soares deCT, Siqueira JO (2010) Effect of fertilizers, lime, and inoculation with rhizobia and mychorrhizal fungi on the growth of four leguminous tree species in a low-fertility soil. Biol Fertil Soils 46:771–779

Diagne O, Ingleby K, Deans JD, Lindley DK, Diaite I, Neyra M (2001) Mycorrhizal inoculum potential of soils from alley cropping plots in Senegal. For Ecol Manag 146:35–43

Diem HG, Gauthier D, Dommergues YR (1982) Isolation of *Frankia* from nodules of *Casuarina equisitifolia*. Canadian J Microbiol 28:526–530

Diem HG, Gauthier D, Dommergues YR (1983) An effective strain of *Frankia* from *Casuarina* sp. Canadian J Botany 61:2815–2821

Diouf A, Diop TA, Ndoye I, Gueye M (2008) Response of *Gliricidia sepium* tree to phosphorus application and inoculations with *Glomus aggregatum* and rhizobial strains in a sub-Saharian sandy soil. Afr J Biotechnol 7(6):766–771

Diouf D, Duponnois R, Tidiane Ba A, Neyra M, Lesueur D (2005) Symbiosis of *Acacia auriculiformis* and *Acacia mangium* with mycorrhizal fungi and *Bradyrhizobium* spp. improves salt tolerance in greenhouse conditions. Funct Plant Biol 32:1143–1152

Dixon RK, Garg VK, Rao MV (1993) Inoculation of *Leucaena* and *Prosopis* seedlings with *Glomus* and *Rhizobium* species in saline soils: rhizosphere relations and seedling growth. Arid Soil Res Rehab 7(2):133–144

Dobereiner J (1984) Nodulation and nitrogen fixation in legume trees. Pesqui Agropecu Bras 19:83–90

Domingo I (1983) Nitrogen fixation in Southeast Asian forestry research and practice. In: Gordon JC, Wheeler CT (eds) Biological Nitrogen Fixation in Forest Ecosystems: Foundations and Applications. Njihoff/Junk, The Hague, The Netherlands, pp 295–313

Dommergues YR (1997) Contribution of actinorhizal plants to tropical soil productivity and rehabilitation. Soil Biol Biochem 29(5/6):931–941

Dreyfus BL, Dommergues YR (1981) Nodulation of *Acacia* species by fast- and slow-growing tropical strains. Appl Environm Microbiol 41:97–99

Dreyfus B, Garcia JL, Gillis M (1988) Characterization of *Azorhizobium caulinodans* gen. nov., sp. nov., a stem-nodulating nitrogen-fixing bacterium isolated from *Sesbania rostrata*. Int J Syst Bacteriol 38:89–98

Duguma B, Tonye J (1994) Screening of multipurpose tree a nd shrub species for agroforestry in the humid lowlands of Cameroon. For Ecol Manag 64(2–3):135–143

Elkan GH (1984) Taxonomy and metabolism of *Rhizobium* and its genetic relationships. In: Alexander M (ed) Biological Nitrogen Fixation, Ecology, Technology, and Physiology. Plenum Press, New York, USA, pp 1–38

Elliott GN, Chen W-M, Chou J-H, Wang H-C, Sheu S-Y, Perin L, Reis VM, Moulin L, Simon MF, Bontemps C, Sutherland JM, Bessi R, De Faria SM, Trinick MJ, Prescott AR, Sprent JI, James EK (2007) *Burkholderia phymatum* is a highly effective nitrogen-fixing symbiont of *Mimosa* spp., and fixes *ex planta*. New Phytol 173(1):168–180

Evans DO, Rotar PP (1987) *Sesbania* in Agriculture. Westview Press, Boulder, CO, USA

Fagbola O, Osonubi O, Mulongoy K, Odunfa SA (2001) Effects of drought stress and arbuscular mycorrhiza on the growth of *Gliricidia sepium* (Jacq). Walp, and *Leucaena leucocephala* (Lam.) de Wit. in simulated eroded soil conditions. Mycorrhiza 11:215–223

Founoune H, Duponnois R, Ba AM, Sall S, Branget I, Lorquin J, Neyra M, Chotte JL (2002) Mycorrhiza Helper Bacteria stimulate ectomycorrhizal symbiosis of *Acacia holoricea* with *Pisolithus alba*. New Phytol 153:81–89

Fried M, Danso SK, Zapata AF (1983) The methodology of measurement of N_2 fixation by nonlegumes as inferred from field experiments with legumes. Can J Microbiol 29:1053–1062

Franco AA (1984) Nitrogen fixation in trees and soil fertility. Pesqui Agropecu Bras 19:253–261

Franco AA, De Faria SM (1997) The contribution of N_2-fixing tree legumes to land reclamation and sustainability in the tropics. Soil Biol Biogeochem 29(5/6):897–903

Ganry F, Diem HG, Wey J, Dommergues YR (1985) Inoculation with *Glomus mosseae* improves N2 fixation by field-grown soybeans. Biol Fertil Soils 1:15–23

Gathumbi SM, Cadisch G, Giller KE (2002) ^{15}N natural abundance as a tool for assessing N_2-fixation of herbaceous, shrub and tree legume in improved fallows. Soil Biol Biogeochem 34(8):1059–1071

Gauthier DL, Diem HG, Dommergues YR (1984) Tropical and subtropical actinorhizal plants. Pesqui Agropec Bras 19:119–136

Gauthier DL, Diem HG, Dommergues YR, Ganry F (1985) Assessment of N_2 fixation by *Casuarina equisitifolia* inoculated with *Frankia* ORS021001 using ^{15}N methods. Soil Biol Biochem 17:375–379

Gehring C, Vlek PLG (2004) Limitations of the 15N natural abundance method for estimating biological nitrogen fixation in Amazonian forest legumes. Basic Appl Ecol 5(6):567–580

Ghosh S, Verma NK (2006) Growth and mycorrhizal dependency of *Acacia mangium* Willd. inoculated with three vesicular arbuscular mycorrhizal fungi in lateritic soils. New Forest 31:75–81

Gibson AH, Dreyfus BL, Dommergues YR (1982) Nitrogen fixation by legumes. In: Dommergues YR, Diem HG (eds) Microbiology of Tropical soils and Plant Productivity. Martinus Njhoff, The Hague, The Netherlands, pp 37–73

Grossman JM, Sheaffer C, Wyse D, Bucciarelli B, Vance C, Graham PH (2006) An assessment of nodulation and nitrogen fixation in inoculated *Inga oerstediana*, a nitrogen-fixing tree shading organically grown coffee in Chiapas, Mexico. Soil Biol Biogeochem 38(4):769–784

Habte M, Aziz T (1985) Response of *Sesbania grandiflora* to inoculation of soil with vesicular-arbuscular mycorrhizal fungi. Appl Environ Microbiol 50(3):701–703

Habte M, Manjunath A (1987) Soil solution phosphorus status and mycorrhizal dependency in *Leucaena leucocephala*. Applied Environ. Microbiol 43:797–801

Habte M, Musoko M (1994) Changes in the vesicular-arbuscular mycorrhizal dependency of *Albizia ferruginea* and *Enterolobium cyclocarpum* in response to soil phosphorus concentration. J Plant Nutr 17(10):1768–1780

Halliday J (1984) Register of nodulation reports for leguminous trees and other arboreal genera with nitrogen-fixing trees. Nitrogen-Fixing Tree Research Reports 4:38–45

Halliday J, Somasegaran P (1983) Nodulation, nitrogen fixation, and *Rhizobium* and strain affinities in the genus *Leucaena*. In: Leucaena Research in the Asia-Pacific region. IDRC, Ottawa, Canada

Haselwandter K, Bowen GD (1996) Mycorrhizal relations in trees for agroforestry and land rehabilitation. For Ecol Manag 81:1–17

Hayman DS (1986) Mycchorhizae of nitrogen-fixing legumes. MIRCEN J 2:121–145

Herridge DF (1982) A whole-system approach to quantifying biological nitrogen fixation by legumes and associated gains and losses of nitrogen in agricultural systems. In: Graham PH, Harris SC (eds) Biological nitrogen fixation technology for tropical agriculture. CIAT, Cali, Colombia

Herridge DF, Bergersen FJ, Peoples MB (1990) Measurement of nitrogen fixation by soybean in the field using the ureide and natural ^{15}N abundance methods. Plant Physiol 93:708–716

Higo M, Isobe K, Kang D-J, Maekawa T, Ishii R (2011) Molecular diversity and spore density of indigenous arbuscular mycorrhizal fungi in acid sulfate soil in Thailand. Ann Microbiol 61:383–389

Högberg P, Kvarnström M (1982) Nitrogen fixation by the woody legume *Leucaena leucocephala*. Plant Soil 66:21–28

Hu TW, Kiang T (1983) *Leucaena* research in Taiwan. In: Leucaena Research in the Asian-Pacific Region. IDRC, Ottawa, Canada

Huang RS, Smith WK, Yost RS (1985) Influence of vesicular-arbuscular mycorrhiza on growth, water relations and leaf orientation in *Leucaena leucocephala* (LAM.) de Wit. New Phytol 99:229–243

Hutton EM (1984) Breeding and selecting Leucaena for acid tropical soils. Pesqu Agropecu Bras 19:263–274

Iglesias L, Salas E, Leblanc HA, Nygren P (2011) Response of *Theobroma cacao* and *Inga edulis* seedlings to cross-inoculated populations of arbuscular mycorrhizal fungi. Agrofor Syst. doi: 10.1007/s10457–011-9400–9

Ingleby K, Diagne O, Deans JD, Lindley DK, Neyra M, Ducousso M (1997) Distribution of roots, arbuscular mycorrhizal colonisation and spores around fast-growing tree species in Senegal. For Ecol Manag 90:19–27

Ingleby K, Wilson J, Munro RC, Cavers S (2007) Mycorrhizas in agroforestry: spread and sharing of arbuscular mycorrhizal fungi between trees and crops: complementary use of molecular and microscopic approaches. Plant Soil 294:125–136

IRRI (1988) Green manure in rice farming: Proceedings of a symposium on sustainable agriculture. International Rice Research Institute, Los Banos, The Philippines

Jalonen R, Nygren P, Sierra J (2009) Transfer of nitrogen from a tropical legume tree to an associated fodder grass via root exudation and common mycelial networks. Plant. Cell Environ 32:1366–1376

Jefwa JM, Sinclair R, Maghembe JA (2006) Diversity of glomalean mycorrhizal fungi in maize/sesbania intercrops and maize monocrop in southern Malawi. Agrofor Syst 67:107–114

Kernaghan G, Hambling B, Fung M, Khasa DP (2002) *In vitro* selection of boreal ectomycorrhizal fungi for use in reclamation of saline-alkaline habitats. Restor Ecol 10:43–51

Kessel C, Naskao P (1986) The use of nitrogen-15-depleted ammonium sulfate for estimating nitrogen fixation by leguminous trees. Agron J 78:549–551

Khasa DP, Furlan V, Fortin JA (1992) Response of some tropical plant species to endomycorrhizal fungi under field condition. Trop Agr 69:279–283

Khasa DP, Vallée G, Bousquet J (1994) Biological considerations in the utilization of *Racosperma auriculiforme* and *R. mangium* in tropical countries with emphasis on Zaire. J Trop Forest Res 6:422–443

Khurma UR, Mangotra A (2004) Screening of some Leguminosae seeds for nematicidal activity. The South Pacific J Nat Sci 22(1):51–53

Knowles R (1983) Nitrogen fixation in natural plant communities and soils. In: Bergersen FJ (ed) Methods for evaluating biological nitrogen fixation. John Wiley, New York, USA

Ladha JK, Peoples MB, Garrity DP, Capuno VT, Dart PJ (1993) Estimating dinitrogen fixation of hedgerow vegetation using the nitrogen-15 natural abundance method. Soil Sci Soc Am J 57(3):732–737

La Malfa S, Tribulato E, Ventura M, Gioacchini P, Tagliavini M (2010) ^{15}N natural abundance technique does not reveal the presence of nitrogen from biological fixation in field grown Carob (*Ceratonia siliqua* L.) trees. Acta Hort. (ISHS) 868:191–196

Lambers H (2003) Dryland salinity: a key environmental issue in southern Australia. Plant Soil 257:v–vii

LaRue TA, Patterson TG (1981) How much nitrogen do legumes fix? Adv Agron 34:15–38

Lesueur D, Ingleby K, Odee D, Chamberlain J, Wilson J, Tiki-Manga T, Sarrailh J-M, Pottinger A (2001) Improvement of forage production in *Calliandra calothyrsus*: methodology for the identification of an effective inoculum containing *Rhizobium* strains and arbuscular mychorrizal inoculates. J Biotechnol 91(2–3):269–282

Lesueur D, Sarr A (2008) Effects of single and dual inoculation with selected microsymbionts (rhizobia and arbuscular mycorrhizal fungi) on field growth and nitrogen fixation of *Calliandra calothyrsus* Meissn. Agrofor Syst 73:37–45

Liang Z (1986) Vegetative propagation and selection of *Casuarina* for resistance to bacterial wilt. Tropical Forestry (Science and Technology) Guangazhou 2:1–6

Ligon JM, Nakas JP (1987) Isolation and characterization of *Frankia* sp. Strain FaC1 genes involved in nitrogen fixation. Appl Environ Microbiol 53(10):2321–2327

Lindbald P, Russo R (1986) C_2H_2-reduction by *Erythrina poeppigiana* in a Costa Rican coffee plantation. Agrofor Syst 4:33–37

Lulandala LLL, Hall JB (1986) *Leucaena's leucocephala*'s biological nitrogen fixation: A promising substitute for inorganic nitrogen fertilization in agroforestry systems. In: Shamsuddin ZH, Othman WMW, Marziah M, Sundram J (eds) Biotechnology of nitrogen fixation in the tropics (BIOnifT), proceedings of UNESCO regional symposium and workshop. Universiti Pertanian Malaysia, Serdang, Malaysia

Maia J, Scotti MR (2010) Growth of *Inga vera* Willd. Subsp. *Affinis* under rizobia inoculation. R. C. Suelo Nutr. Veg 10(2):139–149

Mansour SR, Baker DD (1994) Selection trials for effective N_2-fixing *Casuarina-Frankia* combinations in Egypt. Soil Biol Biogeochem 26(5):655–658

Martinez-Romero E, Segovia L, Martins Mercante F, Antonio Franco A, Graham P, Pardo MA (1991) *Rhizobium tropici*, a novel species nodulating *Phaseolus vulgaris* L. beans and *Leucaena leucocephala* trees. Int J Syst Evol Microbiol 41:417–426

Michelsen A, Rosendahl S (1990) The effect of VA mycorrhizal fungi, phosphorus and drought stress on the growth of *Acacia nilotica* and *Leucaena leucocephala* seedlings. Plant Soil 124:7–13

Milnystki F, Frioni L, Agius F (1997) Characterization of rhizobia that nodulate native legume trees from Uruguay. Soil Biol Biogeochem 29(5–6):989–992

Moreira FMS, Cruz L, Faria SM, Marsh T, Martinez-Romero E, Pedrosa FO, Pitard RM, Young JPW (2006) *Azorhizobium doebereinerae* sp. Nov. Microsymbiont of *Sesbania virgata* (Caz.) Pers Syst Appl Microbiol 29:197–206

Muleta D, Assefa F, Nemomissa S, Granhall U (2008) Distribution of arbuscular mycorrhizal fungi spores in soils of smallholder agroforestry and monocultural coffee systems in southwestern Ethiopia. Biol Fertil Soil 44:653–659

Muthukumar T, Senthilkumar M, Rajangam M, Udaiyan K (2006) Arbuscular mycorrhizal morphology and dark septate fungal associations in medicinal and aromatic plants of Western Ghats, Southern India. Mycorrhiza 17:11–24

Nair PKR (1993) An introduction to agroforestry. Kluwer, The Hague

NAS (1984) Casuarinas: Nitrogen-Fixing Trees for Adverse Sites. National Academy of Sciences, Washington, DC, USA

Ndiaye F, Manga A, Diagne-Leye G, Ndiaye Samba SA, Diop TA (2009) Effects of rockphosphate and arbuscular mycorrhiza fungi on growth and nutrition of *Sesbania sesban* and *Gliricidia sepium*. Afr J Microbiol Res 3(5):305–309

Ndoye I, Dreyfus B (1988) N_2 fixation by *Sesbania rostrata* and *Sesbania sesban* using ^{15}N and total N difference methods. Soil Biol Biogeochem 20(2):209–213

Ndoye I, Gueye M, Danso SKA, Dreyfus B (1995) Nitrogen fixation in *Faidherbia albida*, *Acacia raddiana*, *Acacia Senegal* and *Acacia seyal* estimated using the ^{15}N isotope dilution technique. Plant Soil 172:175–180

Nichols JD, Carpentier FL (2006) Interplanting *Inga edulis* yields nitrogen benefits to *Terminalia amazonia*. For Ecol Manag 233(2–3):344–351

Nick G, Paget E, Simonet P, Moiroud A, Normand P (1992) The nodular endophytes of *Coriaria* spp. form a distinct lineage within the genus *Frankia*. Mol Ecol 1:175–181

Nygren P, Ramirez C (1995) Production and turnover of N_2 fixing nodules in relation to foliage development in periodically pruned *Erythrina poeppigiana* (Leguminosae) tress. For Ecol Manag 73(1–3):59–73

Okon IE (2011) Field response of two cassava genotypes inoculated with arbuscular mycorrhizal fungus to *Gliricidia sepium* mulch in a tropical Alfisol. Bot Res Int 4(1):04–08

Okon IE, Osonubi O, Sanginga N (1996) Vesicular-arbuscular mycorrhiza effects on *Gliricidia sepium* and *Senna siamea* in a fallowed alley cropping system. Agrofor Syst 33:165–175

Oliveira RS, Castro PML, Dodd JC, Vosatka M (2005) Synergistic effects of *Glomus intraradices* and *Frankia* spp on the growth and stress recovery of *Alnus glutinosa* in an alkaline anthropogenic sediment. Chemosphere 60(10):1462–1470

Orchard ER, Darby GD (1956) Fertility changes under continued wattle culture with special reference to nitrogen fixation and base status of the soil. In: Comptes Rendus du sixième Congrès. International Science du Sol, Paris, France

Osonubi O, Atayese MO, Mulongoy K (1995) The effect of vesicular-arbuscular mycorrhizal inoculation on nutrient uptake and yield of alley-cropped in a degraded Alfisol of southwestern Nigeria. Biol Fertil Soil 20:70–76

Osonubi O, Bakare ON, Mulongoy K (1992) Interactions between drought stress and vesicular-arbuscular mycorrhiza on the growth of *Faidherbia albida* (syn. *Acacia albida*) and *Acacia nilotica* in sterile and non-sterile soils. Biol Fertil Soil 14:159–165

Osonubi O, Mulongoy K, Awotoye OO, Atayese MO, Okali DUU (1991) Effects of ectomycorrhizal and vesicular-arbuscular mycorrhizal fungi on drought tolerance of four leguminous woody seedlings. Plant Soil 136:131–143

Peoples MB, Herridge DF (1990) Nitrogen fixation by legumes in tropical and subtropical agriculture. Adv Agron 44:155–223

Peoples MB, Faizah AW, Rerkasem B, Herridge DF (1989) Methods for Evaluating Nitrogen Fixation by Nodulated Legumes in the Field. Monograph No 11, ACIAR, Canberra, Australia

Pindi KP (2011a) Influence of bioinoculants on the growth of *Albizia lebbeck* in nursery conditions. Res J Agr Sci 2(2):265–268

Pindi KP (2011b) Influence of bioinoculants on establishment of micropropagated plantlets of *Sesbania grandiflora*. Res J Agr Sci 2(1):114–118

Pindi KP, Reddy SR, Reddy SM (2000) Mycorrhizal dependency of some agroforestry tree species. Indian Forest 126(4):397–402

Puppo A, Dimitrevic L, Diem HG, Dommergues YR (1985) Homogeneity of superoxide dismutase patterns in *Frankia* strains from Casuarinaceae. FEMS Microbiol Lett 30:43–46

Purcino AAC, Lurlarp C, Lynd JQ (1986) Mycorrhizal and soil fertility effects with growth, nodulation and nitrogen fixation of leucaena grown on a typic eutrustox. Commun Soil Sci Plan 17(5):473–489

Purwantari ND, Date RA, Dart PJ (1995) Nodulation and N_2-fixation by *Calliandra calothyrsus* and *Sesbania sesban* grown at different room temperatures. Soil Biol Biogeochem 27(4–5):421–425

Rahman MK, Parsons JW (1997) Effects of inoculation with *Glomus mosseae*, *Azorhizobium caulinodans* and rock phosphate on the growth of and nitrogen and phosphorus accumulation in *Sesbania rostrata*. Biol Fertil Soil 25:47–52

Reddell PW (1986) Management of nitrogen fixation by *Casuarina*. ACIAR Forest Newslett 2:1–3

Reddell PW, Bowen GD, Robson AD (1986) Nodulation of Casuarinaceae in relation to host species and soil properties. Aust J Bot 34:435–444

Reed SC, Cleveland CC, Townsend AR (2007) Controls over leaf litter and soil nitrogen fixation in two lowland tropical rainforests. Biotropica 39(5):585–592

Reena J, Bagyaraj DJ (1990) Response of *Acacia nilotica* and *Calliandra calothyrsus* to different VA mycorrhizal fungi. Arid Soil Res Rehab 4:261–268

Rodriguez-Echeverria S, Crisostomo JA, Freitas H (2007) Genetic diversity of rhizobia associated with *Acacia longifolia* in two-stages of invasion of coastal sand dunes. Appl. Environ. Microbiol. doi: 10.1128/AEM.00613–07

Roskoski JP (1981) Nodulation and N_2 fixation by *Inga jinicuil*, a woody legume in coffee plantations. I. Measurements of nodule biomass and field C_2H_2 reduction rates. Plant Soil 59:201–206

Roskoski JP (1982) Nitrogen fixation in a Mexican coffee plantation. In: Robertson GP, Herrera R, Roswall T (eds) Nitrogen Cycling in Ecosystems of Latin America and the Caribbean. Martinus Njihoff, The Hague, The Netherlands, pp 283–292

Roskoski JP (1986) Future directions in biological nitrogen fixation research. In: Shamsuddin ZH, Othman WMW, Marziah M, Sundram J (eds) Biotechnology of Nitrogen Fixation in the Tropics. Universiti Pertanian Malaysia, Serdang, Malaysia

Roskoski JP, Kessel C van, Castilleja G (1982) Nitrogen fixation by tropical woody legumes: Potential source of soil enrichment. In: Graham PH (ed) Biological Nitrogen Fixation Technology for Tropical Agriculture. CIAT, Cali, Colombia, pp 447–454

Rouvier C, Prin Y, Reddell P, Normand P, Simonet P (1996) Genetic diversity among Frankia strains nodulating members of the family Casuarinaceae in Australia revealed by PCR and restriction fragment length polymorphism analysis with crushed root nodules. Appl Environ Microbiol 62(3):979–985

Roy S, Khasa DP, Greer CW (2007) Combining alders, frankiae, and mycorrhizae for the revegetation and remediation of contaminated ecosystems. Can J Bot 85:237–251

Russo RO (1989) Evaluating alder-endophyte (*Alnus acuminata*-*Frankia*-Mycorrhizae) interactions. I. Acetylene reduction in seedlings inoculated with *Frankia* strain ArI3 and *Glomus intra-radices*, under three phosphorus levels. Plant Soil 118:151–155

Sanginga P, Danso SKA, Bowen GD (1992) Variation in growth, sources of nitrogen and N-use efficiency by provenances of *Gliricidia sepium*. Soil Biol Biogeochem 24(10):1021–1026

Sanginga N, Mulongoy K, Ayanaba A (1985) Effect of inoculation and mineral nutrients on nodulation and growth of *Leucaena leucocephala*. In: Ssali H, and Keys SO (eds) Biological Nitrogen Fixation in Africa. MIRCEN, Nairobi, Kenya

Sanginga N, Mulongoy K, Ayanaba A (1986) Inoculation of *Leucaena leucocephala* (Lam.) de Wet with *Rhizobium* and its nitrogen contribution to a subsequent maize crop. Biol agric Hortic 3:347–352

Sanginga N, Mulongoy K, Ayanaba A (1988) N contribution of leucaena/rhizobium symbiosis to soil and and a subsequent maize crop. Plant Soil 112:137–141

Sanginga N, Mulongoy K, Ayanaba A (1989) Nitrogen fixation of field-inoculated *Leucaena leucocephala* (Lam) de Wit estimated by the ^{15}N and difference methods. Plant Soil 117:269–274

Sanginga N, Zapata F, Danso SKA, Bowen GD (1990) Effects of successive cuttings on uptake and partitioning of 15N among plant parts of *Leucaena leucocephala*. Biol Fertil Soil 9:37–42

Schenck NC (1982) Methods and principles of mycorrhizal research. American Phytopathological Society, Saint-Paul, Minnesota, USA

Sebuliba E, Nyeko P, Majaliwa JGM, Kizza LC, Eilu G, Adipala E (2010) Effect of selected arbuscular mycorrhiza fungi on the growth of *Calliandra calothyrsus* and *Sorghum bicolor* in eastern Uganda. Second RUFORUM Biennial Meeting 20–24 September 2010, Entebbe, Uganda

Shepherd KD, Jefwa J, Wilson J, Ndufa JK, Ingleby K, Mbuthia KW (1996) Infection potential of farm soils as mycorrhizal inocula for *Leucaena leucocephala*. Biol Fertil Soil 22:16–21

Shepherd M, Nguyen L, Jones ME, Nichols JD, Carpenter FL (2007) A method for assessing arbuscular mycorrhizal fungi group distribution in tree roots by intergenic transcribed sequence variation. Plant Soil 290:259–268

Sogo A, Setoguchi H, Noguchi J, Jaffre T, Tobe H (2001) Molecular phylogeny of Casuarinaceae based on *rbcL* and *matK* gene sequences. J Plant Res 114:459–464

Subhan S, Sharmila P, Pardha Saradhi P (1998) *Glomus fasciculatum* alleviates transplantation shock of micropropagated *Sesbania sesban*. Plant Cell Rep 17:268–272

Staal M, te Lintel-HekkertS, Harren F, Stal L (2001) Nitrogenase activity in cyanobacteria measured by the acetylene reduction assay: a comparison between batch incubation and on-line monitoring. Environ Monit 3(5):343–351

Stamford NP, Ortega AD, Temprano F, Santos DR (1997) Effects of phosphorus fertilization and inoculation of *Bradyrhizobium* and mycorrhizal fungi on the growth of *Mimosa caesalpiniaefolia* in acid soil. Soil Biol Biochem 9(5/6):959–964

Virginia RA, Jarrell WM, Kohl DH, Shearer GB (1981) Symbiotic nitrogen fixation in *Prosopis* (Leguminosae) dominated ecosystems. In: Gibbson AH, Newton WE (eds) Current Perspectives in Nitrogen Fixation. Australian Academy of Sciences, Canberra, Australia

Waring HD (1985) Chemical fertilization and its economic aspects. In: Burnley J, Stewart JL (eds) Increasing Productivity of Multipurpose Species. IUFRO, Vienna, Austria

Wang ET, Martinez-Romero J, Martinez-Romero E (1999) Genetic diversity of rhizobia from *Leucaena leucocephala* nodules in Mexican soils. Mol Ecol 8(5):711–724

Wang ET, Tan ZY, Willems AY, Fernandez-Lopez M, Reinhold-Hurek B, Martinez-Romero E (2002) *Sinorhizobium morelense sp. Nov.*, a *Leucaena leucocephala*-associated bacterium that is highly resistant to multiple antibiotics. Int J Syst Evol Micr 52:1687–1693

Warembourg FR (1993) Nitrogen fixation in soil and plant systems. In: Knowles R, Blackburn TH (eds) Nitrogen isotope techniques. Academic Press, San Diego, USA, pp 127–156

West JB, HilleRisLambers J, Lee TD, Hobbie SE, Reich PB (2005) Legume species identity and soil nitrogen supply determine symbiotic nitrogen-fixation responses to elevated atmospheric [CO_2]. New Phytol 167:523–530

Whitbeck JL (2001) Effects of light environment on vesicular-arbuscular mycorrhiza development in *Inga leiocalycina*, a tropical wet forest tree. Biotropica 33(2):303–311

Williams PK (1984) Current use of legume inoculant technology. In: Alexander M (ed) Biological Nitrogen Fixation: ecology, technology and physiology. Plenum Press, New York, USA, pp 173–200

Witty JF, Minchin FR (1988) Measurements of nitrogen fixation by acetylene reductions assay: myths and mysteries. In: Beck DP, Materon LA (eds) Nitrogen fixation: achievements and objectives. Chapman and Hall, New York, pp 285–292
Woldemeskel E, Singlair FL (1998) Variation in seedling growth, nodulation and nitrogen fixation of *Acacia nilotica* inoculated with eight rhizobial strains. For Ecol Manag 104(1–3):239–247
Wubet T, Kottke I, Teketay D, Oberwinkler F (2003) Mycorrhizal status of indigenous trees in dry Afromontane forests of Ethiopia. For Ecol Manag 179:387–399
Yadav JSP (1983) Soil limitations for successful establishment and growth of Casuarina plantations. In: Midgley SJ, Turnbull JW, Johnson RD (eds) Casuarina Ecology: Management and Utilization. CSIRO, Melbourne, Australia, pp 138–157
Zangaro W, Nishidate FR, Camargo FRS, Romagnoli GG, Vandressen J (2005) Relationships among arbuscular mycorrhizas, root morphology and seedling growth of tropical native woody species in southern Brazil. J Trop Ecol 21:529–540
Zhang Z, Lopez MF, Torrey JG (1984) A comparison of cultural characteristics and infectivity of *Frankia* isolates from root nodules of *Casuarina* species. Plant Soil 78:79–90

Chapter 9
Agroforestry for Soil Conservation

Abstract Soil conservation refers to the maintenance of soil fertility and erosion control. Soil fertility maintenance in agroforestry systems is achieved through the addition of organic matter, typically through litterfall and mulching, while erosion control is achieved through the mitigation of soil losses. Agroforestry systems have been proven to be efficient in soil erosion control. In semiarid Kenya, *Senna siamea* mulch cover reduced soil losses to only 13 % of the standard average loss and barrier hedgerows reduced the loss to 2 %. *Leucaenea leucocephala*-maize plots in a subhumid climate in Malawi reduced soil loss on a steep (44 % gradient) to 2 tons ha^{-1} $year^{-1}$, compared with a loss of 80 tons ha^{-1} $year^{-1}$ in maize plots without agroforestry on a similar slope. On a less steep slope (4 %) in the Himalayan valley of India, runoff was reduced by 27 %, and soil loss by 45 % using a contour cultivation of maize. Contour tree rows or Leucaena hedges reduced runoff and soil loss by 40 % and 48 %, respectively, compared to the maize plot. Soil loss was reduced to about 12.5 Mg (or tons) ha^{-1} $year^{-1}$, a significant improvement over fallow plots, which lost about 39 tons ha^{-1} $year^{-1}$. Agroforestry practices that are widely used in the tropics for erosion control include crop combinations, multi-storey tree gardens, alley cropping, and windbreaks or shelterbelts. Effective windbreaks provide semi-permeable barriers to wind over their full size, from the base of the windbreak up to the top of the tallest trees.

9.1 Introduction

Agroforestry contributes to the increase and maintenance of soil productivity by improving soil conservation. Agroforestry's potential in soil conservation, i.e., in erosion control and maintenance of soil fertility, has been reviewed by Young (1989, 1997). According to Young (1989; p. 9), soil conservation is defined as soil fertility maintenance through control of erosion together with maintenance of organic matter, soil physical properties and nutrients, and avoidance of toxicities.

Soil erosion is an age-old environmental problem. Efforts have long been undertaken to address this problem, and soil conservation was originally synonymous with soil erosion control. During the 1970s, soil conservation became a broader issue that not only involved keeping soils in place, but also included maintaining or improving soil productivity. Soil conservation allows the sustainability of crop

A. Atangana et al., *Tropical Agroforestry*, DOI 10.1007/978-94-007-7723-1_9,

production. Sustainability was defined by Young (1997, p. 6) simply as productivity plus the conservation of resources. Soil fertility maintenance is the most important criterion for land use to be characterized as sustainable. This chapter will present briefly some concepts and trends in soil conservation, followed by a discussion of the measurement of soil erosion, the control of erosion using agroforestry practices, and finally conclude by presenting windbreaks which are agroforestry systems designed for wind erosion control.

9.2 Concepts and Trends in Soil Conservation

Young (1989, p. 23–24) summarized the concepts and trends in soil conservation at the end of the 1980s. The trends were highly focused on soil erosion control, and the drawbacks of land possibility classification (Kingebiel and Montgomery 1961). The same author in 1997 proposed a general soil-agroforestry hypothesis, stating that: "appropriate agroforestry systems have the potential to control runoff and erosion, maintain soil organic matter and physical properties, and promote nutrient cycling and efficient nutrient use". Twelve other hypotheses relating to agroforestry processes, agents and systems were designed (Young 1997, p. 20). These hypotheses constitute the main academic topics in soil conservation research. These research subjects are summarized here (Young 1997):

- Tree-crop competition in agroforestry
- Agroforestry and runoff, soil erosion on sloping lands, and reclamation of degraded and eroded lands, particularly saline or polluted lands
- Agroforestry and water availability in land-use systems
- Agroforestry, soil fertility (maintenance of organic matter and biological activity) and soil physical properties
- Nitrogen-fixing trees and nutrient recycling in agroforestry
- Agroforestry and soil toxicity reduction
- Agroforestry components, croplands, and environmental resources acquisition
- Tree roots in agroforestry systems and soil nutrients.

Maintenance of soil organic matter is possible through litterfall and mulching in agroforestry systems. Erosion control requires the quantification and mitigation of soil losses.

9.3 Measurement of Soil Erosion

It has been proven difficult to measure soil erosion loss. Soil erosion amounts and rates are estimated using predictive models, such as the USLE (Universal Soil Loss Equation; Wischmeier 1976; Wischmeir and Smith 1978). The USLE, which was developed in the United States of America, can be calibrated under given conditions, and the results extrapolated to similar croplands. More recently developed methodologies include Modified Universal Soil Loss Equation (MUSLE), Revised Universal Soil Loss Equation (RUSLE) and RUSLE2 (Foster and Highfill 1983; Toy et al. 1999; Laflen and Moldenhauer 2003). All of these methods are based on USLE but include additional features. Another method of quantifying soil erosion is the use of ^{137}Cs (Caesium-137) measurement (Collins et al. 2001).

Soil erosion loss using USLE is given by the following formula:

$$A = R \times K \times L \times S \times C \times P$$

where:

A is soil loss in tons ha^{-1} $year^{-1}$
R is the rainfall factor (e. g., 1/2 average annual rainfall in mm)
K is the soil erodibility factor (that varies from 0 to 1)
L is the slope length factor
S is the slope steepness factor
C is the cover factor (that varies from 0 to 1)
P is the support practice factor

For practical reasons, half of the annual rainfall is taken to be a good estimation of the R factor in the tropics. The erodibility factor is the resistance of the soil to erosion. When the soil is completely resistant to erosion, $K=0$. Conversely, $K=1$ for a soil completely non-resistant to erosion. The determination of the K value is done through experimentation. Resistant soils containing aggregates, such as Oxisols, have K values approximately equal to 0.1. A highly erodible soil can have a K value as high as 0.5.

The slope length factor and the slope steepness factor are considered as one unique factor, i. e., the length and steepness of slope factor, in the RUSLE2 formula (Laflen and Moldenhauer 2003). Factors L and S provide soil loss ratios from a standard USLE study site of similar length and slope. These factors are generally expressed as a combined factor, the topographic factor. Young (1989) and Nair (1993) further developed this concept. Standard values for LS are given in Table 9.1.

Cover factor (see Table 9.2) indicates the ratio of soil loss of specific agricultural soil under covered and non-covered fallow conditions. In a non-covered fallow, $C=1$. Crop fields that are covered throughout the year have a C close to 0. Values between these two extremes are expected for soils under varying intensities of coverage (Table 9.2). The support practice factor is the ratio of soil loss from a parcel with a given conservation practice compared to that of a culture performed without conservation practices, typically due to erosion, such as planting rows along slopes. Values range from 0 to 1, and values of 0.3 and 0.4 are common in agricultural fields with no special conservation practices.

The USLE was designed for and implemented in the monoculture wheat farms of the United States of America. In tropical croplands, the USLE can yield erroneous soil loss values. Several of the assumptions on which this model is based are not fully applicable in the tropics, leading to unrealistically high values (Nair

Table 9.1 Topographic factor values in the universal soil loss equation (Young 1989)

Slope		Slope length		
Percent	degrees	50	100	200
2	1	0.2	0.3	0.4
4	2	0.5	0.7	0.9
6	3	0.9	1.2	1.7
8	5	1.3	1.8	2.5
10	6	1.8	2.5	3.5
15	9	3.3	4.6	6.5
20	11	5.2	7.5	10.0
25	14	7.5	11.0	15.0
30	17	10.0	15.0	20.0
40	22	16.0	23.0	34.0
50	27	23.0	36.0	45.0

Table 9.2 Cover factor values and corresponding sediment concentration (Young 1989)

Cover factor	Sediment concentration (kg m^{-3})
1.0 (non-covered soil)	190
0.9	55
0.5	8
0.3	4
0.0 (soil covered at 100%)	1

1993). Simplifications were made to USLE, and many USLE variants can be used under different conditions (Young 1989). Variants include the Soil-Loss Estimator for Southern Africa (Elwell 1981; Elwell and Stocking 1982; Stocking 1981), the FAO model (FAO 1979), and the model designed by C.W. Rose (Rose 1988; Rose et al. 1985a, 1985b; Rose and Freebairn 1985). Alterations have also improved the usefulness of the USLE. Geographic Information Systems (GIS) can improve the efficiency of USLE for predicting . A GIS is a system capable of organizing and presenting alphanumeric data spatially referenced, and producing plans and maps. The representation is generally two-dimensional, but a three-dimensional representation or animation showing changes is possible. The USLE along with the use of GIS technology has been used successfully to predict the risk of soil loss by erosion in tropical Africa (Mati et al. 2000; Lufala et al. 2003). Other methods developed for soil erosion assessment include the agricultural non-point source pollution model. The model measured about half the erosion rate predicted by the USLE in West Java, Indonesia, but gave far more realistic values (Kusumandari and Mitchell 1997). Angima et al. (2003) used RUSLE to predict soil erosion loss in a catchment in Kenya. The value ranged from 134 to 549 Mg ha^{-1} $year^{-1}$, indicating the severity of erosion in tropical Africa.

These predictive models indicate that soil management, especially use of soil cover, has a very high potential for reducing soil erosion (Nair 1993). The R and K values are unlikely to change through direct human intervention. However, K values can change in response to soil management practices. The length and angle of the slope can also be manipulated by conservation measures. Physical or biological barriers can reduce slope length and inclination, thus providing benefits effective in erosion control. Woody perennials with cover crops can substantially reduce soil

Table 9.3 Agroforestry practices with potential for soil erosion control (Young 1989)

Agroforestry practice	Suitable environmental conditions	Notes
Plantation crop combinations	Humid to moist subhumid climates	Densely planted combinations of agricultural plantation crops with multipurpose trees appear to control erosion effectively on at least moderate slopes
Multi-storey tree gardens, including homegardens	Mainly developed in humid and moist subhumid climates, but possible potential in drier regions	Possess an inherent capacity to control erosion through combination of herbaceous cover with abundant litter
Hedgerow intercropping (alley cropping) and barrier hedges	Humid, subhumid and possibly semi-arid climates	A considerable apparent potential to combine erosion control with stable use on gentle to moderate slopes, more speculative potential on steep slopes; experimental data is sparse
Trees on erosion-control structures	Any	Supplementary use of trees stabilizes earth structures and gives production from land they occupy
Windbreaks and shelterbelts	Semi-arid zone	Proven potential to reduce wind erosion
Sylvopastoral practices	Semi-arid and subhumid climates, plus some humid (esp. South America)	Opportunities for inclusion of trees and shrubs as part of overall programme of pasture improvement
Reclamation forestry leading to multiple use	Any	Potential for planned design and development
Combination of the above in integrated watershed management	Any	Substantial opportunities include agroforestry with other major kinds of land use in integrated planning and management

erosion, but the rate of erosion control varies greatly depending on the management of twigs and leaves used as mulch.

9.4 Erosion Control Using Agroforestry Practices

Trees, shrubs, and bamboos or palms have always been used to control erosion. The methods used include the direct utilization of trees to reduce erosion rates and the stabilization of physical structures. Direct utilization refers to agroforestry combinations that increase soil cover, establish fences, and maintain or provide soil organic matter. Trees are also used in dikes, embankments, and pipes, and for land reclamation. Agroforestry practices are widely used in the tropics for erosion control, and include crop combinations, multi-storey tree gardens, alley cropping, windbreaks and shelterbelts. These agroforestry practices were evaluated by Young (1989; Table 9.3). However, erosion control is usually a secondary objective, or only one of multiple targeted objectives when designing and implementing an agroforestry practice. Site specificity is also an important factor to consider. Several agroforestry practices have good potential for erosion control; many of which are used worldwide. Best results can be obtained if agroforestry practices are combined with other technologies relevant to land use, in accordance with the biophysical conditions of the farms and the crop production objectives of the farmer.

9.5 Effects of Agroforestry Practices on Erosion Factors

Agroforestry systems are used to control soil erosion. Covering soil surface with *Senna siamea* mulch reduced soil loss to only 13 % of the standard average loss and barrier hedgerows reduces the loss to 2 % in semi-arid Kenya (Kiepe 1996). Banda et al. (1994) reported that *Leucaena leucocephala*-maize plots in Malawi reduced soil loss on a steep (44 % gradient) to 2 tons ha^{-1} $year^{-1}$, compared with a loss of 80 tons ha^{-1} $year^{-1}$ with the agroforestry component on a similar slope in a subhumid climate. On a less steep slope (4 %) in the Himalayan valley of India, runoff was reduced by 27 %, and soil loss by 45 % through contour cultivation of maize. Contour tree rows or *Leucaena* hedges further reduced runoff by 40 %, and soil loss by 48 %, compared to the maize plot. Soil loss was reduced to about 12.5 tons ha^{-1} $year^{-1}$, compared to fallow plots, which lost 39 tons of soil ha^{-1} $year^{-1}$ over the study period (Narain et al. 1998). The effect of agroforestry on soil erosion is the result of the impacts of agroforestry on erosion factors. Agroforestry systems contribute to soil erosion control through the effects of canopy cover, litter, ground vegetation, and the soil stabilizing effect of roots.

Erosivity of rainfall: erosivity indicates the R-factor of USLE, and is designated as the EI_{30} index. This index is obtained using the following formula:

EI_{30} = storm energy x maximum intensity of the storms in 30 min for all storms with precipitation greater than 12.5 mm

There is a widespread assumption that agroforestry practices can reduce the erosivity of the soil, which is not true for all agroforestry systems. High canopies with large leaves can increase the kinetic energy of raindrops. Raindrops may merge into large drops falling from as high as 30 m (Nair 1993). The large drops can reach a high velocity and cause splash erosion when they impact the soil. However, studies have shown that runoff and soil erosion decrease exponentially with an increase in canopy cover (Bochet and Rubio 2006). It is likely that lower and denser canopies reduce erosivity, but very few field measurements have been taken in such an agroforestry system. The relationship between erosion and canopy cover is given by the following equations (Gyssels et al. 2005):

$$Sr = 1^{-aC} \tag{9.1}$$

$$Sr = e^{-bC} \tag{9.2}$$

$$Er = e^{-dC} \tag{9.3}$$

Sr is the relative splash detachment or the relative loss by runoff, Er is the relative soil loss, C is the vegetation cover (in %), and a, b and d are constants varying from 0.0052 to 0.0910, 0.00251 to 0.4770 and 0.0168 to 0.0816 for equations (i), (ii) and (iii), respectively. Although the canopy in an alley cropping system is low, it is not directly above the cultivated land. Well-managed agroforestry systems are known to reduce losses due to erosion, but the extent of reductions caused by changes in rainfall erosivity is not well known.

9.5.1 Soil erodibility

The effects of soil organic matter on soil physical properties constitute the main influence of agroforestry practices on K. Soil structure is better in forests than in cultivated soils, demonstrating high stability, low detachability, and high infiltration capacity in forest soils. Ground cover of the surface soil in alley cropping protects the soil from rainfall detachment and runoff, reducing soil erosion loss (Paningbatan et al. 1995). In the hilly land of the Philippines and in semiarid Kenya, mulching with plant residues and the presence of densely planted hedgerows was observed to increase infiltration rates and reduce runoff (Paningbatan et al. 1995; Kiepe 1996). Shifting cultivation decreases soil organic matter and increases erodibilty. In a *Taungya* system, there is usually a decrease in organic matter content and infiltration capacity, and there is severe erosion during the culture period compared to a young forest plantation without introduced crops. Alley cropping has the potential to maintain or at least limit the rate of decline of soil organic matter, in contrast to the decrease observed in a pure culture. Homegardens and tree-based systems which mimic natural forests, such as cocoa or coffee plantations, also maintain soil organic matter by providing litter through leaves, fruit and plant residues. Roots from perennial plants also help stabilize the soil, reducing soil erodibility.

9.5.2 Runoff reduction

Soil cover reduces runoff (Paningbatan et al. 1995; Kiepe 1996). Runoff reduction uses the 'barrier' approach to control soil erosion. The planting of trees, and strip grasses, or using terraces reduces runoff. Trees and strips of grasses have been found to reduce runoff between 1 % and 10 % in Konx County, Missouri, USA (Udawatta et al. 2002). Living hedges using plants like *Leucaena, Calliandra* and *Setaria*, also reduce soil erosion (Roose and Ndayizigiye 1997).

9.5.3 Soil cover

The impact of raindrops can effectively be limited using living and dead plant materials (Nair 1993). Soil cover has greater potential to reduce soil erosion than the 'barrier' approach. However, it is recommended to use both, as the combined treatment of hedgerows and mulch was proven to be the most effective to reduce soil erosion in a study undertaken in semiarid Kenya (Kiepe 1996). The study also revealed that hedgerows reduced losses more than soil cover. Hedgerows reduced soil loss to between 23 % and 7 %, whereas mulch alone reduced losses to between 41 % and 17 % (Kiepe 1996). More study is needed on the effects of soil cover and hedgerows on runoff reduction and soil erosion control.

Agroforestry can contribute to the maintenance of land cover for long periods of time in several ways. In addition to the supply of living and dead materials on the ground surface, the presence of multiple canopy layers can significantly reduce raindrop velocity, reducing the severity of their impact. Alley cropping, which provides soil cover through mulching and hedgerows, may also be an effective way to control soil erosion.

9.6 Erosion Rates in Agroforestry Systems

It is important to consider the rates of *tolerable* and *acceptable* erosion. Erosion is, to some extent, inevitable. Under practical conditions, it is impossible to achieve a zero rate of erosion. In some areas of the Philippines, intense rainfall, steep slopes, and a lack of soil protection cause severe soil erosion to occur. Farmer's practices in these regions of the Philippines can increase the rate of soil erosion up to 100 or 200 tons ha^{-1} $year^{-1}$ (Paningbatan et al. 1995), far higher than the tolerable limits set by the soil conservation service in the USA, which range from 2.2 to 11.2 tons ha^{-1} $year^{-1}$. However, these values were reduced to 5 tons ha^{-1} $year^{-1}$ using an alley cropping system (Paningbatan et al. 1995). Young (1989) stated that the allowable limits for soil erosion "should be established on the basis of the sustainability of agricultural crops, translated in terms of maintaining organic matter and nutrients. Specifically, the ability of agricultural practices to provide organic matter and recycle nutrients must be integrated with the loss of these nutrients through erosion, to determine if the system is stable".

Little research and field measurement has been done relating to soil loss in agroforestry systems. However, systematic monitoring of soil erosion under different agroforestry practices is increasingly being conducted in the tropics. Roose and Ndayizigiye (1997) observed a reduced risk of erosion in Rwanda when applying agroforestry practices and Angima et al. (2003) reported on the severity of erosion in tropical highlands of Kenya, where values of erosion loss are 2.2 to 10 tons ha^{-1} $year^{-1}$, higher than soil tolerance levels. Erosion rates in agricultural plots managed using farmers' practices were estimated at 100–200 tons ha^{-1} $year^{-1}$ in the Philippines. These rates were reduced by 51 tons ha^{-1} $year^{-1}$ using alley cropping (Paningbatan et al. 1995). A major difficulty in estimating the erosion rate in the tropics is the erratic nature of rainfall from year to year (Lal 1989). Long-term studies are recommended.

Table 9.4 Soil erosion rates in tropical ecosystems (Wiersum 1984; p. 332)

Land use system	Erosion (tons ha^{-1} $year^{-1}$)		
	Minimum	Median	Maximum
Multistory tree gardens	0.01	0.06	0.14
Natural rainforest	0.03	0.30	6.16
Shifting cultivation, fallow period	0.05	0.15	7.40
Forest plantation, undisturbed	0.02	0.58	6.20
Tree crops with cover crop or mulch	0.10	0.75	5.60
Shifting cultivation, cropping period	0.40	2.78	5.60
Taungya, cultivation period	0.63	5.23	17.37
Tree crops, clean weeded	1.20	47.60	182.90
Forest plantations, litter removed or burned	5.92	53.40	104.80

Some erosion rates recorded under agroforestry practices and other relevant land use systems in the tropics are presented in Table 9.4. Erosion rates are classified as low (less than 2 tons ha^{-1} $year^{-1}$), moderate (between 2 and 10 tons ha^{-1} $year^{-1}$), and high (greater than 10 tons ha^{-1} $year^{-1}$). Low erosion rates occur in natural rainforests, forest fallow in shifting cultivation, multi-storey tree gardens, intact forest plantations and tree-crop plantations with mulching, whereas moderate or high erosion rates occur during periods of shifting cultivation and in forest plantations with litter removed or burned (Nair 1993, pp. 331–332).

Management practices are more important than the intrinsic properties of a system for erosion control, and trees alone do not necessarily lead to erosion control. What is important is how agroforestry systems are designed and managed. When designing a tree fallow or an agroforestry system for erosion control, the primary objective should be to establish and maintain plant litter covering the ground. From this perspective, maintaining soil organic matter, and therefore the soil physical properties and resistance to erosion, is the goal. Generally, it is not possible to perform erosion control through the protection of a tree canopy unless the systems below have a low, dense cover.

In summary, maintaining a litter layer to cover the ground is by far the most important management practice to reduce soil erosion. All improved fallows systems have the ability to carry out litter maintenance on the soil surface. Tree gardens, alley cropping, crop combinations, multi-purpose woodlots, and recovery forestry systems also have substantial potential to reduce soil erosion to acceptable levels.

9.7 Windbreaks for Erosion Control

Each windbreak design is unique, and should be chosen depending on the objectives and site conditions (Wright and Stuhr 2002). The height of the windbreak is important to the design, and it should be proportional to leeward distance of the wind protection needed (Figs. 9.1 and 9.2). The length of windbreaks and distance between them can vary considerably. It is common in the dry savannahs and steppes

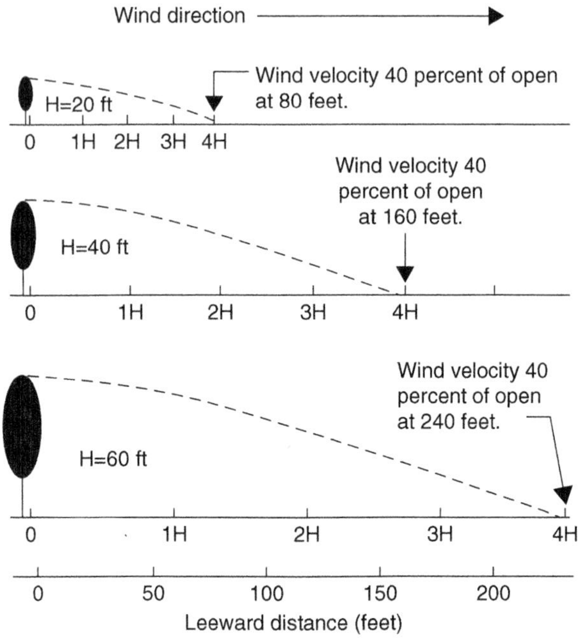

Fig. 9.1 The relationship between wind protection and height of windbreak (Wright and Stuhr 2002)

of Africa to plant windbreaks 100 m long or longer, with a maximum height of 10 m. The density of windbreaks is also an important factor in reducing wind erosion. The more solid or dense a windbreak, the greater the wind speed reduction. However, very dense windbreaks do more harm than good because they tend to create turbulence that will leach the soil on the windward side and damage the crops on the leeward side. Gaps between trees can harness the wind, increasing wind velocity in the leeward side, promoting erosion and damaging crops.

Small living fences and hedgerows may also act as windbreaks for small areas such as homegardens. However, windbreaks are distinguished from plantations by their orientation (facing the wind), their multi-layered structure, and semi-permeable nature.

The protected area created by a windbreak is defined as the area on both the leeward and windward sides, where the wind speed is reduced by 20 % below the incident wind speed. The effective distance of protection is expressed as multiples of the height (H) of the row of the tallest trees (Fig. 9.1). Theoretically, the practical effects of windbreaks extend to a distance of 15–20 H on the leeward side and 2–5 H on the windward side, but a calculation of the normal scope of protection is 10 H on the leeward side. This means that if the tallest trees are 10 m tall, crops up to 100 m downwind of the windbreak will benefit from the windbreak. However, a study on windbreak characterization in Australia indicated that changes in wind speed and microclimate as a result of wind shelter vary spatially and temporally (Sudmeyer

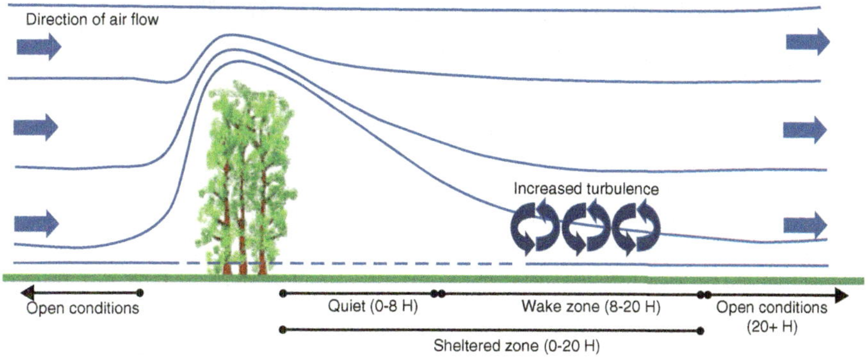

Fig. 9.2 Extent of various microclimate zones of a windbreak (H = tree height; Sudmeyer et al. 2007)

and Scott 2002). The study reported that when the wind direction was perpendicular to windbreaks, wind-run reductions greater than 20 % extended downwind 18 times the height of the windbreak. A permeable windbreak will also host a longer band of cultivated plants than a dense windbreak.

Species composition within the windbreak also affects erosion rates and the effectiveness of the windbreak. Michels et al. (1998) found that in the African Sahel, strips of *Andropogon gayanus* Kunth (a grass) reduced total annual soil flux within a distance of 20 m by 6 ± 55 % SE, and hedges of *Bauhinia rufescens* Lam. 2 m in height reduced soil flux by 47 ± 77 % SE compared with unsheltered pearl millet (*Pennisetum glaucum*) control plots in plantations on-station. This protective influence diminishes with increasing distance from the windbreak. Most effective windbreaks provide semi-permeable barriers to wind over their full size, from the base to the top of tallest trees. An ideal windbreak would consist of a central core with a double-row planting of fast-growing tall species, and another two rows of short species with strong root propagation, such as *Senna siamea*, on both sides of the core. Because trees change their form as they grow, it is necessary to mix several species with different growth rates, shapes, and sizes in multiple rows. Fast-growing species should be used to establish the desired effect as quickly as possible. Species used in windbreaks in the Sahel include *Acacia nilotica, Acacia holosericea* A.Cun. ex G.Don and *Azadirachta indica* (Smith et al. 1998), *Andropogon gayanus*, *Bauhinia rufescens* (Michels et al. 1998; Mayus et al. 1999) *Acacia senegal,* and *Faidherbia albida* (Lamers et al. 1994). Sugar gum (*Eucalyptus cladocalyx* F.Muell.), acacias, tuart (*Eucalyptus gomphocephala* DC., Bird 1998), and *Pinus pinaster* Aiton (Sudmeyer and Scott 2002) are common windbreak species in Australia. *Casuarinas pp.* and *Anacardium occidentale* L. are also used in windbreaks in the tropics of Africa (Nair 1993). Leguminous species such as leucaena, calliandra, setaria and *Senna siamea* are used as living hedges and in hedgerows in mountains of Malawi and Rwanda (Banda et al. 1994; Roose and Ndayizigiye 1997; Kiepe 1996).

9.8 Anticipated Costs and Benefits of Windbreaks for Soil Conservation

The costs and benefits of windbreaks should be considered before planting. Besides the direct costs associated with labor and planting materials, windbreaks occupy part of the land that could be used for cultivation, and compete with crops for water, light and nutrients. In order for a windbreak to be feasible, the increase in crop production, soil improvement and by-products should produce benefits above and beyond all costs. A program to calculate the economics of field shelterbelts, Windbreak economics (WBECON 2.0), has been developed by Kort and Brandle (1996).

In arid and semiarid areas, firewood is a substantive benefit obtained from windbreaks. Roose and Ndayizigiye (1997) found that leguminous living hedges not only reduced soil erosion rates, but also produced 3–8 kg m^{-1} high quality firewood, produced forage, and restored soil fertility in the tropical mountains of Rwanda. The reported effects of windbreaks on crop production vary considerably. In some cases, grain production increased significantly, while in other cases the competition for water and light, the land "lost" to the trees, or changes in microclimate were significantly harmful. The effect of a windbreak on crops is largely dependent on windbreak design, the crop used, and the environment involved. For these reasons, products from trees and long-term soil conservation should be considered as the primary benefits of windbreaks.

Bibliography

Angima SD, Stott DE, O'Neill MK, Ong CK, Weesies GA (2003) Soil erosion prediction using RUSLE for central Kenyan highland conditions. Agr Ecosyst Environ 97(1–3):295–308

Banda AZ, Maghembe JA, Ngugi DN, Chome VA (1994) Effect of intercropping maize and closely spaced *Leucaena* hedgerows on soil conservation and maize yield on a steep slope at Ntcheu, Malawi. Agrofor Syst 27:17–22

Bird PR (1998) Tree windbreaks and shelter benefits to pasture in temperate grazing systems. Agrofor Syst 41:35–54

Bochet PJ, Rubio JL (2006) Runoff and soil loss under individual plants of a semi-arid Mediterranean shrubland: influence of plant morphology and rainfall intensity. Earth Surf Proc Land 31:536–549

Brandle JR, Hodges L, Zhou XH (2004) Windbreaks in North American agricultural systems. Agrofor Syst 61:65–78

Cleugh HA (1998) Effects of windbreaks on airflow, microclimates and crop yields. Agrofor Syst 41:55–84

Collins AL, Walling DE, Sichingabula HM, Leeks GJL (2001) Using ^{137}Cs measurements to quantify soil erosion and redistribution rates for areas under different land use in the Upper Kaleya River Basin, southern Zambia. Geoderma 104:299–323

Elwell HA (1981) A soil loss estimation model for southern Africa. In: Morgan RPC (ed) Soil conservation: problems and prospects. Wiley, Chichester, UK pp 281–292

Elwell HA, Stocking MA (1982) Developing a simple yet practical method of soil-loss estimation. Trop Agric 59(1): 43–48

FAO (1979) A provisional methodology for soil degradation assessment. Fao, Rome, Italy, p 84

Foster GR, Highfill RE (1983) Effect of terraces on soil loss: USLE P factor values for terraces. J Soil Water Conserv 38(1):48–51

Foster GR, Toy TE, Renard KG (2003) Comparison of the USLE, RUSLE1.06c, and RUSLE2 for application to highly disturbed lands. In: Proceedings of the 1st Interagency Conference on Research in the Watersheds, Benson, Arizona, October 27–30, 2003, USDA Agricultural Research Service, Washington, DC, pp 154–160

Gyssels G, Poesen J, Bochet E et al. (2005) Impact of plant roots on the resistance of soils to erosion by water: a review. Prog Phys Geog 29:189–217

Laflen JM, Moldenhauer WC (2003) Pioneering Soil Erosion Prediction: the USLE Story. In: Zoebisch MA (ed) World Association of Soil and Water Conservation Special Publication No. 1

Kang BT, Grimme H, Lawson TL (1985) Alley cropping sequentially cropped maize cowpea with *Leucaena* on a sandy soil in southern Nigeria. Plant Soil 85:267–277

Kiepe P (1996) Cover and barrier effects of Cassia siamea hedgerows on soil conservation in semi-arid Kenya. Soil Technol 9:161–171

Klingebiel LAA, Montgomery PH (1961) Land Capability Classification. USDA Handbook 210. USDA, Washington, DC, USA

Kort J, Brandle JR (1996) Windbreak economics (WBECON 2.0): a program to calculate the economics of field shelterbelts. PFRA Shelterbelt Centre. Indian Head, Sask., Canada

Kusumandari A, Mitchell B (1997) Soil erosion and sediment yield in forest and agroforestry areas in West Java, Indonesia. J Soil Water Conserv 52(5):376–380

Lal R (1989) Agroforestry systems and soil surface management of a tropical Alfisol, Paris I-VI and Summary. Agrofor Syst 8:1–6; 7–29; 97–111; 113–132; 197–215; 217–238; and 239–242

Lamers JPA, Michels K, Vandenbeldt RJ (1994) Trees and windbreaks in the Sahel: establishment, growth, nutritive and calorific values. Agrofor Syst 26:171–184

Lufala A, Tenywa MM, Isabirye M, Majaliwa MJG, Woomer PL (2003) Prediction of soil erosion in a Lake Victoria basin catchment using a GIS-based universal soil loss model. Agr Syst 76(3):883–894

Mati BM, Morgan RPC, Gichuki FN, Quinton JN, Brewer TR, Liniger HP (2000) Assessment of erosion hazard with the USLE and GIS: A case study of the Upper Ewaso Ng'iro North Basin of Kenya. International J Appl Earth Obs Geoinform 2(2):78–86

Mayus M, Keulen H van, Stroosnijder L (1999) A model of tree-crop competition for windbreak systems in the Sahel: description and evaluation. Agrofor Syst 43:183–201

Michels K, Lamers JPK, Buerkert A (1998) Effects of windbreaks species and mulching on wind erosion and millet yield in the Sahel. Exp Agr 34:449–464

Nair PKR (1993) An Introduction to Agroforestry. Kluwer Academic Publishers, Dordrecht, The Netherlands, p 520

Narain P, Singh RK, Sindhwal RS, Joshie P (1998) Agroforestry for soil and water conservation in the western Himalayan Valley Region of India. 1. Runoff, soil, and nutrient losses. Agrofor Syst 39:175–189

Paningbatan EP, Ciesiolka CA, Coughlan KJ, Rose CW (1995) Alley cropping for managing soil erosion of hilly lands in the Philippines. Soil Technol 8(3):193–204

Rose CW (1988) Research progress on soil erosion processes and a basis for soil conservation practices. In: Lal R (ed) Soil erosion research methods. Soil and Water Conservation Society, Ankeny, Iowa, USA. pp 119–139

Roose E, Ndayizigiye F (1997) Agroforestry, water and soil fertility management to fight erosion in tropical mountains of Rwanda. Soil Technol 11(1):109–119

Rose CW (1985a) Developments in soil erosion and deposition models. Adv Soil Sci 2:1–63

Rose CW (1985b) Progress in research in soil erosion processes and a basis for soil conservation practices. In: Craswell ET, Remenyi JV, Nallana LG (eds) Soil erosion management. ACIAR Proceedings series 6. ACIAR, Canberra, pp 32–41

Rose CW, Freebbairn DM (1985) A new mathematical model of soil erosion and deposition processes with application to field data. In: El-Swaify SA, Moldenhauer WC, Lo A (eds) Soil erosion and conservation. Soil Conservation Society of North America. Ankeny, pp 549–557

Smith DM, Jarvis PG, Odongo JCW (1998) Management of windbreaks in the Sahel: the strategic implications of tree water use. Agroforest Syst 40: 83–96

Stocking M (1981) A working model for the estimation of soil loss suitable for underdeveloped areas. Occasional Paper 15. School of Development Studies, University of East Anglia, Norwich, UK

Sudmeyer RA, Scott PR (2002) Characterisation of a windbreak system on the south coast of Western Australia. 1. Microclimate and wind erosion. Aust J Exp Agr 42:703–715

Sudmeyer RA, Bicknell D, Coles N (2007) Tree windbreaks in the wheatbelt. Department of Agriculture and Food, Government of Western Australia, Bulletin 4723.ISSN 1833–7236

Toy TJ, Foster GR, Renard KG (1999) RUSLE for mining, construction and reclamation lands. Journal of Soil and Water Conservation 54(2): 462–467

Udawatta RP, Krstansky JJ, Henderson GS, Garrett HE (2002) Agroforestry practices, runoff, and nutrient loss: A paired watershed comparison. J Environ Qual 31(4):1214–1225

Vandenbeldt RJ (1990) Agroforestry in the semi-arid tropics. In: MacDicken KG, Vergara NT (eds) Agroforestry: Classification and Management. Wiley, New York, USA, pp 50–194

Van Emiren J, Karschorn R, Razumova LA, Robertson GW (1964) Windbreaks and Shelterbelts. WMP Technical Note No 59. World Meteorological Organization No 147. TP. 70. WMO, Geneva, Switzerland

Wiersum KF (1984) Surface erosion under various tropical agroforestry systems. In O'Loughlin CL, Pearce AJ (eds) Symposium on effects of forest land use on erosion and slope stability. University of Hawaii, Honolulu

Wischmeier WH (1976) Use and misuse of the universal soil loss equation. J Soil Water Conserv 31:5–9

Wischmeir QH, Smith DD (1978) Predicting Rainfall Erosion Losses: a Guide to Conservation Planning. USDA Handbook 537, USDA, Washington, DC, USA

Wright B, Stuhr K (2002) Windbreaks: an agroforestry practice. USDA Agroforestry Note 25, USDA Forest Service NAC, University of Nebraska, Lincoln, pp 4

Young A (1989) Agroforestry for soil conservation. CAB International, Wallingford, Oxford, UK, pp 276 (with index)

Young A (1997) Agroforestry for soil management. CABI, ICRAF, pp 320 (with index)

Chapter 10
Carbon Sequestration in Agroforestry Systems

Abstract Agroforestry systems have great potential as carbon (C) sinks, through C sequestration both above- and belowground. The C-sequestration potentials of tropical agroforestry systems are highly variable. The variation may be caused by (i) the estimates of C-sequestration potential of agroforestry systems that are not rigorous, (ii) lack of widely and easily adoptable methodologies for estimating the soil C potential under different conditions, and (iii) the natural variability of soil C stock in agroforestry systems in different agroecological zones. Reported data on soil C sequestration are also highly variable, partly because the term "carbon sequestration potential" can have different meanings depending on the context. Various agroforestry practices and technologies such as alley cropping/intercroping, silvopasture, riparian buffers, parklands, forest framing, homegardens, woodlots, windbreaks, and other similar land-use systems can be valued as carbon sinks in both tropical and temperate regions. The C sequestration potential of agroforestry systems justifies the plea made for its inclusion in the United Nations-based REDD (Reducing Emissions from Deforestation and Forest Degradation) programme for tropical developing regions, aimed at reducing emissions of greenhouse gases. An accurate estimation of C changes is necessary to improve the implementation of REDD + (i. e., conservation and sustainable management of forests, and enhancement of C stocks, on top of REDD) mechanisms, which use financial incentives to promote and popularize the use of any method that would reduce emissions of greenhouse gases.

10.1 Introduction

The potential of agroforestry systems to act as carbon (C) sinks is becoming more and more emphasized, especially with the advent of REDD (Reducing Emissions from Deforestation and Forest Degradation), which is a United Nations-based collaborative initiative to reduce greenhouse gas emission through reduced deforestation and forest degradation in the tropical regions. For more details on REDD and agroforestry, please see chapter 20. Atmospheric C sequestration involves C uptake through photosynthesis and storage in long-living pools such as timber and the soil. Agroforestry systems store C in plant biomass and in the soils. Carbon sequestration has long been an underexploited benefit of agroforestry (Montagnini and

A. Atangana et al., *Tropical Agroforestry*, DOI 10.1007/978-94-007-7723-1_10,

Nair 2004), and this environmental service is becoming increasingly recognized and valued (Schroeder 1994; Albrecht and Kandji 2003; Nair et al. 2009a; Kumar and Nair 2011).

However, some agroforestry systems may act as C sources, because of the complexity of C trade-offs between agroforestry components, and because of the interactions between major greenhouse gases (GHG), e. g., methane, carbon dioxide nitrogen dioxide, in agroforestry systems. Agroforestry systems also release carbon dioxide and nitrogen into the atmosphere, making it possible to have a negative C-balance (i. e., the difference between C sequestered and C released to the atmosphere). Indeed, agrisilvicultural systems are usually C sinks, while ruminant-based silvopastoral systems are largely sources of greenhouse gases (Dixon 1995). The difference in the C sink–source relationship between agroforestry systems reflects the difference in the practices that are carried out between systems, and in the species composition of these systems. The difference in the C sink-source relationship is also a reflection of complexity of agroforestry systems (e.g., soil properties, system components, climate, land cultivation history, farming practices, and socioeconomic context). This chapter will discuss the potential of agroforestry systems to sequester atmospheric C, including brief discussion on methods for estimating C stocks in agroforestry systems and specific agroforestry practices that could help establish agroforestry systems to be C sources, and end with a brief introduction of REDD in agroforestry, which is discussed in more detail in Chapter 20.

10.2 The Potential for C Sequestration in Agroforestry Systems

Agroforestry systems have the potential to sequester C while maintaining crop production (Schoeneberger 2009; Kumar and Nair 2011), and constitute a promising option for environmental management. Potential secondary environmental benefits include food security and secure land tenure in developing countries, increasing farm income, restoring and maintaining aboveground and belowground biodiversity, creating corridors between protected forests, maintaining watershed hydrology, and soil conservation (Pandey 2002). Agroforestry systems can sequester C in standing biomass, in wood products and in the soil, mostly through increased soil organic matter content.

The C that is captured by plants through photosynthesis is stored in long-living C pools including the aboveground plant standing biomass such as timber, and belowground plant biomass such as roots. Carbon pools in agricultural systems also include fruits, soil microorganisms, and any form of organic and inorganic C in soils. Because plants store C in their biomass, it is obvious that the greater the plant size, the greater the amount of C sequestered. For that reason, the introduction of trees in agricultural landscapes will likely increase the C sequestration potential in agricultural systems. Consequently, efforts are increasingly made to bring to light the C sequestration potential of agroforestry systems, as they are tree-based agricultural systems.

Agroforestry systems can sequester large quantities of C in the soil, plant biomass, and wood products (Albrecht and Kandji 2003). Various estimations of C sequestration potential in the tropics have given different figures. In humid areas, agroforestry systems can sequester up to 50 Mg C ha^{-1}, and smallholder agroforestry in the tropics can sequester between 1.5 and 3.5 Mg C ha^{-1} $year^{-1}$ (Montagnini and Nair 2004). Albrecht and Kandji (2003) reported that tropical agroforestry systems have C sequestration potential ranging between 12 and 228 Mg C ha^{-1} with a median value of 95 Mg C ha^{-1}. According to Mutuo et al. (2005), the potential of agroforestry systems in the humid tropics to sequester C in plant biomass may be over 70 Mg C ha^{-1}, plus up to 25 Mg C ha^{-1} can be sequestered in the top 20 cm of the soil. In a review by Oelbermann et al. (2004), it was indicated that the potential for aboveground components in agroforestry systems in Costa Rica to sequester C is estimated to be 2.1×10^9 Mg C $year^{-1}$. The study revealed that a 10-year-old *Erythrina poeppigiana* system can sequester C at a rate of 0.4 Mg C ha^{-1} $year^{-1}$ in coarse roots, and 0.3 Mg C ha^{-1} $year^{-1}$ in tree trunks (Oelbermann et al. 2004).

C sequestration rates are highly variable between agroforestry systems in the tropics. Indeed, the rate of C sequestration ranged between 0.22 Mg C ha^{-1} $year^{-1}$ in *Faidherbia albida*-based plantations in Senegal and 5.8 Mg C ha^{-1} $year^{-1}$ in rotational woodlots in Tanzania (Luedeling et al. 2011). Estimation of C stocks in agroforestry systems varies with the area under consideration. Total C stocks in coffee agroforestry systems amounted to 127 ± 6.6 (SE) and to 93 ± 7.75 (SE) Mg C ha^{-1} in the western highlands of Guatemala (Schmitt-Harsh et al. 2012) and in the central valley of Costa Rica (Häger 2012), respectively. Annual organic C input to the soil from branches and leaves was estimated to be 1.4 Mg C ha^{-1}, and approximately 3.0 Mg C ha^{-1} from crop residues (Oelbermann et al. 2004). Takimoto et al. (2008) estimated that biomass C stock in agroforestry systems in the West African Sahel ranged from 0.7 to 54 Mg C ha^{-1} using allometric equations. The same study revealed that soil C stock determined in three layers (0–10, 10–40 and 40–100 cm) ranged from 28.7 to 87.3 Mg C ha^{-1}, indicating that more C is stored in the soil than in aboveground plant biomass. Shaded-perennial-crop-based agroforestry systems have great potential for soil C sequestration. In the weathered Oxisols of Bahia, Brazil, soil C stock in shaded cocoa (*Theobroma cacao* L.) -based agroforestry systems was estimated as 302 Mg ha^{-1} to 1 m depth in 2009 (Gama-Rodrigues et al. 2010).

Agroforestry practices affect the amount of potential C captured in a system. Traditional parkland agroforestry systems have larger C stock than the improved systems (Takimoto et al. 2008). In sub-Saharan Africa, Vagen et al. (2005) reported that an improved fallow system can increase attainable soil C sequestration rates from 0.1 to 5.3 Mg ha^{-1} $year^{-1}$. Also, organic coffee agroforestry systems (i. e., on which farmers applied between 0 and 10,500 kg of organic fertilizers ha^{-1} annually) stored more C (109.1 ± 29 (SD) Mg ha^{-1}) than conventional (i. e., on which farmers applied between 600 and 3,300 kg of synthetic nitrogen-phosphorus-potassium fertilizers ha^{-1} $year^{-1}$) coffee agroforestry systems (76.1 ± 18 Mg ha^{-1}) between November 2008 and April 2011 on Alfisols in Costa Rica (Häger 2012).

Large quantities of C can also be stored belowground. In India, Lal (2004) estimated the organic C pool sequestered in the soil at 21 billion tons in the first 30 cm,

Table 10.1 Summary of soil carbon stock under tropical agroforestry systems (Adapted from Nair et al. 2009 and Raji and Ogunwole 2006)

Agroecological zones	Major Agroforestry systems	Reported values of C stock (Mg ha^{-1})[a]
Humid lowlands	Shaded perennial systems	21–35
	Alley cropping	10–25
	Improved fallows	135–149
	Homegardens	108–119
	Tree intercropping	27–78
	Silvopasture	96–173
	Woodlots	61–75
Tropical highlands	Shaded perennial systems	21–97
	Silvopastoral systems	132–173
Arid and semiarid lowlands	Silvopastoral systems	
	Fodder banks	33
	Live fences	5.20–24
	Grazing systems	12.64–33
	Tree intercropping systems	
	Crop dominated	20–70
	Fodder dominated	25–80
	Fuelwood dominated	30–90
	Shelterbelts	39.09

[a] The soil depths and the methods used for different studies were highly variable. The listed range of values for each system is compiled from multiple literature sources. Specific literature references are therefore not given for each; literature citations can be found in Nair et al. (2009b) and Raji and Ogunwole (2006)

and 63 billion tons in the first one and a half meters. In the same study, organic C was estimated to be 196 billion tons in the first m of the soil. India's total potential of soil C sequestration was estimated to be between 39 and 49 Tg (1 Tg ~ 10^6 Mg) C $year^{-1}$ (Lal 2004).

Differences in C sequestration estimates may also exist because the estimates are not rigorous or standardized, as "the extent of C sequestered in any agroforestry system will depend on a number of site-specific biological, climatic, soil, and management factors" (Nair et al. 2009a). Generalizations based on such data are unrealistic and widely and easily adoptable methodologies are not available for estimating soil carbon sequestration potential under different conditions (Nair et al. 2009b). Amounts of soil carbon C stock will also vary by agroforestry system and agroecological zone (Tables 10.1 and 10.2). An emphasis should be placed on the development of widely acceptable and rigorous standard methods for estimating carbon sequestration in agroforestry systems (Nair 2012). Several difficulties hampered the accuracy of past estimates. The estimation of tree biomass used to quantify C sequestration is based on species-specific allometric equations (Tamang et al. 2012), which, when available, were developed for trees in natural forests, and may be location-specific. In addition, the size of the tree canopy in agroforestry systems can be different from that of the same tree species in a natural forest, creating a bias in C sequestration estimates.

Table 10.2 Indicative values of soil carbon stock and sequestration potential under major agroforestry systems in the tropics[a] (Nair et al. 2009b)

Major ecological regions and agroforestry systems	System characteristics *E*: existing; *N*: new plantings; *TD*: tree density (trees ha^{-1}); age: years (yr)	Soil carbon (Mg C ha^{-1})[b]		Time frame for realizing the potential (yr)[c]
		Stock to 50 cm depth	Potential for sequestering additional C to 100 cm soil depth	
Humid lowlands				
Shaded perennial systems	E > 15 yr	100–200	20–30	10
	N/young < 5-yr-old	70–150	100–200	
Alley cropping	E > 5 yr	20–45	25–75	> 5
	N or young < 5 yr	20–70	30–120	> 10
Improved fallows		60–100	80–150	
Homegardens	Low TD < 750 trees ha^{-1}	60–90	70–150	> 20
	Medium TD > 750 trees ha^{-1}	70–120	100–180	> 20
Silvopasture (grazing systems)	E, TD low < 50 ha^{-1}	20–80	50–100	> 20
Silvopasture (fodder bank)	E > 10-yr-old	60–95	30–60	
	N or young < 10 yr	75–95	50–100	
Woodlots	E > 10 yr	80–100	40–60	> 20
	N young or < 8 yr	50–80	50–150	
Tropical highlands				
Shaded perennial systems	E > 15-yr-old	100–200	20–50	10
Alley cropping	E > 5 yr	30–60	40–70	
	N or young < 5 yr	20–70	40–120	> 10
Homegardens	Low TD < 250 trees ha^{-1}	50–80	70–150	> 20
	Medium TD > 250 trees ha^{-1}	70–150	100–200	
Silvopasture (grazing systems)	E, TD low, > 20 trees ha^{-1}	70–120	80–150	> 20
	E, TD high	80–150	90–160	
Silvopasture (fodder bank)	E > 10 yr	60–100	30–70	> 20
	N young or < 8 yr	75–110	60–150	
Woodlots	E > 10 y-r-old	80–100	40–70	> 20
	N young or < 5 yr	50–80	60–170	
Arid and semiarid lands (mostly lowlands)				
Intercropping systems	Parklands, W Afr Sahel E ~ 50 trees ha^{-1}	30–40	5–10	> 25
?	Parklands, enrichment planting	20–30	30–50	> 25
Silvopasture, semiarid regions	E ~ 50 trees ha^{-1}	30–40	5–10	> 15
Grazing systems	N: planting trees in existing grazing lands	20–30	30–50	> 10
Fodder bank	N		30–100	
Fuelwood	N			

[a] The values are "best guess" estimates based on literature data (from nearly 150 reviewed papers and reports) and the authors' experience. Detailed literature citations are included in Nair et al. (2009)

[b] The soil stock values are reported mostly from the upper soil layers, to less than 50 cm depth. Therefore the estimates are for 0–50 cm soil depth. These, as well as the values for sequestration, will vary enormously depending on a large number of site- and system-specific factors[b]

[c] The values proposed as potential for sequestering additional C (column 4) are for up to 1 m depth considering the substantial amounts of the roots and the SOC (Soil Organic Carbon) in deeper soil layers. It is assumed that the existing systems have only limited potential in SCS (Soil Carbon Sequestration) unless they are significantly modified by management interventions such as (new) tree planting and fertilization; but the potential could be substantial in new agroforestry initiatives. It is also recognized that fairly long periods of time (column 5) are required to realize the potential for additional C sequestration in soils

Methodological issues involved in the direct and indirect estimation of C sequestration include: the accuracy of direct estimation of C stock, remote sensing and modeling, the influence of stand age on C accumulation, the influence of tree species and management practices on soil C sequestration, and information on stocks of organic C in deep soil layers. Many studies on belowground C sequestration focused on soil surface layers, but C sequestration in deep soil layers may be more important than in surface layers. Nair et al. (2009c) reported that tree-based systems store more C than treeless systems in soil layers as deep as 1 m. Issues related to C sequestration estimates in agroforestry systems are further discussed in Nair et al. (2009b).

The composition of an agroforestry system also influences the net soil C gains or losses. When estimating C stocks in a nitrogen-fixing trees and crop intercropping system, particular attention should be given to the effects of nitrous oxide, carbon dioxide, and methane emissions on net C gain and on the mitigation of GHG. Indeed, Kim (2012) reported that previous estimates of C stocks in a gliricidia-maize intercropping system in Malawi were incorrect, as the authors overlooked soil C loss as carbon dioxide emissions and the beneficial impacts of the reduction of nitrous oxide emissions from this agroforestry system on GHG mitigation. The C loss as soil carbon dioxide emissions amounted to be 64 % of the sequestered soil C (76 ± 8.6 Mg C ha^{-1} in the 0–2 m soil layer) for 7 years in the gliricidia-maize intercropping system, and the annual net gain of soil C was estimated to be 3.5 Mg C ha^{-1} $year^{-1}$ (Kim 2012). Also, the gliricidia-maize intercropping system reduced nitrous oxide emissions, thereby mitigating GHG by an equivalent of 3.5–4.1 Mg CO_2 ha^{-1} $year^{-1}$ (Kim 2012).

Another reason for the differences in the soil C sequestration data is that the term *carbon sequestration potential* has different meanings depending on its usage. Ingram and Fernandes (2001) drew on existing agroecosystem research concepts to define three levels of production, namely "potential", "attainable" and "actual". The authors suggested that the term "$attainable_{max}$" be used as the preferred term for carbon sequestration in mineral soils.

Systems also have an indirect effect by helping decrease pressure on natural forests by reducing the amount of land cleared for agricultural purposes. Dixon (1995) reported that 1 ha of sustainable agroforestry could provide enough goods and services to potentially offset 5–20 ha of deforestation. Proper design and management of agroforestry systems can lead to C sequestration (Montagnini and Nair 2004), whether it is sequestered in the ground biomass of plants, the soil or in wood products.

10.3 Agroforestry and REDD

Technologies for soil conservation using agroforestry practices can increase the storage of C in the soil, while adoption of agroforestry systems may reduce pressure on forests, indirectly increasing C sequestration (Montagnini and Nair 2004).

Agroforestry practices also contribute to C sequestration in wood products, standing biomass, roots, and soil organic matter. This role of agroforestry in C sequestration justifies the calls made for its inclusion in programs to reduce emissions of greenhouse gases (Schoeneberger 2009). Deforestation and forest degradation account for a large part of the emissions of greenhouse gases.

The United Nations-based REDD Programme, supported by The World Bank, uses financial incentives to promote and popularize the use of any method that would reduce emissions of greenhouse gases. Any strategy for reducing deforestation or promoting afforestation is eligible to be included in this program. The REDD initiative is based on the concept that developed countries should pay developing countries not to cause deforestation, and these payments should be based on the amount of C emitted by developed countries, and on the amount of C sequestered in developing countries, most of which are located in the tropics. The REDD initiative could be an opportunity for poor farmers in the tropics who practice sustainable land-use systems such as agroforestry to benefit from carbon payments. However, the financial evaluation of environmental services needs to be refined in order to be accepted by the international community (Schoeneberger 2009). As pointed out by Melick (2008), several practical questions are associated with the implementation of REDD: (i) Can forest changes and degradation be measured and monitored? Rigorous and standardized methods to estimate carbon sequestration potential in some land use systems, such as agroforestry, are not yet available (Nair et al. 2009b, c; Nair 2012); (ii) Can REDD schemes be implemented in the social, economic, and political climates of forested developing countries, most of which have problems with poor governance, low transparency, and corruption?; (iii) Will benefits from carbon payments reach forest communities? This last problem could be overcome by dealing directly with communities owning forests. For example, a system of community forests is being implemented in the Congo basin. Village communities organize themselves as legal entities, and are allowed by forest authorities to manage the surrounding forests delineated on the basis of a sustainable management plan of forest resources.

Agroforestry and other tree-based systems (woodlots, afforestation) can contribute to REDD + under certain forest definitions and for achieving REDD + in landscapes (Minang et al. 2011). In the context of REDD + agroforestry has the potential for reducing degradation by supplying timber and fuelwood that would otherwise be sourced from adjacent or distant forests, thereby reducing deforestation and pressure on natural forests. As pointed out by Minang et al. (2011), enabling market infrastructure, policies on tree rights and ownership, and safeguards would be necessary for agroforestry and other tree-based systems in the landscape to effectively contribute to the goals of REDD + and Nationally Appropriate Mitigation Actions (NAMAs).

The development of a carbon market in the tropics would certainly have an effect on the adoption of land use systems promoting afforestation and reforestation. Antle et al. (2007) developed a model to simulate the impact of carbon contracts for the adoption of agroforestry in the tropical highlands of Peru. The analysis of this model indicated that participation in carbon contracts could increase the adoption of terraces and other agroforestry practices, and the rate of adoption depends on the accumulation of C and other key factors affecting land productivity, such as the slope of the land. However, an accurate estimation of C change is necessary for better implementation of REDD mechanisms. Also needed are internationally accepted REDD standards and national and international policies on climate change (Melick 2008). Agreements for C emissions and forest protection have begun in Indonesia, as part of the REDD process (Akiefwanati et al. 2010). For REDD implementation to be successful, more awareness of REDD mechanisms is needed on the part of policy-makers, with increased support from scientists, Non-Governmental Organizations, as well as stakeholders involved in the Clean Development Mechanism (CDM) implementation.

Bibliography

Akiefwanati R, Villamor GB, Zulfikar F, Budisetiawan I, Mulyoutami E, Ayat A, van Noordwijk M (2010) Stewardship agreement to reduce emissions from deforestation and degradation (REDD): Lubuk Beringin's hutan desa as the first village forest in Indonesia. ICRAF Working Paper No 102

Antle JM, Stoorvogel JJ, Valdivia RO (2007) Assessing the economic impacts of agricultural carbon sequestration: Terraces and agroforestry in the Peruvian Andes. Agric Ecosyst Environ 122(4):435–445

Albrecht A, Kandji ST (2003) Carbon sequestration in tropical agroforestry systems. Agric Ecosyst Environ 99:15–27

Dixon RK (1995) Agroforestry systems: sources or sinks of greenhouse gases? Agrofor Syst 31:99–116

Gama-Rodrigues EF, Nair PKR, Nair VD, Gama-Rodrigues AC, Baligar V, Machado RCR (2010) Carbon storage in soil size fractions under two cacao agroforestry systems in Bahia, Brazil. Environ Manage 45:274–283

Häger A (2012) The effects of management and plant diversity on carbon storage in coffee agroforestry systems in Costa Rica. Agrofor Syst 86:159–174

Ingram JSI, Fernandes ECM (2001) Managing carbon sequestration in soils: concepts and terminology. Agric Ecosyst Environ 87:111–117

Lal R (2004) Soil carbon sequestration in India. Climatic Change 65:277–296

Melick D (2008) Credibility of REDD and experiences from Papua New Guinea. Conserv Biol 24(2):359–361

Kim DG (2012) Estimation of net gain of soil carbon in a nitrogen fixing-tree and crop intercropping system in sub-Saharan Africa: results from re-examining a study. Agrofor Syst 86:175–184

Kumar BM, Nair PKR (2011) Carbon sequestration potential of agroforestry systems: opportunities and challenges. Advances in Agroforestry 8, Springer Science and Business Media. pp 307

Luedeling E, Sileshi G, Beedy T, Dietz J (2011) Carbon sequestration potential of agroforestry systems in Africa. In: Kumar BM, Nair PKR (eds) Carbon sequestration potential of agroforestry systems: opportunities and challenges. Advances in Agroforestry. Springer Science and Business Media, vol 8, + Business Media. pp 61–83

Minang PA, Bernard F, van Noordwijk M, Kahurani E (2011) Agroforestry in REDD + : Opportunities and Challenges. ASB Policy Brief No. 26, ASB Partnership for the Tropical Forest Margins, Nairobi, Kenya (http://www.asb.cgiar.org/PDFwebdocs/ASB_PB26.pdf)

Montagnini F, Nair PKR (2004) Carbon sequestration: an underexploited environmental benefit of agroforestry systems. Agrofor Syst 61:281–295
Mutuo PK, Cadisch G, Albrecht A, Palm CA, Verchot L (2005) Potential of agroforestry for carbon sequestration and mitigation of greenhouse gas emissions from soils in the tropics. Nutr Cycl Agroecosys 71:43–54
Nair PKR (2012) Carbon sequestration studies in agroforestry systems: A reality-check. Agrofor Syst 86:243–253
Nair PKR, Kumar BM, Nair VD (2009a) Agroforestry as a strategy for carbon sequestration. J Plant Nutr Soil Sc 172(1):10–23
Nair PKR, Nair VD, Kumar BM, Haile SG (2009b) Soil carbon sequestration in tropical agroforestry systems: a feasibility appraisal. Environ Sci Policy 12(8):1099–1111
Nair PK, Nair VD, Gama-Rodrigues EF, Garcia R, Haile SG, Howlett DS, Mohan Kumar B, Mosquera-Losada MR, Saha SK, Takimoto ANG, Tonucci RG (2009c) Soil carbon in agroforestry systems: An unexplored treasure? Nature Precedings hdl:10101/npre.2009.4061.1
Oelbermann M, Voroney RP, Gordon AM (2004) Carbon sequestration in tropical and temperate agroforestry systems: a review with examples from Costa Rica and southern Canada. Agric Ecosyst Environ 104:359–377
Pandey DN (2002) Carbon sequestration in agroforestry systems. Clim Policy 2:367–377
Raji BA, Ogunwole JO (2006) Potential of soil carbon sequestration undervarious landuse in the sub-humid and semi-arid savanna of Nigeria: Lessons from long-term experiments. Int J Soil Sci 1(1):33–43
Schmitt-Harsh M, Evans TP, Castellanos E, Randolph JC (2012) Carbon stocks in coffee agroforests and mixed dry tropical forests in the Western Highlands of Guatemala. Agrofor Syst 86:141–157
Schoeneberger MM (2009) Agroforestry: working trees for sequestering carbon on agricultural lands. Agrofor Syst 75:27–37
Schroeder P (1994) Carbon storage benefits of agroforestry systems. Agrofor Syst 27:89–97
Takimoto A, Nair PKR, Nair VD (2008) Carbon stock and sequestration potential of traditional and improved agroforestry systems in the West African Sahel. Agric Ecosyst Environ 125:159–166
Tamang B, Andreu MG, Staudhammer CL, Rockwood DL, Jose S (2012) Equations for estimating aboveground biomass of cadaghi (*Corymbia torelliana*) trees in farm windbreaks. Agrofor Syst 86:255–266
Vågen T-G, Lal R, Singh BR (2005) Soil carbon sequestration in Sub-Saharan Africa: a review. Land Degrad Dev 16:53–71

Chapter 11
Agroforestry and Biodiversity Conservation in Tropical Landscapes

Abstract Agroforestry systems contribute to the maintenance of biodiversity in tropical landscapes. In the East Usambara Mountains of Tanzania, conservational functional group richness was significantly higher in an agrosilvopastoral system than that of an agrisilvicultural or aquasilvicultural system or monoculture plantation. A compilation of nine studies undertaken in Africa, Latin America and Asia indicated that animal diversity is highest in cocoa agroforests that have high plant diversity, structurally complex canopies, and abundant surrounding forest cover. Three assumptions support expected agroforestry effects on biodiversity conservation. First, it is assumed that the adoption of agroforestry practices by farmers induces a reduction of the pressure of deforestation on additional land. Second, agroforestry systems provide new habitats and resources for local plant and animal species that are in part dependent on the forest for survival, and could not survive in a purely agricultural landscape. Third, the value of the conservation of remnants of natural vegetation is greater if the remains are embedded in a landscape dominated by agroforestry elements. This only holds true if the surrounding matrix consists of crop fields or pastureland largely enriched with tree cover.

11.1 Introduction

Agroforestry systems, through the maintenance and diversification of specific trees on farms, help maintain biodiversity in tropical landscapes exploited by men. Agroforestry systems can have species richness equivalent to more than 60 % of that of the natural forest (Bhagwat et al. 2008). Land-use transformation in the rainforests of Sulawesi, Indonesia, from near-primary forest to agroforestry had little effect on overall species richness (Steffan-Dewenter et al. 2007). Intensive farming exposes the soil and fragments vegetation cover in the landscape. In such landscapes, agroforestry can help maintain biodiversity and provide ground cover similar to that of the natural ecosystem. Integrating germplasm of different multi-purpose tree species on farms and homegardens also contributes to the creation and conservation of biodiversity, and is promoted by the tree domestication program implemented by ICRAF. Revegetation of the soil using agroforestry practices can also promote biodiversity conservation. Various land use systems impact the conservation of biodiversity. The impact of functional groups (i.e., sets of species with similar impacts on

A. Atangana et al., *Tropical Agroforestry*, DOI 10.1007/978-94-007-7723-1_11,

ecosystem processes) of agroforestry components on biodiversity has been reported by Huang et al. (2002). Their study was undertaken in the East Usambara Mountains of North-Eastern Tanzania and demonstrated that the richness of conservational functional groups in an agrosilvopastoral system was significantly higher than that of an agrisilvicultural or aquasilvicultural system, or monoculture plantation, suggesting that functional groups are more diverse in agroforestry systems than in tree monoculture. Global priority rankings focusing on conservational, livelihood, and ecological functional groups, identified agrosilvopastoral systems as the best management alternative for biodiversity conservation, followed by agrosilvicultural, aquasilvicultural, and plantation systems (Huang et al. 2002). Schroth and Harvey (2007) reported that animal diversity is highest in cocoa agroforests that have high plant diversity, structurally complex canopies, and abundant surrounding forest cover, indicating that cocoa agroforests can sustain a significant level of biodiversity. The expected effects of agroforestry on biodiversity conservation are supported by three postulates presented by Schroth et al. (2004), given here in the form of research questions to be discussed throughout this chapter:

- Agroforestry and deforestation: What are the links between agroforestry and deforestation? Can agroforestry help mitigate deforestation, thereby preserving species richness in natural forests?
- Agroforestry and habitat: Can agroforestry provide for species occurring in natural forests?
- Agroforestry-matrix: Is agroforestry likely to contribute effectively to biodiversity conservation in a landscape mosaic of natural vegetation and agricultural areas?

11.2 The Agroforestry and Deforestation Hypothesis

The deforestation hypothesis is based on the assumption that the adoption of agroforestry practices by farmers reduces the deforestation pressure on additional land, and allows farmers to cope with a limited farming area and limited resources. Although the hypothesis is yet to be proven, it is widely used to justify the development and popularization of agroforestry technologies. The participatory domestication of agroforestry species is heavily reliant on this argument (Tchoundjeu et al. 2006). It relies on the assumption that the needs and resources of farmers remain stable over time. If the farmer's needs could be met by the optimal management of a small area, the farmer would not seek to operate an additional parcel of forest. However, it should be noted that population growth is generally strong in the tropics, and this growth alone is enough to increase the need for cultivated areas. The needs of a farmer can also increase over time, and if resources permit, he could extend a profitable technology over a greater area, increasing his farm size or the number of crop fields.

Another issue is the increasing involvement of agro-industrial companies in Non-Timber Forest Product (NTFP) market chains. Some NTFPs, such as *Allanblackia floribunda* nuts, *Irvingia gabonensis* nuts, and *Dacryodes edulis* fruits, are becoming more commonly used in food and cosmetic industry, and seeing increased financial value. There is a risk that agro-industrial companies may promote the establishment of large mono-specific plantations of these species when domesticated, thus contributing to deforestation by land conversion from forest to agriculture.

The agroforestry-deforestation hypothesis is heavily based on the study of Sanchez and Benites (1987), which reported that 1 ha of "low-input system" (i. e., an improved cropping system with selected varieties of rice, cowpea, and chemical weed control, which would not necessarily be considered agroforestry) is likely to save 5 ha of natural forests. The study of Sanchez and Benites (1987), which was based on a single study from Yurimaguas, Peru, is valid only in situations similar to that of Yurimaguas. Consequently, the agroforestry-deforestation link does not always hold.

Angelsen and Kaimowitz (2004) discussed the conditions (that heavily depend on farmers' constraints, including labor, land and capital) under which agroforestry systems are likely to reduce the conversion of forests. For instance, land tenure, market conditions, and labor constraints for a particular agricultural innovation likely influence deforestation activities, especially in the tropics. It is obvious that poor farmers without property rights for the land on which they grow crops would not convert more natural forests to agriculture or agroforestry, even if the system were highly profitable, since they would not reap the benefits of their labor. Also, any agricultural innovation that requires intensive labor would be hardly adopted by poor farmers. Conversely, agroforestry can be an incentive for deforestation in conditions of tenure security, no labor constraints and profitability of the system. Profitable agroforestry systems may give farmers an incentive to convert more natural forests to agroforestry, indicating that agroforestry adoption can have contradictory effects on the conservation of natural forests (Angelsen and Kaimowitz 2004).

On the other hand, agroforestry can be considered as a land-use system that can revert marginal land, including tropical savannas, into native and wild areas with increased biodiversity. An example of successful transformation of such a landscape is the Mampu Agroforestry village in the Democratic Republic of the Congo where a pilot phase of 8,000 ha reforestation project of *Acacia auriculiformis* on sandy soil of the Bateke plateau has been allocated to 320 farming families with 25 hectare plots each to provide sustainable green renewable fuelwood energy and improve food security by employing improved fallow sequential agroforestry technologies (Peltier et al. 2010; Schure et al 2010). Total charcoal production from the plantation varies from 8,000 to 12,000 tons year^{-1}, in addition to 10,000 tons year^{-1} of cassava, 1,200 tons year^{-1} of maize, and 6 tons year^{-1} of honey. Other agroforestry systems deserve to be tested or developed for different ecological and social or economic conditions, such as managing the natural regrowth of local multiple-use species as applied to the traditional system of fallow enrichment (Nkunku) in the Bas-Congo.

11.3 The Agroforestry—Habitat Hypothesis

The agroforestry-habitat hypothesis states that agroforestry systems can provide a habitat and resources for local plant and animal species that are in part dependent on the forest for survival, and could not survive in a purely agricultural landscape

(Schroth et al. 2004). Agroforestry systems, in their complexity, are generally more species diverse than intensive agricultural systems. Tropical homegardens act as a reservoir of tree and crop germplasms (Torquebiau 1992). Schroth and Harvey (2007) reported the contribution of cocoa agroforests to the conservation of biodiversity in Africa, Latin America and Asia. Indeed, both plant and animal diversity within cocoa farms was greatest than those of other agricultural land use as demonstrated by analysis of nine studies that document the contribution of cocoa farms to biodiversity conservation in Latin America, Africa and Asia (Schroth and Harvey 2007). A possible explanation of species richness in the cocoa farms is that a large number of bird species present in the forest are found in cocoa plantations. In addition, the diversity of birds in cocoa farms increases with species richness of trees used for shading and the reduction of the intensity of the management practices (Schroth and Harvey 2007). Two hundred and six tree species have been identified in cocoa plantations in Cameroon, with an average of 21 species per agroforest (Sonwa et al. 2007). Bird species not commonly found in agroforests are mainly various insectivore species. Their reduced numbers in agroforests can be explained by the reduced number of insects in cocoa farms due to chemical treatments against capsids. This demonstrates the importance of the richness of tree species in agroforests for the conservation of biodiversity.

Agroforestry practices contribute to biodiversity through enrichment planting of cropping systems and connectivity between tree populations. Participatory tree domestication of indigenous fruit species, which promotes the integration of various species into existing land-use systems, would help increase and maintain biodiversity in agricultural landscapes (Leakey 1998). Germplasm of indigenous tree, shrub and liana species with desired characteristics for fruit, nut, and leaf production in Latin America, Africa, and Southeast Asia can be grown in complex multistrata agroforests, depending on the farmer's objectives, by enrichment planting (Leakey 1998). Agroforestry practices such as mulching contribute to soil humus formation and reduce tillage frequency, thereby contributing effectively to the conservation of soil microfauna. The intensity of land management is important because soil humus that is undisturbed by tillage is better at conserving moisture, hence improving conditions for soil microfauna. Agroforestry systems are a refuge for forest species in human-exploited areas (Bhagwat et al. 2008). Farmers deliberately retain high-value trees when clearing land for agricultural purposes. This likely explains the genetic connectivity between tree populations, thus minimizing the risk of genetic drift that results in a loss of genetic diversity, as postulated by Atangana et al. (2010), who reported the neutral genetic diversity of *Allanblackia floribunda* using microsatellite DNA markers.

Agroforestry systems are also a refuge for biodiversity following fire. Indeed, 83 bird species were recorded in two agroforestry farms that had suffered severe fire in 1998 at Tikal, Petén region, Guatemala (Griffith 2000); 11 of these 83 bird species were found to be obligates of mature forests, and another 15 were considered to be forest generalists of Tikal (Whitacre 1995; Griffith 2000).

11.4 The Agroforestry-Matrix Hypothesis

In landscapes that are mosaics of agricultural areas and natural vegetation, the value of the conservation of remnants of natural vegetation is greater if the remains are embedded in a landscape dominated by agroforestry elements. This only holds true if the surrounding matrix consists of crop fields or pasture largely devoid of tree cover (Schroth et al. 2004).

The importance of having a mosaic of vegetation for the conservation of biodiversity has been emphasized by Bennett et al. (2006). The composition of the mosaic, based on the proportion of elements present, has a strong influence on the composition of assemblages of fauna. The diversity of elements is often positively correlated with the richness of taxonomic assemblages. Agroforestry, which has more components than pure agriculture, would have a higher potential for biodiversity conservation in a landscape mosaic of natural vegetation and agricultural areas. Kanowski et al. (2005) identified a smaller positive impact on biodiversity in plantation monocultures after comparing several scenarios of monoculture tree plantations, mixed tree plantations, mosaics of monoculture tree plantations, and mosaics of tree planting and ecological restoration cultures. The mosaic planting and ecological restoration areas had the highest positive impact on biodiversity in the landscape of cleared rainforests. Diverse, low-input systems using agroecological principles and constructed by smallholders are probably the best option for a high-quality matrix for tropical biodiversity conservation (Perfecto and Vandermeer 2008). The 2008 study conducted by Perfecto and Vandermeer concluded that biodiversity conservation should incorporate a landscape approach in order to create a landscape matrix dominated by productive agroecological systems that facilitate interpatch migration while promoting a sustainable livelihood for rural communities.

Agroforestry systems also serve as a buffer zone in the form of strips of forests against winds and fires. Agroforests located between areas of natural vegetation serve as an excellent wildlife corridor, facilitating the dispersion of wildlife from trees. More exploration of the importance of this aspect in the conservation of biodiversity is required. The book entitled "Agroforestry and Biodiversity Conservation in Tropical Landscapes" that was edited by Schroth et al (2004) presents the potential role of agroforestry in conserving tropical biodiversity, increasing connectivity in mosaic landscapes where natural habitat has been highly fragmented, and forming boundaries with agricultural areas. The book explores the potential of agroforestry for landscape-scale tropical biodiversity conservation, discusses the benefits related to the biodiversity of agroforestry systems and the landscapes of which they are part. It also identifies some of the ecological, socio-economic, and political constraints on biodiversity-friendly land use systems, presents some practical examples of the use of agroforestry in biodiversity conservation projects in the tropical nations, and identifies knowledge gaps that warrant further research.

Bibliography

Angelsen A, Kaimowitz D (2004) Is agroforestry likely to reduce deforestation? In: Schroth G, da Fonseca GAB, Harvey CA, Vasconcelos HL, Gascon C, Izac AM (eds) Agroforestry and Biodiversity Conservation in Tropical Landscapes. Island Press, Washington DC, USA, pp 87–106

Atangana AR, Beaulieu J, Khasa DP (2010) Wild genetic diversity preservation in a small-sized first-generation breeding population of *Allanblackia floribunda* (Clusiaceae). Tree Genet Genomes 6(1):127–136

Bennett AF, Radford JQ, Haslem A (2006) Properties of land mosaics: Implications for nature conservation in agricultural environments. Biol Conser 133(2):250–264

Bhagwat SA, Willis KJ, Birks HJB, Whittaker RJ (2008) Agroforestry: a refuge for tropical diversity? Trends Ecol Evol 23(5):261–267

Griffith DM (2000) Agroforestry: a refuge for tropical biodiversity after fire. Conserv Biol 14(1):325–326

Huang W, Luukkanen O, Johanson S, Kaarakka V, Räisänen S, Vihemäki H (2002) Agroforestry for biodiversity conservation of nature reserves: functional group identification and analyses. Agrofor Syst 55:65–72

Kanowski J, Catterall CP, Wardell-Johnson GW (2005) Consequences of broadscale timber plantations for biodiversity in cleared rainforest landscapes of tropical and subtropical Australia. Forest Ecol Manag 208(1-3): 359–372

Leakey RRB (1998) Agroforestry for biodiversity in farming systems. In: Collins WW, Qualset Co (eds) Biodiversity in agroecosystems. Advances in Agroecology, CRC Press, pp 127–145

Peltier R, Bisiaux F, Dubiez E, Marien J-N, Muliele J-C, Proces P, et Vermeulen C (2010) De la culture itinérante sur brûlis aux jachères enrichies productrices de charbon de bois, en Rep. Dem. Congo. ISDA, Montpellier, France

Perfecto I, Vandermeer J (2008) Biodiversity conservation in tropical agroecosystems. Ann NY Acad Sci. 1134:173–200

Sanchez PA, Benites JR (1987) Low-input cropping for acid soils of the humid tropics. Science 238(4833):1521–1527

Schroth G, da Fonseca GAB, Harvey CA, Vasconcelos HL, Gascon C, Izac AM (2004) Introduction: The role of agroforestry in biodiversity conservation in tropical landscapes. In: Schroth G, da Fonseca GAB, Harvey CA, Vasconcelos HL, Gascon C, Izac AM (eds) Agroforestry and Biodiversity Conservation in Tropical Landscapes. Island Press, Washington DC, USA, pp 1–14

Schroth G, Harvey CA (2007) Biodiversity conservation in cocoa production landscapes: an overview. Biodivers Conserv 16:2237–2244

Sonwa DJ, Nkongmeneck BA, Weise SF, Tchatat M, Adesina AA, Janssens MJJ (2007) Diversity of plants in cocoa agroforests in the humid forest zone of Southern Cameroon. Biodivers Conserv 16:2385–2400

Schure J, Assembe Mvondo S, Awono A, Ingram V, Lescuyer G, Sonwa D, Somorin O (2010). L'état de l'art du bois énergie en RDC: Analyse institutionnelle et socio économique de la filière bois énergie. Projet Makala—EuropeAid DCI-ENV/2008/151–384

Steffan-Dewenter I, Kessler M, Barkmann J, Bos MM, Buchori D, Erasmi S, Faust H, Gerold G, Glenk K et al. (2007) Tradeoffs between income, biodiversity and ecosystem functioning during tropical rainforest conversion and agroforestry intensification. PNAS 104(12):4973–4978

Schroth G, da Fonseca GAB, Harvey CA, Gascon C, Vasconcelos HL, Izac AMN (2004) Agroforestry and Biodiversity Conservation in Tropical Landscape. Island Press, Wasington

Tchoundjeu Z, Asaah EK, Anegbeh PO, Degrande A, Mbile P, Facheux C, Tsobeng A, Atangana AR, Ngo Mpeck ML Simons AJ (2006) Putting participatory domestication into practice in West and Central Africa. Forests, Trees and Livelihoods 16:53–69

Torquebiau E (1992) Are tropical homegardens sustainable? Agric Ecosyst Environ 41:189–207

Whitacre DF, Schulze M, Seavy N (1995) Habitat affinities of a central American forest avifauna: implications for conservation in Neotropical slash-and-burn farming landscapes. Technical report to the U.S. Man and the Biosphere Program. The Peregrine Fund, Boise, Idaho

Chapter 12
Integrated Pest Management in Tropical Agroforestry

Abstract Agroforestry systems most often harbor more pests than monocrop systems. Parasitism is increased by management practices that increase light availability in agroforestry systems, together with connectivity between these agroforestry systems and the forest. In cocoa farms, the trees used to provide shade also host the fungus *Phytophthora megakarya*, which causes brown rot in cocoa plants. In Malawi, some insects such as *Brachyplatys testudonigro*, *Mesoplatys ochroptera*, *Exosoma* sp. and *Ootheca* sp. feed on the sap of *Sesbania sesban*, and are associated with other agroforestry species. Four types of caterpillar that defoliate *Ricinodendron heudelotii* were identified in the Democratic Republic of Congo, namely *Lobobunaea phaedusa, Imbrasia petiveri, Imbrasia epimethea, Imbrasia obscura* and probably *Imbrasia melanops*. Parasitism in agroforestry systems is also due to the integration of germplasms from species with high nutritional and commercial value in local agroforests, and that germsplasms' integration can be problematic if the genetic base of the species being domesticated is reduced. The integration of trees in agricultural landscapes can also contribute to pest control. Trees can serve as a barrier to insect movement, thereby reducing crop infestation. Integrated pest management in agroforestry can be achieved by: (i) the identification and use of host plants that are resistant to pests and pathogens, (ii) crop rotation between host plants and plants that do not harbor pests, (iii) the biological control of pest abundance, and (iv) the use of farming practices that do not increase light intensity in agroforestry systems.

12.1 Pests in Tropical Agroforestry Systems

Integrated pest management is defined as a set of techniques that aim at maintaining pathogen, pest and weed populations at levels below those that cause economic loss (Dix et al. 1998). Agroforests are more likely to harbor pests and insects than agricultural systems. This difference could be attributed to the intensity of cultivation of non-domesticated species, the taxonomic relatedness of trees and crops, the amount of light intensity, the distance from natural forest, and the degree of maturity of the agroforest. In Central Indonesia, parasitism has been found to increase with management practices that amplify light availability in agroforestry systems and improve connectivity between these systems and the forests (Klein et al. 2006).

A. Atangana et al., *Tropical Agroforestry*, DOI 10.1007/978-94-007-7723-1_12,

Epila (1986) postulated that the behavior of insects is determined by plant species diversity, cultivated trees, age of agroforest, cultivation method and the degree of relationship between associated crops.

Agroforestry practices can be a source of pest infestation on crops. The integration of germplasms from species with high nutritional and commercial value into local agroforests can be problematic if the genetic base is reduced. Such conditions can occur if the germplasm is obtained from vegetative propagules. The strategy of participatory domestication of high-value agroforestry species in the humid tropics of West and Central Africa is based on the vegetative propagation of individuals with desirable characteristics (Tchoundjeu et al. 2006). Some species enrolled in this program already face problems with insects and pests. Four types of caterpillar defoliating *Ricinodendron heudelotii*, a priority species for domestication in the region, have been identified in the Democratic Republic of Congo (Latham 2003). The species are *Lobobunaea phaedusa* (Drury), *Imbrasia petiveri* (Guerin-Meneville), *I. obscura* (Butler) and probably *I. melanops* (Bouvier) and *I. epimethea* (Drury) as well. The caterpillars cause significant damage to the leaves of *R. heudelotii* between December and January (*I. petiveri*), October and May (*L. phaedusa*), or from October to February (*I. obscura* and *I. melanops*). Biological control of these caterpillars is possible, as these pests are edible and constitute a source of protein for local people. In Cameroon, a psyllid, *Dichlidophlebia xuani* (subfamily Paurocephalinae) attacks the buds of *R. heudelotii* seedlings in the nursery (Messi and Tamesse 1999). Predators of this psyllid were identified in a study conducted around Yaoundé, Cameroon. Predators are among the Anthocoridae, Coccinellidae, Miridae, and Syrphidae species, and include an unidentified species of spiders belonging to the Encyrtidae, but their impact on the psyllid is low. Other species included in this domestication program face pest and diseases problems. Safou (*Dacryodes edulis*) suffers from sagging of flowers, and the nuts of *Cola acuminata* and *C. anomala* are attacked by *Rhincophorus* sp.

Insects that feed on the sap of *Sesbania sesban* and other agroforestry species were identified by Sileshi et al. (2000). The insect species are mainly *Brachyplatys testudonigro*, *Mesoplatys ochroptera*, *Exosoma* sp. and *Ootheca* sp. A study by Mchowa and Ngugi (1994) reported that the leaf beetle *M. ochroptera* (Stal.) causes severe defoliation of *Sesbania* spp., and that an unidentified lepidopteran caterpillar species also fed on *Sesbania* spp., causing considerable defoliation in Malawi. The study also presented some insects associated with multipurpose species in Malawi (Table 12.1). A cropping system can control Sesbania defoliation from pest infestation, by modifying the microclimate of agroecosystems, which can influence the abundance of pests or natural enemies (Mchowa and Ngugi 1994). The study postulated that agroforestry practices, such as increasing the diversity of multipurpose trees incorporated in agroforestry systems, would certainly reduce pest incidence and damage to both the tree and crop species. However, an increase in tree diversity may not be associated with an increase in tree density, as the trees used to provide shade often host plant pathogens. For example, in cocoa farms in Ghana, the trees used to provide shade also host the fungus *Phytophthora megakarya*, which causes brown rot in cocoa plants (Opoku et al. 2002).

Table 12.1 Species of insects associated with multipurpose trees for agroforestry at Makoka, Malawi (Mchowa and Ngugi 1994)

Insect species	Plant species (site)	Relative abundance	Habitat, status	Reference
ORDER COLEOPTERA				
Family Anthicidae				
Formicorus sp.	*Sesbania* sp. (nursery)	Rare (one specimen)	Anthicid adults are said to sometimes occur in flowers and larvae in detritus	Unpublished manual for identification of Coleoptera (IIE[a])
Family Chrysomelidae				
Subfamily Chrysomelinae				
Mesoplatys orchoptera (Stal.)	*Sesbania* sp. (nursery and field)	Abundant, outbreak population	Both larvae and adults caused serious defoliation on *Sesbania* sp. They are known to occur throughout East Africa on *Sesbania*, *Aechynomene* and *Erythrina* sp.	M.L. Cox, Identification report (IIE)
Subfamily Galerucinae				
Medythia quaterna (Fairmaire)	*Sesbania* sp. (nursery)	Rare	Adult members of this subfamily usually feed on lower surface parenchyma, causing a lace-like effect by leaving veins intact. This species is widely distributed throughout Africa and is a pest of various pulse crops	Unpublished identification manual for Coleoptera. M.L. Cox, Identification report
Family Coccinellidae				
Cheilomenes sulphurea (Olivier)	*Sesbania* sp. (nursery and field)	Rare	Some members of this genus are reported as predators of aphids	Kalra (1988)
Chilocorus angolensis (Crotch)	6 MPT[b] spp.	Many	Preying on *Macropulvinaria inopheron*	R.G. Booth, Identification report (IIE)
Platynaspis capicela (Crotch)	*Sesbania* sp. (nursery and filed)	Scarce	Not known	–
Family Phalacridae				
Genus and species indeterminate	6 MPT spp.	Many	In association with *M. inopheron*. Family is mostly fungal feeders. Not conclusively shown as predators	R.G. Booth, Identification report (IIE)
ORDER DIPTERA				
Diopsis sp. (*apicalis* (Dalman gp.))	*Sesbania* sp. (nursery)	One	Members of this family are miners or shoot pests of the Graminae plant family as larvae or else saprophytic	Hill (1975)

Table 12.1 (continued)

Insect species	Plant species (site)	Relative abundance	Habitat, status	Reference
Family Sciomyzidae				
Sepedon ap.	*Sesbania* sp. (nursery)	One	Larvae of spp. Of known biology develop as predators of snails in damp or aquatic conditions. Adult *Sepedon* are usually found near margins of acidic pools	Det. I.M. White, White, Identification report (IIE)
ORDER HOMOPTERA				
Family Aphididae				
Aphis fabae (Scopoli)	*Phaseolus vulgaris* (L.) (field)	Many small colonies	A pest of beans and other pulses	Hill (1975)
Family Coccidae				
Macropulvinaria inopheron (Laing)	6 MPT spp. *Cajanus* sp.	Outbreak populations	Was described on *Erythrina* sp. (Uganda) and is reported on cotton (Nigeria), *Salvia* sp. (Kenya), Tung and *Hibiscus* sp. (Malawi) and *Ziziphus macronata* (Zimbabwe)	G.W. Watson, Identification report
Family Tettigometridae				
Hilda patruelis (Stal.)	*Sesbania* sp. (field)	Numerous	A polyphagous pest causing local and occasional damage to *Arachis hypogaea*	Weaving (1980)
ORDER HYMENOPTERA				
Family Encytidae				
Psyllechthrus oophagus (Gliesquiere)	*Sesbania* sp. (field)	Many	Parasitizing eggs of *Hilda patruelis*	Weaving (1980)
Family Ichneumonidae				
Subfamily Ichneumonidae				
Genus and species indeterminate	*Sesbania* sp. (nursery)	One	The subfamily comprises solitary endophagous parasites of lepidopteran pupae. Species are often host-specific	Unpublished manual for identification of hymenoptera (IIE)

[a] IIE: CAB International Institute of Entomology; [b] MPT: multipurpose tree

The damage caused by insects on agroforestry species was also reported in India. Singh et al. (2004) reported on insects that cause damage to poplars (mainly *Populus deltoides*) in agroforestry systems in India. Insects include stem borers (*Apriona cinerea*), shoot borers (*Eucosma glaciate*), defoliators (*Clostera cupreata, C. fulgurita,*

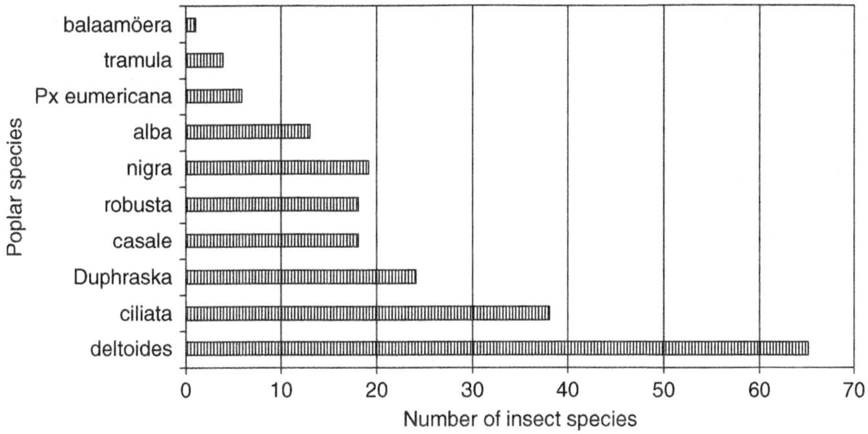

Fig. 12.1 Relative susceptibility of *Populus deltoides* to insects in relation to other poplarspecies in northwestern India (in terms of insect species infestation) (Singh et al. 2004)

C. restitura and the Common Leopard Butterfly, *Phalantha phalantha*). Other parasitical insects found were leaf beetles (*Nodostoma waterhousie*) sap suckers, gall formers (*Aphids, Eriosoma spp.*), root feeders (*Holotrichia* spp., *Granida* spp., and *Brhamina* sp., known collectively as white grubs), and termites (*Odontotermes* spp. and *Copototermes* spp.). Lepidoptera constitutes the most damaging insect order on *P. deltoides* in northwestern India (Fig. 12.1).

12.2 Agroforestry Practices for Pest Management

Insects, especially herbivorous arthropods, have always been a pest for crops. In Malawi, insects are the main cause of tree mortality, and are an obstacle to the expansion of agroforestry (Sileshi et al. 2008). However, trees can also serve as a barrier to the movement of insects, thus reducing infestations of plants. Integrated approaches used to fight pests in agroforestry include combining host plants resistant to pests, exploitation of alternative tree species, and biological controls promoting the development of insects that prey on the pest in question. Various approaches could be used to deal with the proliferation of pests in agroforestry systems (Rao et al. 2000). Exploitation of biological controls against these pests requires the identification and regulation of the abundance of natural enemies of insects causing damage to plants. Parasitic wasps, ants, beetles, birds, rodents and spiders have for years controlled the abundance of herbivorous arthropods in natural ecosystems by keeping them below the epidemic threshold (Mason 1987; Crawford and Jennings 1989). In agroforestry systems, the presence or absence of fences may influence the level of infestation of maize by stem borers (*Busseola fusca* and *Chilo* spp.), aphids (*Rhophalosiphum maidis*) and beetles (Girma et al. 2000). In addition, Girma et al. (2000) concluded that the effects of hedges on the infestation of crops by pests and their role as a refuge from predators depend on the specific arthropods in question (Girma et al. 2000). Integrated pest management in agroforestry is also directed towards the empowerment of small farmers in the management of agroecosystems. Farmers decide on the methods used to control pests below non-epidemic thresholds, such as the use of resistant species (van Huis 2009). Integrated pest management in agroforestry is exercised by:

- Identification and use of host plants resistant to pests and pathogens;
- Crop rotation between host plants and plants that do not harbor pests;
- Biological control of pest abundance;
- Use of farming practices that do not increase light intensity in agroforestry systems;
- Use of species that are tolerant or resistant to insects and pathogens;
- Training of farmers and empowering them to practice agroforestry system management to maintain pest populations below non-epidemic thresholds;
- Use of farmers' local knowledge to fight against insects and pathogens.

12.3 Research Areas on Pest Management Using Agroforestry Techniques

Traditional practices for integrated pest management include: site selection, soil management, timing of planting and harvesting, crop resistance, intercropping, weed management, harvest residue management, post-harvest management, natural enemies management, mechanical control, and use of repellents and traps (Morales 2002). Agroforestry practices involving the timing of planting, intercropping, crop resistance, and weed management are suited for pest management in any landscape. Integrated pest management practices using agroforestry techniques are based on three assumptions:

- The specific diversification of crops, both herbaceous and woody, induces a reduction of pests.
- Crop rotation in an agroforestry system reduces the risk of pest and disease infestation.
- Increasing the incident light intensity in an agroforestry system reduces moisture and therefore reduces the risk of crop infestation by pathogens.

Microclimate and light intensity are important factors for pest management in agroforestry systems. Their effects on pest and pathogen control need to be investigated further, as contradictory results have often been reported. When investigating pests and diseases in agroforestry systems in the humid tropics, Schroth et al. (2000), found that:

- The risk of diseases and pests does not automatically decrease through the introduction of perennial plants or the increase in plant diversity in the system. This is contradictory to the assumption that the diversification of crop species causes a reduction of pests in the system.
- If the introduced plants host pests or diseases of other species in the system, the risk of infection is increased.
- To evaluate such risks, one should consider the range of hosts and pathogens.
- In addition to the selection of compatible plant species, spatial arrangement is important for reducing the spread of pathogens in the system. An increase in pests and diseases has often been identified at the tree-crop interface. This is probably because of the humid microclimate, physical protection of mammals and bird pests by trees, and possibly the tolerance of pests and diseases by plants stressed by competition. Optimum shade conditions obtained with a well-designed, multi-strata crop system, minimizes the entire pest complex and maximizes the effects of beneficial microflora and fauna acting against it (Staver et al. 2001).
- Linear planting of trees and hedgerows can affect the transport of small insects and pathogens by wind, and the active immigration and emigration of pests as populations of natural enemies change.
- Higher shading has a major effect on the microclimate in which pathogens, pests, populations of natural predators, and the crop itself grow. Optimization of shade is an effective strategy to control pests and diseases.
- On infertile lands, crop susceptibility to pests and diseases is strongly influenced by nutrient availability. Agroforestry techniques can influence nutrient availability in various ways.
- Soil management practices such as mulching and the use of cover crops affect plant health by improving soil fertility and directly affect the populations of pests and diseases.

The importance of a more systematic collection of information concerning pests and diseases in agroforestry, preferably in a centralized database, was highlighted by Schroth et al. (2000). The study further underlined the importance of the developing strategies to reduce the risks of pests and diseases in agroforestry in cooperation with farmers (Schroth et al. 2000). Strategies for insect control in agroforestry involve the practice of cultural methods, the use of tolerant varieties or clones, natural enemies, and bio-pesticides (e.g., secondary cycling of the fungal pathogen *Metharizium flavoviride* after single application on crops effectively controlled locust and grasshopper infestation) (Thomas et al. 1995). Further recommendations include avoiding alternate host plants, as indicated by Singh et al. (2004) based on a study on poplar agroforestry in northwestern India. The architectural complexity of Eucalyptus plantations was found to have an effect on communities of wasp parasitoids (Steinbauer et al. 2006). Further investigation needs to be done on the effects of complexity and species richness in agroforests on insect management, and the identification of the suitable non-epidemic threshold.

Bibliography

Crawford HS, Jennings DT (1989) Predation by birds on spruce budworm *Choristoneura fumiferana*: functional, numerical, and total responses. Ecology 70:152–163

Dix ME, Bishaw B, Workman SW, Barnhart MR, Klopfenstein NB, Dix AM (1998) Pest management in energy- and labor-intensive agroforestry systems. In: Buck LE, Lassoie JP, Fernandes ECM (eds) Agroforestry in Sustainable Agricultural Systems. CRC Press, pp 131–156

Epila JSO (1986) The case for insect pest management in agroforestry research. Agric Syst 19:37–54

Girma H, Rao MR, Sithanantham S (2000) Insect pests and beneficial arthropods populations under different hedgerow intercropping systems in semiarid Kenya. Agrofor Syst 50:279–292

Klein A-M, Steffan-Dewenter I, Tscharntke T (2006) Rain forest promotes trophic interactions and diversity of trap-nesting Hymenoptera in adjacent agroforestry. J Animal Ecolo 75:315–323

Latham P (2003) Edible caterpillars and their food plants in Bas-Congo. Mystole Publications, Canterbury

Mason RR (1987) Non-outbreak species of forest Lepidoptera. In: Barbosa P, Schultz JC (eds) Insect Outbreaks, pp 31–57. Academic Press, New York

Mchowa JW, Ngugi DN (1994) Pest complex in agroforestry systems: the Malawi experience. Forest Ecol manag 64:277–284

Messi J, Tamesse JL (1999) Numerical fluctuations of populations of *Trioza erytreae* (Del Guercio) (Hemiptera: Psylloidea: Triozidae), the citrus psyllid, in an orchard at Yaounde (Cameroon). Annales de la Societe Entomologique de France, 35: Supp. 238–241

Morales H (2002) Pest management in traditional tropical agroecosystems: Lessons for pest prevention research and extension. Intergrated Pest Manag Rev 7:145–163

Opoku IY, Akrofi AY, Appiah AA (2002) Shade trees are alternative hosts of the cocoa pathogen *Phytophthora megakarya*. Crop Protection 21:629–634

Rao MR, Singh MP, Day R (2000) Insect pests problems in tropical agroforestry systems: Contributory factors and strategies for management. Agrofor Syst 50:243–277

Schroth G, Krauss U, Gasparotto L, Duarte Aguilar JA, Vohland K (2000) Pests and diseases in agroforestry systems of the humid tropics. Agrofor Syst 50:199–241

Sileshi G, Maghembe JA, Rao MR, Ogol CKPO, Sithanantham S (2000) Insects feeding on *Sesbania* species in natural stands and agroforestry systems in southern Malawi. Agrofor Syst 49:41–52

Sileshi GW, Kuntashula E, Matakala P, Nkunika PO (2008) Farmers' perception of tree mortalitypests and pest management practices in agroforestry in Malawi, Mozambique and Zambia. Agroforest Sys 72:87–101

Singh AP, Bhandari RS, Verma TD (2004) Important insect pests of poplar agroforestry and strategies for their management in northwestern India. Agrofor Syst 63:15–26

Staver C, Guharay F, Monterroso D, Muschler RG (2001) Designing pest-suppressive multi-strata perennial crop systems: shade-grown coffee in Central America. Agrofor Syst 53:151–170

Steinbauer MJ, Short MW, Schmidt S (2006) The influence of architectural and vegetational complexity in eucalypt plantations on communities of native wasp parasitoids: Towards silviculture for sustainable pest management. For Ecol Manag 233:153–164

Tchoundjeu Z, Asaah EK, Anegbeh PO, Degrande A, Mbile P, Facheux C, Tsobeng A, Atangana AR, Ngo Mpeck ML, Simons AJ (2006) Putting participatory domestication into practice in West and Central Africa. Forests, Trees and Livelihoods 16:53–69

Van Huis A (2009) Challenges of Integrated pest management in Sub-saharan Africa. In: Peshin R, Dhawan AK (eds) Integrated Pest Management: Dissemination and Impact: DOI 10.1007/978-1-4020–8990-9-12

Thomas MB, Wood SN, Lomer CJ (1995) Biological control of locusts and grasshoppers using a fungal pathogen: the importance of secondary cycling. Proceedings of the Royal Society London Biological Sciences 259(1356):265–270

Part III
Research Methods in Agroforestry

Chapter 13
Diagnosis and Design (D & D) Approach and Participatory Rural Appraisal (PRA)

Abstract The design of agroforestry systems should be developed in response to specific needs. The basic premise of Diagnosis and Design (D & D), a method of diagnosing land management problems and designing agroforestry solutions, is the identification of a problem and determination of the appropriate type of research to solve the identified problem. The goals of D & D are: (i) to describe and analyze existing land use systems, (ii) to develop appropriate agroforestry technologies for the alleviation of constraints, and (iii) to develop appropriate research goals and a method of efficient examination. The D & D method is applied at the household (micro-level), community or watershed (meso-level), and regional or country (macro-level). On the other hand, Participatory Rural Appraisal (PRA) is a broader approach involving all the stakeholders, especially the farmers, from the beginning to the end of the diagnosis. Participatory Rural Appraisal can be defined as "an approach and method for learning about rural life and conditions from, with, and by rural people". Participatory Rural Appraisal, which evolved from RRA, or Rapid Rural Appraisal, i.e., a series of techniques used for quick identification, appraisal and evaluation of information on rural resources relevant for planning action, is the most popular participatory research method, and has been used in research since the 1990s.

13.1 Introduction

Agroforestry, as a science, was born from concern over the sustainable management of natural resources. Any development of an agroforestry system should be made in response to specific needs. Problem diagnosis and the proposal of a solution are closely related. Research methods, the social aspects as well as the biophysical aspects of agroforestry have been inventoried. Methods mainly consist of procedures for the full evaluation of constraints and problems of land use, the identification of specific intervention points for improvement, and the adaptation of methods and procedures that are already available in other agricultural sciences to fulfill the needs of agroforestry. Before developing agroforestry experiments, it is necessary to determine the nature of the problem, and decide what type of research is appropriate to find a solution. This is the basis of Diagnosis and Design (D & D). However, D & D focuses on problem-solving, whereas agroforestry research needs a broader

A. Atangana et al., *Tropical Agroforestry*, DOI 10.1007/978-94-007-7723-1_13,

approach (Rocheleau 1985a). For that reason, participatory rural appraisal (PRA) has become more and more important in agroforestry research. The PRA involves all stakeholders, especially farmers, who are the first beneficiaries of the project, from the beginning to the end of the diagnosis. Therefore, PRA is a tool that is used in participatory research to identify problems and propose solutions that then will be tested/evaluated by farmers. Fischer and Vasseur (2002) reported farmers' perceptions of agroforestry projects in Panama, and recommended the use of participatory techniques to involve farmers in all stages of project design, implementation, and monitoring. The authors also recommend that all groups within a community should be involved in an agroforestry project (Fischer and Vasseur 2002). The most important element of PRA is the participation of all the stakeholders. Problem identification is time-consuming, and the perceptions, beliefs, attitudes and values of rural people should be known and taken into account in any agroforestry system design. Specificities of each site should also be taken into account, as the nature, conditions, and solutions to rural poverty can vary across households, groups, villages, and regions (Mukherjee 1993).

13.2 The D & D method

The D & D method was developed by ICRAF scientists in the early 1980s (Raintree 1983 a,b) to meet the requirements of research for development. The main goals of this approach are to:

- Describe and analyze existing land use systems;
- Develop appropriate agroforestry technologies for the alleviation of constraints; and
- Develop appropriate research work, and efficient methods of examination.

Diagnosis and Design is a method used to initiate, monitor and evaluate agroforestry programs, based on the land use system, regardless of the farming system, ecozone, or country. The structure and function of the land use system are determined by climatic, biological, physical, technological, socio-economic and political factors. Diagnosis and Design then, is an iterative method allowing the fine-tuning of appropriate agroforestry systems for any region (Fig. 13.1). It should be noted that several analysis and evaluation methods for land use systems existed long before the D & D method was developed, the best known methods being the Farming Systems Research/Extension (FSR/E) method (Shaner et al. 1982; Hildebrand 1986), and the Land Evaluation Methodology (FAO 1976). Each of these methods was developed for specific purposes. The FSR/E method was developed to overcome the shortcomings of the traditional methods of technology transfer that had been initiated to spread green revolution technology from researchers to poor farmers. The FSR/E was an interdisciplinary and global approach requiring farmers' participation at the beginning of a project.

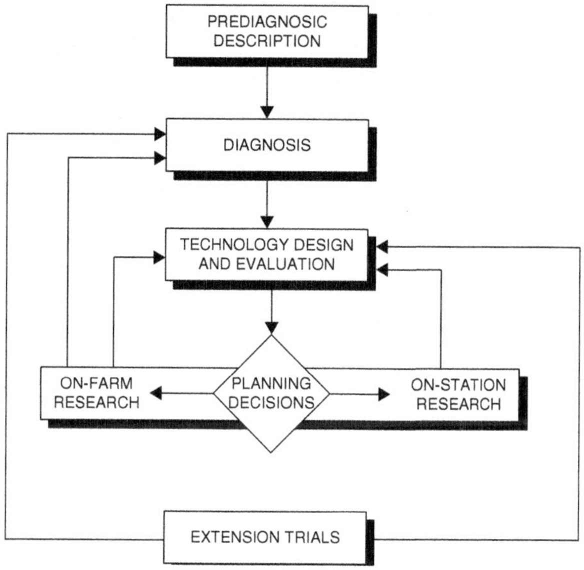

Fig. 13.1 Components of project design incorporating the iterative process of D & D (Raintree 1987a)

The D & D method comes from the "exigencies of the situation of agroforestry. It attempts to develop a *special focus on agroforestry-related constraints and opportunities* within existing land-use systems and to highlight agroforestry potentials that might be overlooked by other methodologies" (Raintree 1987a). As Muller and Scherr (1990) put it, "development of sustainable land use systems, including agroforestry, requires systematic diagnosis of constraints and potentials, design of relevant technologies, and continuous 're-design' until locally appropriate and validated technology designs are achieved". An example of the efficiency of this iterative method in developing appropriate agroforestry systems in the tropics was studied by Pinners and Balasubramanian (1991). The study reported the use of feedback from farmers and extensionists for the periodic evaluation of research assumptions and technology designs in Rwanda. In an area under increasing population pressure, a project was implemented to develop appropriate food production systems. Each family in the area owned an average of 1.2 ha, and the main crops grown were banana (*Musa* spp.), sweet potato (*Ipomoea batatas*), cassava (*Manihot esculenta*), sorghum (*Sorghum* spp.), maize (*Zea mays*), beans (*Phaseolus vulgaris*), groundnuts (*Arachis hypogaea*), peas (*Pisum sativum*), and coffee (*Coffea arabica*). Goats (*Capra aegagrus hircus* L.) and chicken (*Gallus gallus domesticus* L.) were also found on each farm. Using surveys, traditional wisdom was identified, documented, analyzed, and a list of researchable topics was established to address the issues of a lack of arable and pasture land, soil degradation, and scarcity of fodder, and to identify most appropriate interventions. However, farmers were not actively involved in the initial selection of technologies for on-station testing. The study concluded that the farmers' needs, circumstances, available resources and management capacity regarding tree planting are important points to consider when identifying suitable agroforestry techniques for any region (Pinners and Balasubramanian 1991).

The analysis of agroecosystems is a simple and rapid assessment in rural areas (Conway 1985) that shares the same philosophy as D & D, which has been very popular due to its focus on agroforestry. Another approach to land use system management, focusing on ranch management, puts a greater emphasis on design compared to diagnostics (Savory 1988).

13.3 Main Concepts and Procedures of the D & D Method

Diagnosis and Design can be defined as a method for diagnosing land management problems and designing agroforestry-based solutions. The clear identification of the problem is often all that is needed to suggest the solution. Diagnosis and Design is simply a systematic approach to this principle in agroforestry (Nair 1993). The nature of the data and information that needs to be collected, and the type of questions or surveys to be conducted at each stage of the process are given in Tables 13.1 and 13.2. This iterative process allows adjustments when necessary, and continues until the system is optimized (Pinners and Balasubramanian 1991, Fig. 13.1). The D & D method is characterized by flexibility, speed and repeatability. It assumes that by including farmers in the research and extension, any subsequent interventions and recommendations are easily adopted. During the diagnostic phase, researchers communicate with farmers to characterize agroforestry practices, identify constraints for production, and discuss management strategies and production alternatives. Farmers' perceptions are considered, and used as a guide for extension if the proposed agroforestry technologies are already available, or else as a basis for the design of suggested technologies. The D & D method is applied several times in development-oriented projects.

13.4 Scales of D & D

The D & D method can be applied at the household level (micro-scale) to identify problems of individual plantations, at the community or watershed level (meso-scale) to identify problems like slope erosion, and at the regional or country level (macro-scale) to solve problems such as pest infestation. Methods used at the household level are intended to identify constraints (i.e., the basic needs approach), and to help design solutions (i.e., the troubleshooting procedure) so as to address the diagnosed problems. Using troubleshooting logic, D & D attempts to identify *what is causing the problems* that farmers face in meeting their every day needs for food, firewood, shelter, and fodder, and *why* the problem exists (Nair 1993). Although it has been proven to be efficient at the micro-scale level, the D & D method cannot be used for problem identification in large-scale enterprises. Another drawback of the

Table 13.1 Procedures of the diagnosis and design (D & D) methodology (Raintree 1987a)

D & D stage	Basic questions to ask	Key factors to consider	Mode of inquiry
Prediagnosis	Definition of the land use system and site selection (which system to focus on?)	Distinctive combinations of resources, technology and land objectives	Seeing and comparing the different land use systems
	How does the system work? (how is it organized, how does it function to achieve its objectives?)	Production objectives and strategies, arrangement of components	Analyzing and describing the system
Diagnosis	How well does the system work? (what are its problems, limiting constraints, problem-generating syndromes and intervention points?)	Problems in meeting system objectives (production shortfalls, sustainability problems)	Diagnostic interviews and direct field observations
		Causal factors, constraints and intervention points	Troubleshooting the problem subsystems
Design and evaluation	How to improve the system? (what is needed to improve system performance?)	Specification for problem solving or performance enhancing interventions	Iterative design and evaluation of alternatives
Planning	What to do to develop and disseminate the improved system?	Research and development needs; extension needs	Research design; project planning
Implementation	How to adjust to new information?	Feedback from on-station research, on-farm trials and special studies	Rediagnosis and redesign in the light of new information

micro-scale procedure is that the household is not a homogeneous unit. Women and men have different perceptions of problems. For this reason, meso-scale methods are used when dealing with units larger than farms, such as landscapes or watersheds, or areas covering a hundred or a thousand farms (Rocheleau 1985b), such as pest infestation in a region. However, it should be noted that there is no "single best *fixed agroforestry package for any given region or group of people*, but rather a vast array of agroforestry principles and components that can be recombined, tested and modified to suit changing social, economic and ecological conditions for individuals, households, communities and nation-states" (Rocheleau 1989).

Methods used to solve identified problems should be developed or adapted according to the scale of the problem. For example, analysis of aerial photos could help identify topographical features of a landscape affected by soil erosion, as well as be used to determine checkpoints and make decisions to support the development of hedges of trees. Global positioning system (GPS), geographic information systems (GIS) and remote sensing are powerful and efficient tools to monitor erosion risk and other problems of land use management at the regional level. These tools are now being used in precision agroforestry planning, participatory mapping in the

Table 13.2 Information needs and sources for agroforestry (AF) diagnosis and design (Raintree 1987a)

Design decisions	Questions and sources of information	
	External knowledge base	Diagnostic field survey
Potentially relevant AF prototypes (provisional identification)	What kind of a system is it? (environment, land use system type, land use intensity, sources of production increase, typical problems, potentials and functional needs, adoptability consideration)	What are the identifying characteristics of the system? (what are its component parts, how is it organized, how does it work? (from brief reconnaissance survey)
Site-specific development algorithm Development strategy	What kinds and rate of change is this type of system able to absorb? What is the optimal pathway of intensification?	What is the best overall development strategy for the system? (incremental improvement vs. complete transformation; phase approach to introduction of changes)
What *problems* and *potentials* should the design address?	What are the typical problems and potentials of this type of system at its present stage of development?	What are the actual problems and potentials of the system? (How do local people normally cope with these problems?)
What *functions* should the design perform?	What functional needs and constraints are typical of such systems?	What are the actual *functional needs* of the system? (as perceived by both farmers and researchers?)
Which functions should be performed *separately* and which in *combination*?	What are the needs and possibilities for functional combinations in such systems?	How does the land user perceive the relative advantages of different possibilities
At what *locations* within the landscape should these functions be performed?	What landscape niches are usually found in such systems?	What landscape niches are actually available, which offer the best choice, what are the land user's preferences?
What species *components* or component combinations are best used to perform the desired functions?	What exotic components are thought be suitable for these functions in such systems?	What indigenous components could perform these functions? (local or ethnobotanical knowledge)
How many of each are required to achieve the objectives of the design?	What is the expected yield of the chosen components in this environment? (If for service role, how much impact are they likely to have?)	Is it possible to fit the required number of components into available spaces? (If not, how can the supply gap be filled? Review local strategies for coping with supply shortages and other problems to suggest additional approaches.)What precise arrangement of the plant and animal components is envisaged?
What *arrangements* are possible? (simultaneously in space and/or sequentially in time)	Which arrangements are preferred by the land users?	

Table 13.2 (continued)

Design decisions	Questions and sources of information	
	External knowledge base	Diagnostic field survey
What *management practices* are envisaged to achieve the performance objectives?	What are the management options?	Which management options are preferable to local users? (check compatibility with local skills, availability of labor and other inputs)

design and implementation of development projects; and for extension professionals (Di Gessa et al. 2008). Meso-level analysis is also used to examine differences between land use management in different landscapes of a region, and to determine possible opportunities for further production (Rocheleau and van den Hoek 1984).

The macro-level of D & D is used to address problems at the eco-region or provincial level, using procedures with a wider application than those of the micro- and meso-levels (Scherr 1989). Moreover, the D & D method is compatible with some environmental investigative techniques, such as those associated with soil evaluation methodology (FAO 1976). However, this level of application is extremely complex.

It should be noted that the suggested procedures should always be tailored to the needs and circumstances of users. The best results are obtained when the procedures are used resourcefully as an aid for sensitive diagnosis and creative design. Some failures observed in the adoption of agroforestry systems have caused the popularization of the PRA since the mid-1990s. These failures were the result of outside experts developing solutions to rural problems on the basis of very weak and poorly consistent diagnoses and without a real and deep consultation of the concerned populations.

13.5 Participatory Rural Appraisal in Agroforestry

Participatory Rural Appraisal (PRA), which evolved from RRA (Rapid Rural Appraisal, i.e., a series of techniques used for quick identification, appraisal and evaluation of information on rural resources relevant for planning action (Chambers 1981)), is the most popular participatory diagnosing method used in agroforestry research, and has been in use since the 1990s. Participatory Rural Appraisal can be defined as "an approach and methods for learning about rural life and conditions from, with, and by rural people" (Chambers 1994a).

The PRA method draws on active participatory research, agroecosystem analysis, applied anthropology, field research on farming systems, and rapid rural appraisal, the last being the most direct source (Chambers 1994a). It is a set of approaches and visual tools used to enable farmers to share, enhance and analyze their knowledge of life, situation, and living conditions, create a plan and act on that plan (Chambers 1994a). Tools used in PRA include interviews with key informants, transect walks, mapping the village and village resources by farmer groups, matrix scoring, seasonal calendars, trend and change analysis, well-being definition by farmers, wealth ranking and grouping, and institutional and analytical diagramming. Exercises are conducted with different community groups to highlight the needs of the community. Men, women, children, young, poor, rich, old, and any other group that might be vulnerable are interviewed and involved. Exercises can be supplemented by surveys and interviews of resource-persons or key informants from the community, and feedback sessions are needed to validate the information received. In this sense, the PRA process is longer, but allows a better understanding of the community. The process allows researchers to trace the history, culture, eating habits, land tenure rights, land use systems practiced in the region, farm size per household, and gender issues of the community. It also allows them to develop a map of the community, locate the resources in this community, establish with farmers the criteria of well-being, perform a well-being ranking of the community (highlighting the livelihoods and main source of income in the community) as well as the frequency and accessibility of different markets, and understand the daily problems faced by community members, the solutions to these problems and the priorities of these communities.

The PRA method provides a more effective and revealing communication than a questionnaire, making PRA longer and more time-consuming. The PRA tools allow a direct consultation of the main beneficiaries of the project, namely the farmers (Mbosso 1999). For example, Farmers' perception of poverty is much broader than the indicators proposed by organizations such as the United Nations Development Program. Information obtained through PRA has been proven to be highly valuable and more reliable than data acquired through more traditional methods (Chambers 1994b).

Participatory Rural Appraisal is used in four different situations, in correspondence with the cycle of a project: during the diagnostic phase (longer exploratory diagnostic), when analyzing a thematic issue, in the planning stages and evaluation of the action. The PRA method has three fundamental pillars:

- The behaviors and attitude of outsiders; outsiders are there to facilitate, rather than to control or to impose (no top-down approach).
- Methods are open, and group-oriented, although interviews with key informants are sometimes needed, and should be visual and comparative.
- Information and experiences are shared between all the stakeholders.

The application of PRA helps determine an effective diagnosis. Therefore, it is necessary that the implementation of solutions, and evaluation of these solutions, be done with the cooperation of the farmers. A study on the opportunities and constraints related to fruit tree integration by poor farmers in the humid lowlands of West and Central Africa is an excellent example of PRA (Leakey et al. 2003; Degrande et al. 2006). Community-level participatory tools were used in six communities in Cameroon and Nigeria. The research team spent one week working with farmers in each community, and performed semi-structured household interviews and farm fruit tree inventories to understand farmers' fruit tree-planting strategies (Degrande et al. 2006; Schreckenberg et al. 1999). Group exercises were performed using local materials like beans, and twigs of leaves, for matrix ranking. It is important that the exercises are visual and that the material used is familiar to local people. To extract more information from the community, it is important that these group exercises be done with community members from different groups. For instance, in Nko'ovos II, South Cameroon, the women provided a more detailed and meaningful map of the village and existing resources than the men (Degrande et al. 2006).

Another example of application of participatory methods in agroforestry is reported by Cardoso et al. (2001). Their study used both PRA, for assessing core problems and possible solutions, and D & D, for identification of specific local tree knowledge. The application of these two approaches, preceded by a survey of existing systems, led to the initial design of agroforestry experiments. The steps used in the PRA process of this project are given in Table 13.3, and those of the D & D are illustrated in Table 13.4. Once the design had been set up, participatory monitoring and evaluation of the activities were regularly performed to fine-tune and improve these systems.

There are, however, some concerns with regard to PRA. Though PRA appears suited to the understanding of local needs, the "rapid spread of this method has made quality assurance a concern, with dangers from 'instant fashion', rushing formalism, and ruts" (Chambers 1994c). The scientific quality of this method has been questioned, as it may not be a theory-based investigation. This method is more empiric, and develops theory from practice.

13.6 Evaluating Agroforestry Technologies

The adoption of agroforestry technologies by farmers depends on site-specific factors, such as biophysical functionality, economic feasibility, policy issues, and social compatibility (Alavalapati and Nair 2001). Economic and policy issues that may influence the adoption of agroforestry technology include profitability, household benefits, equity, sustainability, soil conservation, environmental services, markets for inputs and outputs, gender, and institutions such as property rights (Alavalapati and Nair 2001; Mercer and Hyde 1991). Pattanayak et al. (2003) identified five factors explaining technology adoption within an economic framework: preferences, market incentives, resource endowments, biophysical factors, and risk and uncertainty.

Technology adoption is a complicated process, and starts with research planning. Methods have been developed for determining criteria in agroforestry research planning. One such method is the Delphi method developed by Ndour et al. (1992). Using a qualitative approach, fourteen criteria were identified to be used when planning agroforestry research in developing countries. These criteria were ranked in four groups. The first group includes local people's needs, sustainability, adoptability and research quality. The second group is comprised of existing systems and economic criteria. The third group consists of biophysical effects, institutional capabilities, partnerships and transferability. The last group of criteria encompasses diversity of products, tree/crop interface, flexibility, and species selection (Ndour et al. 1992). Therefore, it is very important to identify local people's needs

Table 13.3 Steps in the participatory rural appraisal undertaken in Araponga, Minas Gerais, Brazil (Cardoso et al. 2001)

Stage	Method	Outputs
Pre-analyses	Literature study and organization of the available data on environment and development of the municipality	First insights about the region Information about local agroecosystems
Use of some PRA tools with farmer union leaders	Venn diagram Map Seasonal diagram	Information about agroecosystems from farmer union leaders, discussion about this information
Planning of the next stages	Meetings with farmer union leaders	Draft plan community of work Key topics to guide interviews: production, migration, health and education
Semi-structured interviews with farm families	During one day, a selected group of houses in one community (normally a small watershed) were visited and the family interviewed (11 communities in total) Interview guide was based on main topics as suggested by farmer union leaders and by Altieri's (1995) checklist of issues for studying agroecosystems In the evenings, after the interviews, the interview responses were discussed with farmers	Information about local agroecosystems from farmers, and discussion with them
Large meeting	Meeting with 350 farmers involved in the interviews and others interested in joining	Main problems were identified and discussed collectively Choice of community representatives to participate in the next stage
Small meeting	Meeting with representatives of the communities chosen in the large meeting to prioritize the problems to be resolved	Prioritization of land-use problems Creation of the "healthy land" community composed of farmers. Researchers and NGO staff
"Healthy land" committee	Regular meetings	Decision to adopt agroforestry as a practice for sustainable land use

when planning agroforestry research, and this cannot be done after only a few visits to the village. Agroforestry research teams, comprised of biophysical, social, policy and anthropology researchers, should undertake a deep study of the community where research is to be done, to identify the local people's needs, culture, history, gender issues, and land tenure rights.

Table 13.4 Approach for the adapted Diagnosis & Design in two watersheds in Araponga, Minas Gerais, Brazil (Cardoso et al. 2001)

Stage	Method	Outputs
Analysis of conclusions from the PRA process	Meetings	The decision to undertake D & D in two watersheds to discuss agroforestry as an alternative land use in depth with farmers
Use of some PRA methods for specific questions	Maps Seasonal calendar Matrix Causal diagramming	Specific information and discussion about trees and their potential
Discussion about concepts, advantages and disadvantages of agroforestry systems	Two meetings, one in each watershed	Establishment of two agroforestry plots, one in each watershed

Another approach used for planning and evaluating agroforestry systems is Multiple Objective Programming (MOP, Mendoza et al. 1986). The MOP allows the optimal allocation of land to various alternative-cropping systems based on management regimes. It is "a time-based schedule of cropping patterns that can be implemented on a given tract of land over the length of the planning horizon" (Mendoza et al. 1986). Management regimes are decision variables, and the MOP is a mathematical technique in which several functions are optimized simultaneously. The MOP accommodates most of the common features of agroforestry, but does not adequately consider the ecological and economic interactions of woody components and crops (Mendoza et al. 1986).

The evaluation of agroforestry technology adoption could also be done using either ex-ante adoption potential analysis or ex-post adoption analysis (Sirrine et al. 2010). Ex-ante adoption analysis is the evaluation of primarily socioeconomic factors, such as lessons for effective dissemination, feasibility and acceptability, feedback for research and extension, and boundary conditions (e.g., market opportunities) that allow or prohibit the agroforestry practices to be profitable, feasible, and acceptable to farmers (Sirrine 2008; Sirrine et al. 2010). Ex-post adoption analysis consists of interviewing households on adoption. Both ex-ante and ex-post techniques were found to contribute distinct and valuable data (Sirrine et al. 2010).

Household surveys and the monitoring of modifications introduced to agroforestry technology were also used to assess rotational hedgerow intercropping (A technology that evolved from alley cropping thanks to modifications introduced by farmers after evaluating alley cropping in the humid lowlands of West Africa (Kanmegne and Degrande 2002)), and to identify key characteristics likely to influence the adoption of this technology (Degrande and Duguma 2000). Improved fallows are likely to be adopted in areas of high demographic pressure on land. Franzel et al. (2002) reported that adoption of an improved fallow system is influenced by several key elements:

- Effective diagnosis of farmers' problems and screening of potential solutions;
- Farmer participation in the early stages of testing improved fallows;
- Testing of a range of management options by farmers and researchers and encouraging farmers to innovate; and
- Development of an adaptive research and dissemination network.

The effective diagnosis of farmers' problems should take into account gender differences in needs and preferences. Gender issues have been considered by agroforestry research for many years, and are important factors influencing agroforestry technology adoption in Africa (Doss 2001). There is often a division of labor and responsibilities between women and men within the household and in farming and cultivation activities, which influences the adoption of any agroforestry technology. In the humid lowlands of West and Central Africa, women must feed the household, and are therefore much more involved in food cropping. Men provide income, and are more involved in cash crop activities, such as cocoa farming. Women farmers testing shrub fallows in the humid lowlands of Cameroon were found to be more interested in *Cajanus* fallow for several reasons (Degrande and Duguma 2000):

- Clearing of *Cajanus* fallow is much easier to manage compared to natural fallow.
- After slashing *Cajanus* shrubs and burning of woody residues, the field is clean and can easily be ploughed for groundnut (*Arachis hypogaea*) planting, the main crop in the area.
- *Cajanus* shrub is established through direct seeding, and is less labor-intensive.
- Yield response to *Cajanus* fallow is relatively quick compared to tree fallow.

Other factors influencing women's adoption of improved fallows are a lack of land and tree tenure security, low level of education, and limited access to information on new innovations (Degrande 2005).

As pointed out by Mercer (2004), "achieving the full promise of agroforestry requires a fundamental understanding of how and why farmers make long-term land-use decisions and applying this knowledge to the design, development, and 'marketing' of agroforestry innovations". Farmers sometimes create or adapt technologies to suit local conditions and meet their own needs (de Wolf 2010). Most often, agroforestry research has not incorporated farmer's experiences and adaptation of agroforestry technologies into their agenda (de Wolf 2010). The evolution of alley cropping to rotational fallows, and then to shrub fallows in the humid tropics of Cameroon is an exception (Kanmegne and Degrande 2002; Degrande and Duguma 2000). Often, this lack of consideration of farmers' innovations is due to the way agroforestry research is structured (on-station experiments first and then on-farm, a top-down approach) and the decline of social science capacities in international agricultural research (de Wolf 2010), although, there has been an increase in awareness of social and economic understanding of agroforestry systems in agroforestry research since the 1990s (Buck et al. 1998). Key elements for agroforestry system adoption include land tenure, labor, and the marketability of products (Nair 1993). Mercer (2004) reviewed the adoption of agroforestry innovations in the tropics, and suggested that research is needed to develop a better understanding of: the role of risk and uncertainty, insights into how and why farmers adapt and modify adopted systems, factors influencing the intensity of adoption, village-level and spatial analyses of adoption, and the temporal path of adoption.

Participatory rather than prescriptive or consultative approaches are effective in the adoption of agroforestry practices. Additionally, scaling-out and -up can be used to disseminate a technology that has proven to be successful (Franzel et al. 2001). Scaling-up agroforestry technology involves building the institutional capacity in the community to support and replicate improved practices (Franzel et al. 2001). Scaling-up is practiced in the ICRAF participatory tree domestication programme in the humid tropics of West Africa. ICRAF's biophysical and socio-economic researchers and Non-Governmental Organizations (NGO) work with farmers, government extension services, farmer organizations, and farmer network organizations. The main objective of these scaling-up activities is to massively and simultaneously disseminate tree domestication techniques to many farmers in many villages in the region (Tchoundjeu et al. 2006). Farmer groups are trained in rural resource centers, using educational material adapted to their circumstances. Most often tree nurseries are key elements of these resource centers, because tree domestication in West and Central Africa is heavily based on the vegetative propagation of selected "plus-trees" (i.e., trees with superior agronomic traits such as large fruits). Farmer-to-farmer technology transfer as well as information sharing is facilitated through the farmers' network.

Bibliography

Alavalapati JRR, Nair PKR (2001) Socioeconomic and institutional perspectives of agroforestry. In: Palo M, Uusivuori J (eds) World forests, society and environment: markets and policies. Kluwer Academic Publishers, Dordrecht, The Netherlands, pp 71–81

Buck LE, Lassoie JP, Fernandes ECM (1998) Agroforestry in sustainable agricultural systems. CRC Press, Boca Raton

Cardoso IM, Guijt I, Franco FS, Carvalho AF, Ferreira Neto PS (2001) Continual learning for agroforestry system design: university, NGO and farmer partnership in Mibas Gerais, Brazil. Agricultural Systems 69(3): 235–257

Chambers R (1981) Rapid rural appraisal: rationale and repertoire. Public Adm Dev 1:95–106

Chambers R (1994a) The origins and practice of Participatory Rural Appraisal. World Dev 22(7):953–969

Chambers R (1994b) Participatory Rural Appraisal (PRA): analysis of experience. World Dev 22(9):1253–1268

Chambers R (1994c) Participatory Rural Appraisal (PRA): challenges, Potential and Paradigm. World Dev 22(10):1437–1454

Conway GR (1985) Agroecosystems analysis. Agric Admin 20:31–35

Degrande A (2005) Adoption potential of two agroforestry technologies: improved fallows and domestication of indigenous fruit trees in the humid forest and savannah zones of cameroon. PhD dissertation. Faculty of Bioscience Engineering, Ghent University, Belgium, p 247

Degrande A, Duguma B (2000) Adoption potential of rotational hedgerow intercropping in the humid lowlands of Cameroon. Agricultural Research Network Paper No 103, ODI, UK

Degrande A, Schreckenberg K, Mbosso C, Anegbeh P, Okafor V, Kanmegne J (2006) Farmer's fruit growing strategies in the humid forest zone of Cameroon and Nigeria. Agrofor Syst 67:159–175

de Wolf JJ (2010) Innovative farmers, non-adapting institutions: a case study of the Organization of agroforestry research in Malawi. In: German L et al. (eds) Beyond the biophysical: knowledge, culture, and power in agriculture and natural resource management. Springer, Germany, pp 901–910

Di Gessa S, Poole P, Bending T (2008) Participatory mapping as a tool for empowerment: experiences and lessons learned from the ILC network. International Land Coalition Secretariat, Rome, p 45

Doss CR (2001) Designing agricultural technology for African women farmers: Lessons from 25 years of experience. World Dev 29(12):2075–2092

FAO (1976) A framework for land evaluation. FAO Soils Bulletin 32. FAO, Rome

Fischer A, Vasseur L (2002) Smallholder perceptions of agroforestry projects in Panama. Agrofor Syst 54:103–113

Franzel S, Cooper P, Denning GL (2001) Scaling up the benefits of agroforestry research: lessons learned and research challenges. Dev Pract 11(4):524–534

Franzel S, Phiri D, Kwesiga F (2002) Assessing the adoption potential of improved fallows in Eastern Zambia. In: Franzel SC, Scherr SJ (eds) Trees on farms: Assessing the adoption potential of agroforestry practices in Africa, CABI, pp 37–64

Hildebrand PE (1986) Perspectives on farming systems research and extension. Lynne Rienner, Boulder

Kanmegne J, Degrande A (2002) From alley cropping to rotational fallow: Farmer's involvement in the development of fallow management techniques in the humid forest zone of Cameroon. Agrofor Syst 54:115–120

Leakey RRB, Schreckenberg K, Tchoundjeu Z (2003) The participatory domestication of West-African fruits. Int For Rev 5(4):338–347

Mbosso MFPC (1999) Opportunités et contraintes liées à la culture des arbres fruitiers locaux dans les basses terres humides du Cameroun: *Dacryodes edulis* et *Irviniga gabonensis*. Mémoire de fin d'études présenté pour l'obtention du diplôme d'Ingénieur Agronome, Université de Dschang, p 111

Mendoza GA, Campbell GE, Rolfe GL (1986) Multiple Objective Programming: an approach to planning and evaluation of agroforestry systems-Part 1: Model description and development. Agric Syst 22:243–253

Mercer DE (2004) Adoption of agroforestry innovations in the tropics: A review. Agroforestry Sys 61–62:311–328

Mercer DE, Hyde WF (1991) The economics of agroforestry. In: Burch WR Jr, J. K. Parker JK (eds) Social science applications in Asian Agroforestry. New Delhi: Oxford & IBH Publishing, pp 111–143

Mukherjee N (1993) Participatory rural appraisal: Methodology and applications. Ashok Kumar Mittal, New Delhi, p 164

Müller EU, Scherr SJ (1990) Planning agroforestry Interventions in extension projects. Agrofor Syst 11:23–44

Nair PKR (1993) An introduction to agroforestry. Kluwer, The Hague

Ndour B, Force JE, McLaughlin WJ (1992) Using the Delphi method for determining criteria in agroforestry research planning in developing countries. Agrofor Syst 19:119–129

Pattanayak SK, Mercer DE, Sills E, Yang J-C (2003) Taking stock of agroforestry adoption studies. Agrofor Syst 57:173–186

Pinners E, Balasubramanian V (1991) Use of the iterative diagnosis diagnosis and design approach in the development of suitable agroforestry systems for a target area. Agrofor Syst 15:183–201

Raintree JB (1983) Guidelines for agroforestry diagnosis and design. ICRAF Working papers

Raintree JB (1987a) The state-of-the art of agroforestry diagnosis and design. Agrofor Syst 5:219–250

Raintree JB (1987b) D & D user's manual: An Introduction to Agroforestry Diagnosis and Design. ICRAF, Nairobi

Rocheleau DE (1985a) Criteria for Re-appraisal and Re-design: Intra-household and between House-hold aspects of FSRE in three Kenyan Agroforestry projects. ICRAF Working Paper 37. ICRAF, Nairobi

Rocheleau D (1985b) Land use planning with rural farm households and communities: participatory agroforestry research. ICRAF Working paper No 36

Rocheleau DE (1989) Agroforestry as popular science: a land user perspective for research and design in rural landscapes. Paper presented to the Annual Meeting of the American Association for the Advancement of Science, January 14–19, 1989. San Francisco

Rocheleau DE, van den Hoek A (1984) The application of ecosystems and landscapes analysis in agroforestry diagnosis and design: a case study from Kathama Sublocation, Machakos, Kenya. ICRAF Working Paper 11. ICRAF, Nairobi

Savory A (1988) Holistic resource management. Island Press, Covelo

Scherr SJ (1989) The diagnosis and design approach to agroforestry planning and implementation: examples from western Kenya. In Budd WW, Duchhart I, Hardesty LH, Steiner F (eds) Planning for agroforestry. Selected contributions from an international symposium held at Washington State University, Pullman, Washington, on 24-27 April 1989, pp 132–160

Schreckenberg K, Boyd C, Degrande A (1999) Methodological guidelines for socio-economic fieldwork at community and household level. Internal project document. FRP project R7190. Overseas Development Institute, London

Shaner WW, Philip PE, Schmel WR (1982) Farming systems research and development. Westview press, Boulder

Sirrine D (2008) Agroforestry, soils, and food security: interdisciplinary on-farm research linking sustainability and livelihoods. Doctoral dissertation, Department of Environmental Studies. University of California, Santa Cruz

Sirrine D, Shennan C, Sirrine JR (2010) Comparing agroforestry systems' ex ante adoption potential and ex post adoption: on-farm participatory research from southern Malawi. Agrofor Syst 79:253–266

Tchoundjeu Z, Asaah EK, Anegbeh PO, Degrande A, Mbile P, Facheux C, Tsobeng A, Atangana AR, Ngo Mpeck ML, Simons AJ (2006) Putting participatory domestication into practice in West and Central Africa. Forests, Trees and Livelihoods 16:53–69

Chapter 14
Experimental Design in Agroforestry

Abstract Agroforestry research is theme-based, methodological, or experimental in nature. Research is carried out at the organizational level, and can be classified according to the site of research as either *on-station* or *on-farm*. Agroforestry research experiments follow the principles established by Fisher (1947): randomization, replication and blocking or local control. The varied agroforestry research categories and experimental designs are not different from those found in agriculture. The difference between agricultural and agroforestry research is that agroforestry systems are more complex. Particularities of agroforestry experiments include: (i) the presence of multiple components in the system and the treatments applied to each component or the entire system will have a significant impact on the necessary plot size; (ii) the natural longevity of trees and the area on which they extend influence complicates the design of an experimental unit as well as the experimental design itself; and (iii) the difficulty in finding homogeneous sites, along with the big plot size necessary for agroforestry experiments, makes it difficult to delimit blocks that are as homogenous as desired. On-going field-based agroforestry experiments in the tropics include screening and selection tests, system management and component trials, studies on the interaction of components, prototype assessment tests, testing and evaluation of germplasm and provenances, testing of the development and fine-tuning of protocols for the vegetative propagation of agroforestry species, characterization of phenotypic variation of candidate species for domestication and selection of "plus trees" (i.e., trees with desirable characteristics for domestication) for improvement, development of methods for quantification of carbon stored by agroforestry systems, and trials for soil conservation.

14.1 Agroforestry Research

Research can be defined as any investigation or experimentation in order to produce or develop knowledge, discover or interpret facts, review accepted theories in the light of new facts, or practical application of theories or rules. It is obvious that research in agroforestry encompasses many perspectives, owing to the complexity of agroforestry systems. The multiple dimensions and perspectives of agroforestry research are summarized in Table 14.1. Agroforestry research can be done at the organizational level, or based on themes. Agroforestry research can be done at the

A. Atangana et al., *Tropical Agroforestry*, DOI 10.1007/978-94-007-7723-1_14,

Table 14.1 Perspectives of agroforestry research

	Category/Operational unit	Example/Type of research
Organizational level	Ecosystem	Agroforestry-systems design
	Farm/Plot	Field experiments
	Component	MPT evaluation
	Cellular/Molecular	Biotechnology
Stage of technology generation	Exploratory	Survey
	Component/system management	Plant-plant interactions
	Prototype	Alley cropping
Subject	Biophysical	Soil productivity
	Social, economic, political	Economic evaluation
Nature	Methodological	D&D, statistical methods
	Experimental	Plant- and soil management
Application of results	Basic	DNA, N_2-fixation process
	Applied	MPT management
	Strategic	Genotype
	Adaptive	Soil-erosion control
Nature of questions addressed	What	The result of growing crops near tree rows
	Why	Why it happens?
	How	How it happens?
Place of research	On-station	On research stations
	On-farm	On farmers' fields

D&D Diagnosis and design; *MPT* Multipurpose tree

cellular or molecular level (biotechnological applications), at the genotype level, at the farm level, at the ecosystem level or at the landscape level. Theme-based agroforestry research is mostly focused on the investigation of biophysical, socio-economic, or policy questions. Biophysical research involves the investigation of biology and systems management, their effect on soil and other environmental factors, and ways of handling components and systems for better results. Socio-economic research relates to the social acceptability and economic/financial benefits of agroforestry. Policy research focuses on political issues related to the implementation of agroforestry policy and regulation.

Investigations in agroforestry can be divided into specific steps in the process of technology generation. For instance, the pre-diagnosis and diagnosis portions of D&D involve exploratory research or surveys, followed by trials of the management of system components, such as planting methods and the arrangement of plants. Results from these trials are used to develop technology prototypes consisting of research and synthesis products, whose practical performance has not yet been tested. An example of technology that has proven effective in agroforestry with regard to vegetative and massive propagation of agroforestry species is the non-mist poly-propagator, designed by Leakey et al. (1990). For more details on vegetative propagation of agroforestry species, please see Chap. 6.

Agroforestry research is methodological or experimental. Methodological research includes the development of methodologies to be applied in specific areas or subjects, whereas experimental research involves testing hypotheses. Sanchez

(1995) reported on some of the hypotheses in agroforestry that need to be or have been tested. Emphasis should be placed on the elaboration of hypotheses that are innovative and rooted in theory. Scientific hypotheses are intended to explain why or how systems work, and should have testable alternatives. Agroforestry research should be move away from descriptive studies towards having strong inferences, and apply the following steps as detailed by Platt (1964):

1. Devise alternative hypotheses
2. Devise one or several crucial experiments with alternative possible outcomes, each of which should exclude one or more hypotheses
3. Carry out the experiment so as to get accurate results
4. Recycle the procedure, making subhypotheses or sequential hypotheses to define the possibilities that remain.

Agroforestry is a complex discipline, and many factors interact to produce variations in observed patterns and processes in the system. Quinn and Dunham (1983) pointed out that "the objective of investigation is not to determine the single cause of a pattern, as no such cause exists, but rather to assign relative importance to the contributions of, and interactions between, a number of processes, all known or reasonably suspected of operating to some degree." Care should be taken when elaborating a hypothesis and designing experiments in agroforestry to ensure that the various causes of the investigated phenomena have been taken into account. However, hypothesis-driven research could be categorized as a fundamental research method.

Fundamental research and applied research are the two research methods most often used in agroforestry. Fundamental research, which aims to advance knowledge in the field, investigates processes and mechanisms. The results from fundamental research should be widely applicable and have large-scale consequences. Applied research is the application of research results to address specific problems. As pointed out by Nair (1993), two categories of applied research exist in agricultural sciences: the strategic and the adaptive. Strategic research is the innovative application of the results of fundamental research to address medium to long-term problems. An example is the development of alley cropping technology. Adaptive research is the development of technologies in specific locations to address problems of immediate concern, for example, the use of alley cropping for soil conservation in a specific location.

Fundamental research and applied research are also characterized by the nature of the issues addressed ("what", "why" and "how" research types) (Nair 1993). "What" type of research is mostly observational, whereas the "why" type of research aims at understanding why the observed behavior happen. The results of "what" type of experiments are strongly site-specific, and research is of an applied nature. "Why" type of research results are applicable on a larger-scale. "How" type of research investigates how a given phenomenon happens. As "why" and "how" research types tend to be fundamental in nature, it is often difficult to distinguish between these two types of research.

Agroforestry research can also be classified according to the site of the study area, as either *on-station* or *on-farm*. *On-station* indicates that the research is conducted on station, in a controlled area. *On-farm* research is conducted outside of a controlled area, or in a farmer's plantations, with or without the involvement of the farmer or the owner of the land. These two types of agroforestry research have been abandoned since the late 1990s in favor of participatory research, and research and development methods. Research and development, commonly called R&D, has been widely used in the 2000s, and refers to creative work undertaken on a systematic basis to increase the knowledge of humans, culture and society, and use of this stock of knowledge to devise new applications. R&D has an economic approach, because the consumer or user is emphasized. The aim is to more effectively produce what already sells, or has an established market opportunity. This approach is used in the participatory domestication of high-value multi-purpose indigenous tree species, a farmer-driven and market-oriented practice (Simons and Leakey 2004).

The varied agroforestry research categories are not so different from those found in agriculture. What differentiates agroforestry from agricultural research is that agroforestry systems are more complex than those of agriculture and forestry. Agroforestry systems harbor many species including trees, shrubs and herbaceous plants, and often combine different components. In addition, the exploitation of ecological and economic interactions between the components of different combinations is of primary importance in agroforestry. The outputs of agroforestry systems are more numerous, and these systems offer a higher degree of land sustainability than monocrop fields. Agroforestry is considered by farmers and other agricultural stakeholders as a promising system of land management for difficult or fragile ecosystems not suited for conventional agriculture due to technical and social factors.

Donors will support research that can produce results of immediate practical application in the shortest time possible with the least costs. Therefore, applied research involving field trials, mostly carried out in relatively small units, is the preferred form of agroforestry research. However, the ecological approach to research is increasingly being adopted through the integration of domesticated or semi-domesticated trees into farms and landscapes. The objective is to maintain the ecological functions of watersheds using agroforestry practices.

14.2 Experimentation in the Field

In 1947, Fischer pioneered field experimentation, and established its basic principles. Three research procedures are considered cardinal for any field experiment: **randomization**, **replication** and **blocking** or **local control**. **Randomization** is the process of randomly allocating treatments to experimental units, which are the basic objects on which the experiment is conducted, or the smallest units on which treatments are applied. The aim is to reduce or eliminate the inherent effects of uncontrolled factors on the results that may occur in the plot. A bias is an influence, condition, or set of conditions that can affect the results of research and prevent their spread. Any bias on the part of the researcher in the assignment of treatments to plots should be avoided. Randomization is the first postulate of the analysis of data variance. **Replication** refers to the process of repeating the same treatment in several plots. It allows the average response of the same treatment over different plots to be obtained, giving a better sense of the typical response than would be obtained from the response of a single plot. By observing several plots, the researcher can also estimate the variability between plots, which is important for quantifying the reliability of the results through **statistical analysis. Local control** is the process of reducing the variability in the experimental material and plots to ensure that the experimental unit is as homogeneous as possible. One way to do this is to make blocks by the grouping of plots in relatively homogeneous units, and repeating the treatments in each block throughout the experiment. Blocks should be set up taking into account the existing sources of variation in the site where the experiment is conducted. Other control methods include the selection of homogeneous plots for the experiment, the use of seeds and other planting materials of uniform quality, and standardization of management procedures as well as observations and measurements, unless they are to be conditioned differently by the experimental treatments.

Before engaging in field experiments, researchers should be aware of all experimental procedures by referring to manuals and other information sources. Several books contain information on agricultural and biological experimentation. Useful references include Gomez and Gomez (1984), Steel et al. (1997), and Montgomery (2012). As a rule of thumb, researchers should always discuss research objectives and experimental design with a statistician prior to field implementation, because several principles other than the basic ones listed above must be considered when designing a rigorous agroforestry experiment.

14.3 Particularities of Agroforestry Experiments

Agroforestry experiments are uniquely complex due to several factors. First, agroforestry involves the presence of multiple components, including agricultural crops, animals, shrubs and trees, and the treatments applied to each area, or the entire system, as well as the space required to accommodate the woody perennials, all have a significant impact on plot size. Second, the natural longevity of the trees and the area which they influence complicates the design of an experimental unit, the experiment itself, and sampling. Finally, to overcome problems related to soil variability, agroforestry experiments must be established on marginal sites representative of areas that are potential targets of intervention, such as sloping land. The difficulty in finding marginal homogeneous sites, along with the large size of plots required in agroforestry experiments makes it difficult to demarcate blocks that are as homogenous as desired by the researchers. This is one of the biggest issues in agroforestry research. Agroforestry research uses biometrical tools developed for agriculture or ecology studies. Agroforestry researchers should strive to develop acceptable and suitable experimentation and statistical analyses tools specific for agroforestry, that take into account all of its specificities.

A few researchers, namely Huxley (1987, 1990), Rao and Roger (1990), Rao et al. (1991), Mead (1991), and Rao and Coe (1992), have addressed issues related to specific sites in general terms, and suggested some general recommendations.

14.3.1 Size and Arrangement of the Plot

Many factors influence the size of the plot required for an experiment in agroforestry. Factors include the subject of investigation (e.g., tree selection trials), the characteristics of the study area (sloping land or not), the type and nature of measurements to be made, the expected life of trees and their ultimate height, the variability of the site, and the chosen experimental design. Tree selection trials require smaller plot sizes than experiments where specific agroforestry technologies are tested. A tree selection can be done on a plot of 20–30 m^2, while a technology experiment might need a plot ranging from 50–200 m^2. A large tree canopy may influence adjacent crops, hence the need for a guard zone to reduce this influence, further increasing needed plot size. An example of a guard zone is given by Rao and Roger (1990) in an alley cropping study. Alley spacing of 2, 4, and 6 m could be used to create a plot 12 m wide with 3, 4 and 7 hedgerows. A 4 m guard zone could be used between plots, increasing the net plot width to 12 m.

Sites most often display large variations over a small area, including changes in soil fertility, topography, microclimate and past management of the land. Blocking is one of the available means to control site variations. As a block is a subset of homogeneous experimental units, and each block must be as uniform as possible. This is very difficult to accomplish in agroforestry because of the number of components and their complexity, spatial variation of soil fertility, and differences in past management of the land. Small plots can be used to overcome site variability effects, but this can cause excessive use of experimental space for borders. In addition, some types of treatments or experimental designs require large parcels of land, such as studies on soil conservation.

Experimental design type has a major influence of the arrangement of plots. Many experimental designs exist for agricultural experiments, depending on the objective of the study, field conditions, and other variable environment factors (Gomez and Gomez 1984). Plots on terraces or sloping land should be long and linear, and on sloping lands, parcels must be perpendicular to the direction of the slope or the strongest slope.

14.3.2 Experimental Designs

Experimental design refers to the way the various treatments are allocated among the plots. There are several types of experimental designs, ranging from completely randomized designs (CRD; Fig. 14.1) and randomized complete block designs (RCBD; Fig. 14.2), to Latin square designs (Fig. 14.3), balanced (Fig. 14.4) and partially balanced lattice designs, split-plot, split-split-plot (Fig. 14.5), split-block, strip-split plot (Fig. 14.6), or group balanced block (Fig. 14.7) designs (Gomez and Gomez 1984; Steel et al. 1997; Kirk 2009; Montgomery 2012). Experimental designs can

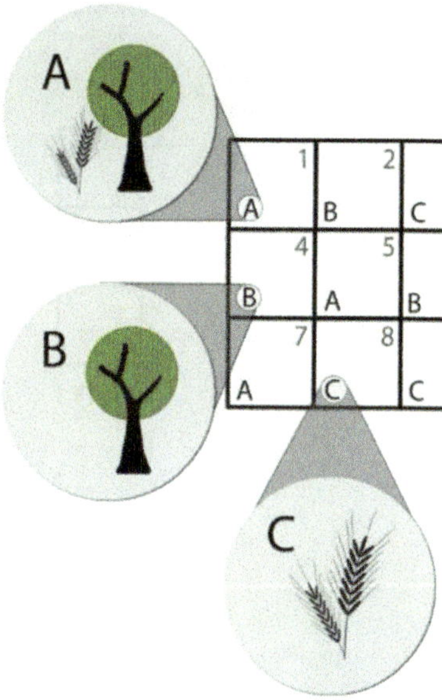

Fig. 14.1 A sample layout of completely randomized design involving 3 treatments (i.e., trees mixed with crops (**a**), trees (**b**) and monocropping (**c**))

also be classified as factorial (i.e., testing more than one factor; Fig. 14.9) or simple (i.e., testing one factor) experiments, or repeatedly measured plans which are split-plot designs in which levels of one or several factors cannot be assigned randomly within one experimental unit (e.g., time-measured experiments) (Steel et al. 1997; Ngo Mpeck et al. 2009).

Completely randomized designs are best suited to lab experiments, where there are little or no external sources of variation. The RCBD is widely used in field experiments, and is far better suited to field studies. The RCBD is characterized by the imposition of a restriction on randomization: the treatments are randomized within each block, allowing some control of the variation. Blocks are established perpendicular to the direction of the variation, and variations should be minimal within the blocks and maximal between the blocks. This design is not suitable when the number of treatments to be used is large, or when considerable variability exists within each block. An RCBD would not be suitable for a study evaluating multipurpose trees, or factorial experiments with three or more factors at each level. Studies in which the number of plots in a block is less than the total number of treatments should use incomplete block designs, like lattice (Fig. 14.4) or "confounded" designs (Montgomery 2012). However, the advice of a statistician is required before such a design is implemented.

Split-plot designs or drawer plans are technically not designs at all. They are a special case of incomplete block design, frequently used for factorial experiments. The design is composed of main plots and subplots in which one or several levels of a factor are applied. If there is a third factor, the design is referred to as a split-split plot. Randomization is done at each level, and the main plots can be arranged in various ways (completely random, complete blocks, Latin square, etc.). Treatments that require more controls are put in subplots, creating "hidden repeti-

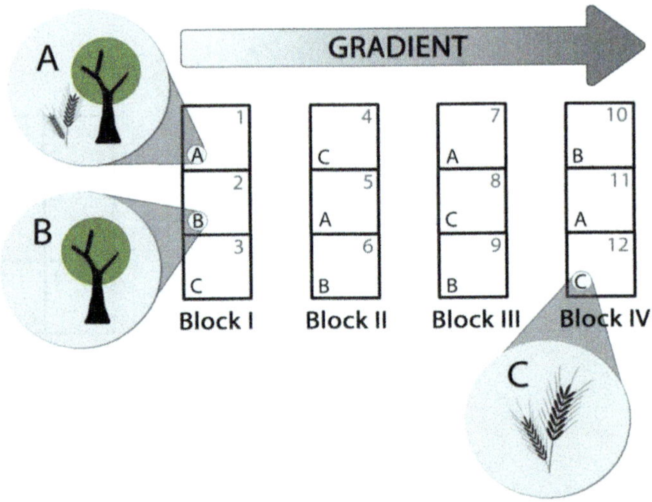

Fig. 14.2 A sample layout of randomized complete block design with 4 replicates (blocks) involving 3 treatments (i.e., trees mixed with crop (**a**), trees (**b**) and monocropping (**c**)); the gradient illustrates the direction of variation

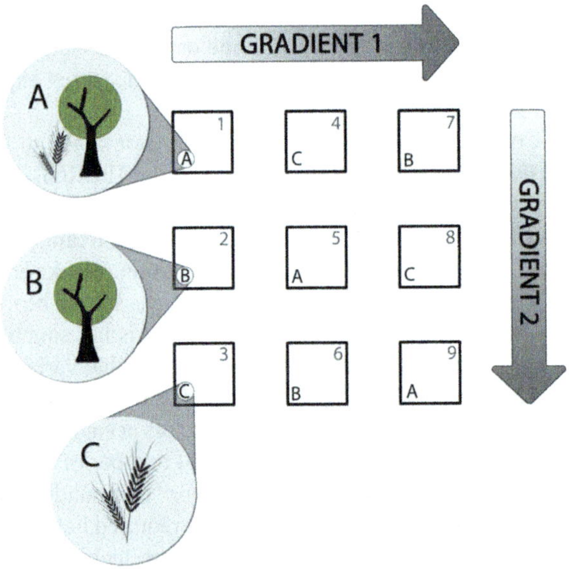

Fig. 14.3 A sample layout of a Latin square design involving 3 treatments (i.e., trees mixed with crop (**a**), trees (**b**) and monocropping (**c**)); the gradients illustrate two perpendicular directions of variation

tions". In alley cropping experiments or in experiments testing species and pruning management, species can be allocated to main plots, and frequency and pruning height allocated to subplots. One common situation where split-plot designs are useful is when there is a soil gradient, such as a slope or fertility difference in one direction. Split-plot designs are particularly useful if treatments are susceptible

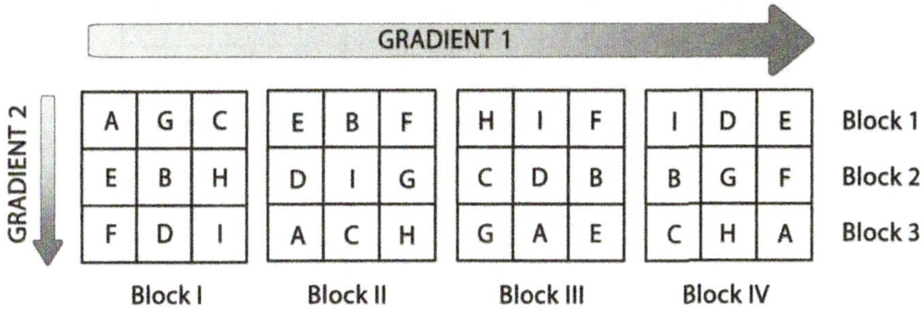

Fig. 14.4 A sample layout of a balanced lattice design involving 6 treatments (A, B, C, D, E, F, G, H and I); the gradients illustrate the directions of variations

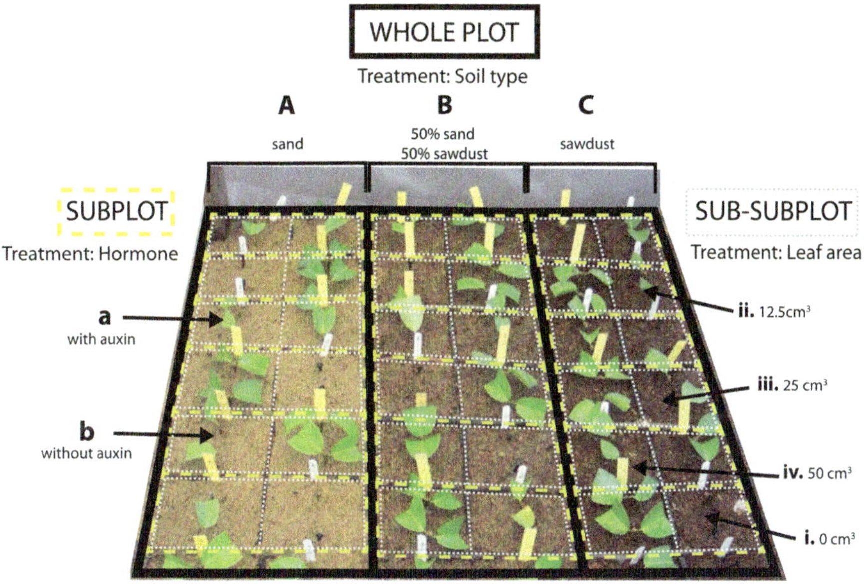

Fig. 14.5 Layout of a split-split-plot design for an experiment that investigated the effects of rooting medium (sand, 50:50 sand/sawdust and sawdust; whole plot), surface area (0, 12.5, 25 and 50 cm^2) and application of auxin on the rooting of *Allanblackia floribunda* leafy stem cuttings (Atangana et al. 2006)

Fig. 14.6 A sample layout of one replicate of a strip-plot design with cropping system at the whole plot level and amount of nitrogen supply at the strip plot level

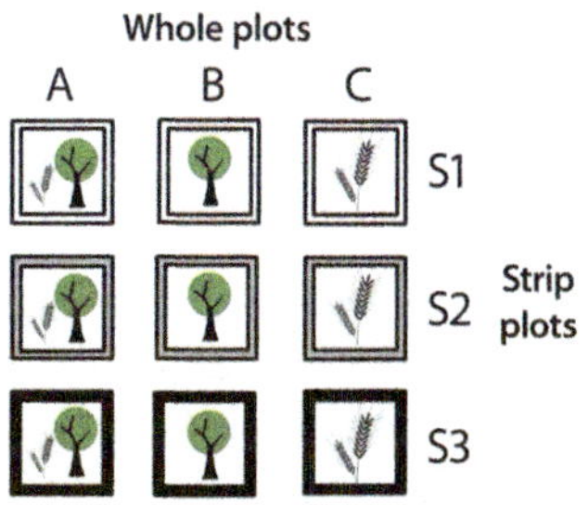

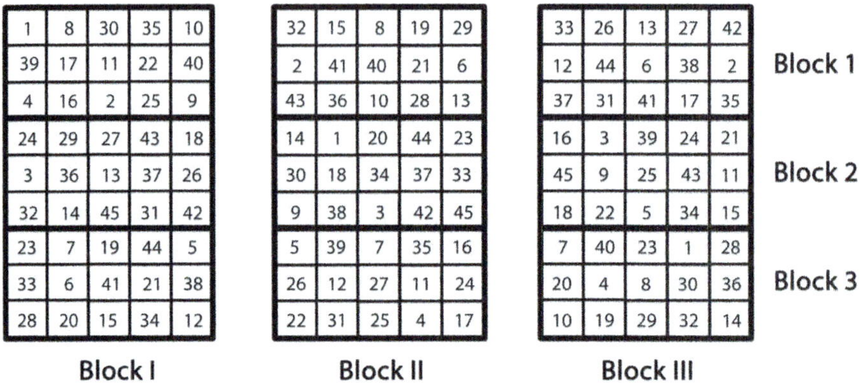

Fig. 14.7 A sample layout of a group balanced block design involving 45 agroforestry shrub species divided into 3 groups, each consisting of 15 shrub-crop distances

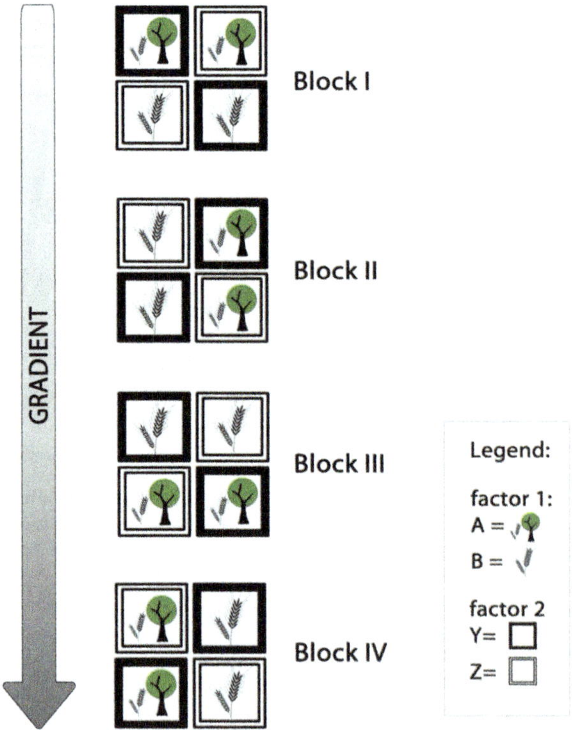

Fig. 14.8 A sample layout of a 2 × 2 factorial complete design experiment; the gradient illustrates the direction of variation

to neighborhood effects, such as in irrigation experiments, or in experiments involving tree species with different growth habits, especially tree height. In most split-plot experiments, it is possible that the treatments in the main plots are less precisely comparable than those in the subplots. In split-split-plots, treatments in sub-subplots are more comparable than those in subplots.

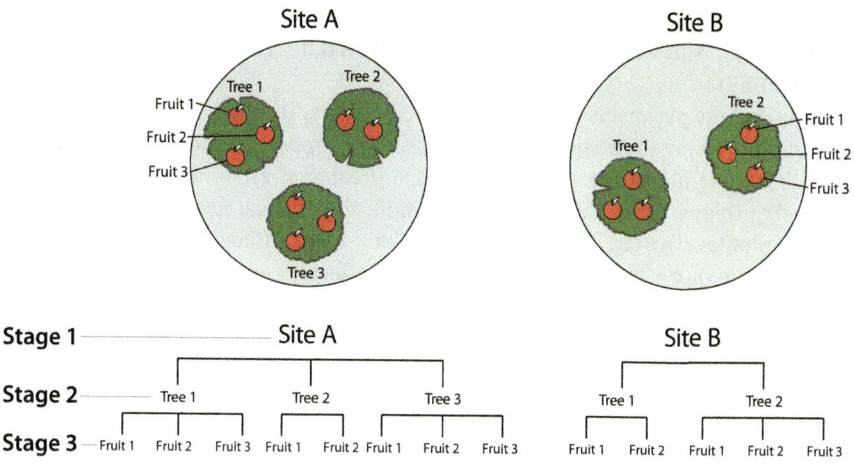

Fig. 14.9 A sample layout of nested design for a field survey aiming at characterizing phenotypic variation in fruit straits of tree species in wild stands

Box 14.1 Example of analysis of data from a split-plot design in agroforestry

Alley cropping experiment testing the effects of agroforestry species (main plots) and pruning frequency (sub-plots) on maize production

Statistical analysis of data is performed following the mathematical model below:

Y=(overall mean)+(block effect)+(species effect)+(species × block effect)+(frequency effect)+(frequency × species effect)+(frequency × block effect)+(frequency × species × block effect)+experimental error, where y is the average value for the dependent variable.

Split-block designs are a type of drawer designs, similar to split-plot, but where there is a constraint on the randomization of subplots. Plots are randomized within blocks, and the subplots constitute bands within the main plots. Split-block designs grant additional flexibility to deal with certain experimental situations where the application of experimental treatments to small units is impractical, and allow maximum precision for the estimation of interactions. Unfortunately, the design can lead to a loss of important information concerning the main factor, and interactions between factors must be present. In addition, statistical analysis of the data obtained is more complex than for ordinary RCBD. Another type of drawer design is strip-plot design (Fig. 14.7). The design is specially adapted for experiments testing two factors, when these factors require large plots as is the case for agroforestry experiments, and when the interaction between the two factors must be measured more accurately than the effects of each factor separately (Fig. 14.8).

Box 14.2 Mathematical model for analysis of data from Split-block design (Fig. 14.7)
Data from experiments that are performed using split-block designs are analyzed following the statistical model, assuming replicates (blocks):

Y = (overall mean) + (block effect) + (treatment effect at whole plot level) + (block × treatment effect at the whole plot) + (treatment effect at the strip plot level) + (treatment at the strip plot × block effects) + (treatments at the strip plot × treatment at the whole plot effects) + error, where y is the average value for the dependent variable.

Non-random systematic designs can be used to overcome problems created by certain circumstances in agroforestry experiments (Huxley 1985). Nair (1993) explained that in alley cropping trials, spacing width can gradually increase through a site rather than having parcels of varying widths located at random. Systematic designs are also suitable for experiments where complete randomization is not possible. However, the statistical analysis of data from systematic designs is complex. The introduction of randomization at some level by the systematic repetition of sets of treatments arranged at different locations will certainly allow a rigorous statistical analysis of the data from such designs.

With the advent of the participatory domestication of multipurpose trees using agroforestry practices, trials are commonly done in the nursery using non-mist poly propagators (Leakey et al. 1990). The most typical designs are RCBD; however, split-plot designs are sometimes used (Atangana et al. 2006). Ecological approaches are used for the introduction of germplasm of trees under domestication into the landscape. Data from experiments that deal with the introduction of trees in agricultural landscapes could be performed effectively using *Tree diversity analysis* software (Kindt and Coe 2005).

It should be noted that agroforestry experiments are unique in terms of sampling, plot selection, management of experimental plots, data collection, analysis and interpretation.

14.4 On-going Field Agroforestry Experiments in the Tropics

Field agroforestry experiments in the tropics are mostly conducted by the World Agroforestry Centre, Winrock International, the International Crops Research Institute for the Semi-Arid Tropics, the International Institute of Tropical Agriculture, the Centre for Tropical Agricultural Research and Higher Education (Costa Rica), national research centers, and various universities. These experiments can be classified into 9 groups:

- Screening and multi-purpose tree and shrub (MPTS) selection tests
- System and component management trials
- Studies on the interaction of components

- Assessment tests of prototypes
- Testing and evaluation of germplasm and provenances
- Tests to develop and fine-tune protocols for the vegetative propagation of agroforestry species
- Characterization of phenotypic variation and selection of plus-trees for improvement
- Development of methods for the quantification of carbon stored by agroforestry systems
- Trials on soil conservation

Soil conservation trials are carried out in areas with high erosion risk such as sloping lands, and mainly consist of an increase in soil cover, the establishment of fences, maintenance or provision of soil organic matter, dikes, embankments, and pipes, and land reclamation. Crop combinations, multi-storey tree gardens, alley cropping, windbreaks and shelterbelts are agroforestry practices that are widely used in the tropics to control soil erosion.

The development of methods to estimate soil carbon sequestration in the tropics constitutes a theme recently highlighted by Nair et al. (2009a). Widely and easily adoptable methodologies are not available for estimating the soil carbon potential under different conditions (Nair et al. 2009b). However, Nair et al. (2009c) conducted a wide-scale study on soil C estimation, and the methods used in this study are promising.

14.4.1 Screening and MPTS Selection Tests

The screening and evaluation of multipurpose trees or shrubs (MPTS), typically through experiments of an exploratory nature, was by far the most common element of agroforestry trials in the 1980s and 1990s. Experimental designs were typically RCBD (Duguma and Tonye 1994; Arredondo et al. 1998) or systematic designs. These experiments are generally designed to screen the germplasm of several promising species, either several species or several varieties of a species. These experiments aim to identify the most promising species and provenances based on early performance in terms of growth, establishment, and other factors, in view of developing specific varieties. A major difficulty encountered in these studies is that the improvement of one trait, such as fruit size, can negatively affect the performance of another important trait, like taste. This problem could be overcome by a thorough screening of wild trees or species for important traits for the improvement of the species (Atangana et al. 2001), and the identification and selection of "ideotypes" (trees or species exhibiting good performances in a set of traits) (Atangana et al. 2002; Leakey et al. 2005). The selected individuals should be evaluated and improved to obtain varieties. A second difficulty is the lack of standard procedures for tree evaluation in agroforestry systems. Traditional forestry research procedures, typically aimed at improving timber characteristics, are not always suitable for multipurpose trees. With the spreading of tree domestication research conducted by

ICRAF in the tropics of Africa, southern Africa, Latin America and Southeast Asia, there is hope that this difficulty will be overcome.

The genetic improvement of multipurpose trees used in agroforestry is underway in several places. Work has been undertaken on *Acacia mangium*, *A. auriculiformis*, *Eucalyptus spp*, *Cordia alliodora*, *Gliricidia*, *Erythrina*, *Sesbania*, *Dalbergia* and *Leucaena leucocephala*. Tree selection is also underway for the improvement of priority species for domestication in the tropics. For details on tree selection in agroforestry tree domestication programs, refer to Chap. 6.

14.4.2 Experiments on Management of Systems and Components

Agroforestry technology improvement is the purpose of experiments on the management of systems and components. These experiments are simultaneously of an experimental and applied nature. Agroforestry research placed more emphasis on alley cropping and soil fertility improvement in the 1980s and 90s, while the domestication of multipurpose indigenous species has been receiving more attention since the end of the 1990s. Examples of system and component management experiments in alley cropping include the application method of the harvested biomass (mulch), the spacing of hedgerows, and the frequency of pruning. Tree domestication examples include the identification of genes and group of genes influencing a desired characteristic of a selected species, the quantification of the effect of the environment on the desired trait, and the effect of trees from improved germplasm on adjacent crops. The characterization of phenotypic variation and the selection of plus-trees, evaluation of provenances, and the development of protocols for rooting and marcotting (air layering) are some other themes under discussion in tree domestication. The non-mist poly propagator (Leakey et al. 1990) is a widely used technology in the domestication of agroforestry species to date. Some work on the development and adaptation of this technology in rural areas (Mbile et al. 2004) falls within component management testing.

Soil conservation trials have also received attention in agroforestry research. These trials aim to identify the best agroforestry practices to control soil erosion, and the best configuration of planting along with management approaches employed in practices such as agroforestry crop combinations or silvopastoral systems (Nair 1993).

14.4.3 Studies on the Interaction of Components

Agroforestry systems have many components, and the relationships between these components need to be quantified and understood (Ong and Huxley 1996). Component interaction studies involve, among other aspects, the sharing of water and resources, and the presence or absence of positive or negative interactions. These studies are primarily of an environmental nature, and have strongly advanced agroforestry soil research. Agroforestry research also use Type I and Type II studies. Type I studies are those where changes in soil properties are tracked over time on the same site. Type II studies involve the sampling of soil from nearby farms or other planting sites with known planting dates at the same time (Sanchez et al. 1985; Sanchez 1987). Type I experiments are preferred, because they can be replicated and characterized. Shading and tree-crop interaction experiments are considered to be type I.

14.4.4 Prototype Evaluations

Prototype evaluations are trials aimed at evaluating specific packages of agroforestry technology under farm conditions. Prototype trials are mostly undertaken either entirely or partially in rural farms or other sites in the field, such as rural nurseries. Prototype evaluations are a good example of the link between research and extension.

14.4.5 Testing and Evaluation of Germplasm and Provenances

These trials assess the performance of individuals or provenances for selection, and also serve as living gene banks. Trials are becoming increasingly common in the domestication of multipurpose tree species. Provenance trials include those of *Adansonia digitata* established in Burkina-Faso, with accessions from Burkina-Faso, Côte d'Ivoire, Senegal, and Kenya (Raebild et al. 2010). Also included are provenance trials of *Parkia biglobosa* conducted in Burkina-Faso (Raebild et al. 2010). Germplasm of *Irvingia gabonensis* and *Irvingia wombolu* were collected in Cameroon, Gabon, and Nigeria, and gene banks established in Cameroon and Nigeria, when ICRAF began its tree domestication program in the 1990s. Provenance-progeny trials also allow the assessment of intra-specific variability, and can be used to obtain estimates of heritability that are useful in breeding programs. More details on provenance trials are given in Chap. 6.

14.4.6 Development and Fine-Tuning of Protocols for Vegetative Propagation of Agroforestry Species

Vegetative propagation using rooting of single-node leafy stem cuttings is a key element in the domestication of multipurpose tree species. The use of non-mist poly propagators (Leakey et al. 1990), which are inexpensive and robust, is promoted in village nurseries. More details on the operation and capacity of these poly propagators are given in Chap. 6. The development of a practical protocol for the rooting of cuttings of an unstudied species involves the identification of an appropriate rooting medium, leaf area and cutting size, and auxin application. Other factors reported to influence rooting of cuttings include the propagation environment, within- and between-shoot factors, pre-severance stock plant nutrient and light environment, stock plant management (pruning, fertilizer use and light management), ontogenic and physiological aging of stock plants, genetic origin, and the interaction of all these factors (Leakey 2004; Atangana et al. 2006; Atangana and Khasa 2008). Air-layering trials mostly investigate the size of the marcott, marcott position, number of marcotts per tree and genetic variation.

14.4.7 Characterization of Phenotypic Variation and Selection of Plus-trees for Improvement

Surveys are carried out in wild stands of priority species for domestication to assess the phenotypic variation in important characteristics for improvement. These studies appeared with the advent of the participatory tree domestication program. Pioneering studies were conducted by Roger Leakey and his colleagues (Leakey et al. 2000, 2002; Atangana et al. 2001). In the humid tropics of Africa, species that have been studied include *Irvingia gabonensis*, *Dacryodes edulis*, *Ricinodendron heudelotii*, and *Allanblackia floribunda* (Leakey et al. 2000, 2002; Ngo Mpeck et al. 2003; Atangana et al. 2001, 2011). Fruit and nut characteristics are most often selected as traits of interest for such studies. The studies usually lead to the selection of individuals to be used in the first-generation breeding population (Atangana et al. 2002, 2011). Common gardens need to be established to confirm the superiority of individuals selected, in order to obtain "elite trees" (i.e., trees that have proven superiority in the desired characteristics for domestication based on genetically controlled traits). Because the phenotype is caused by the interaction of genotype and environment, it is very important to determine the degree of genetic control over the traits of interest, and to be certain that the observed superiority is of genetic origin.

Box 14.3 Mathematical model for the analysis of data from surveys aiming at characterizing the phenotypic variation of fruit traits in trees in wild stands (Fig. 14.9)
Assume *k* fruits (stage 3) from each of j randomly selected trees (stage 2) in *x* sites (stage 1) are measured for phenotypic traits (e.g., fruit mass and size); data analysis will be performed following the statistical model (all effects are random):

y_{xjk} = overall mean + site_x + tree_j nested in site_x + fruit_k nested in tree_j nested in site_x

Bibliography

Arredondo S, Aronson J, Ovalle C, del Pozo A, Avendano J (1998) Screening multipurpose legume trees in central Chile. For Ecol Manag 109:221–229

Atangana AR, Tchoundjeu Z, Asaah EK, Simons AJ, Khasa DP (2006) Domestication of *Allanblackia floribunda*: amenability to vegetative propagation. For Ecol Manag 237:246–251. doi:10.1016/j.foreco.2006.09.081

Atangana AR, Khasa DP (2008) Preliminary survey of clonal variation in rooting of *Allanblackia floribunda* leafy stem cuttings. Can J For Res 38:10–15

Atangana AR, Tchoundjeu Z, Fondoun J-M, Asaah EK, Ndoumbe M, Leakey RRB (2001) Domestication of *Irvingia gabonensis*: 1. Phenotypic variation in fruit traits in 52 trees from two populations in the humid lowlands of Cameroon. Agrofor Syst 53:55–64

Atangana AR, Ukafor V, Anegbeh P, Asaah E, Tchoundjeu Z, Fondoun J-M, Ndoumbe M, Leakey RRB (2002) Domestication of *Irvingia gabonensis*: 2. The selection of multiple traits for potential cultivars from Cameroon and Nigeria. Agrofor Syst 55:221–229

Atangana AR, van der Vlis E, Khasa DP, van Houten D, Beaulieu J, Hendrickx H (2011) Tree-to-tree variation in stearic and oleic acid content in seed fat from *Allanblackia floribunda* from wild stands: potential for tree breeding. Food Chem 126:1579–1585

Duguma B, Tonye J (1994) Screening of multipurpose tree and shrub species for agroforestry in the humid lowlands of Cameroon. For Ecol Manag 64:135–143

Fisher RA (1947) The design of experiments. Oliver and Boyd, Edinburgh

Gomez KA, Gomez A (1984) Statiscal procedures for agricultural research, 2nd edn. Wiley, New York

Huxley PA (1985) Systematic designs for field experimentation for multipurpose trees. Agrofor Syst 3:251–266

Huxley PA (1987) Agroforestry experimentation: separating the wood from the trees? Agrofor Syst 5:251–275

Huxley PA (1990) Experimental agroforestry. In: MacDicken KG, Vergara NT (eds) Agroforestry: classification and Management. Wiley, New York, pp 332–353

Kindt R, Coe R (2005) Tree diversity analysis. A manual and software for common statistical methods for ecological and biodiversity studies. World Agroforestry Centre, Nairobi

Kirk RE (2009) Experimental design. In: Millsap R, Maydeu-Olivares A (eds) Sage handbook of quantitative methods in psychology. Sage, Thousand Oaks, pp 23–45

Leakey RRB (2004) Physiology of vegetative propagation. In: Burley J, Evans J, Youngquist JA (eds) Encyclopaedia of forest sciences. Academic Press, London, pp 1655–1668

Leakey RRB, Mesén JF, Tchoundjeu Z, Longman KA, Dick JMcP, Newton A, Matin A, Grace J, Munro RC, Muthoka PN (1990) Low-technology techniques for the vegetative propagation of tropical trees. Commonw For Rev 69:247–257

Leakey RRB, Fondoun J-M, Atangana A, Tchoundjeu Z (2000) Quantitative descriptors of variation in the fruits and seeds of *Irvingia gabonensis*. Agrofor Syst 50:47–58

Leakey RRB, Atangana AR, Kengni E, Waruhiu AN, Usoro C, Anegbeh PO, Tchoundjeu Z (2002) Domestication of Dacryodes edulis in West and Central Africa: Characterization of genetic variation. Forest Trees Livelihoods 12: 57–71

Leakey RRB, Greenwell P, Hall MN, Atangana AR, Usoro C, Anegbeh PO, Fondoun J-M, Tchoundjeu Z (2005) Domestication of Irvingia gabonensis: 4. Tree-to-tree variation in food-thickening properties and in fat contents of dika nut. Food Chem 90(3): 365–378

Mbile P, Tchoundjeu Z, Degrande A, Avana M-L, Tsobeng A (2004) Non-mist propagation by resource poor, rural farmers of the forest zone of Cameroon: some technology adaptations to enhance practice. For Trees Livelihoods 14:43–52

Mead R (1991) Designing experiments for agroforestry research. In: Avery M, Cannell MGR, Ong CK (eds) Biophysical research for Asian Agroforestry. Winrock International, Arlington. pp 3–20

Montgomery DC (2012) Design and analysis of experiments, 8th edn. Wiley, New York

Nair PKR (1993) An introduction to agroforestry. Kluwer, The Hague

Nair PKR, Kumar BM, Nair VD (2009a) Agroforestry as a strategy for carbon sequestration. J Plant Nutr Soil Sci 172(1):10–23

Nair PKR, Nair VD, Kumar BM, Haile SG (2009b) Soil carbon sequestration in tropical agroforestry systems: a feasibility appraisal. Env Sci Policy 12(8):1099–1111

Nair PK, Nair VD, Gama-Rodrigues EF, Garcia R, Haile SG, Howlett DS, Mohan Kumar B, Mosquera-Losada MR, Saha SK, Takimoto ANG, Tonucci RG (2009c) Soil carbon in agroforestry systems: An unexplored treasure? Nature Precedings hdl:10101/npre.2009.4061.1

Ngo Mpeck ML, Asaah E, Tchoundjeu Z, Atangana AR (2003) Strategies for the domestication of *Ricinodendron heudelotii*: Evaluation of variability in natural populations from Cameroon. Food Agric Env 1:257–262
Ngo Mpeck M-L, Atangana AR, Omgba AM (2009) Temporal variation in the rooting of *Annickia chlorantha*. New For 38:273–283
Ong CK, Huxley PA (1996) Tree-crop interactions: a physiological approach. ICRAF, C.A.B. International, p 386
Platt JR (1964) Strong inference. Science 146:347–353
Quinn JF, Dunham AE (1983) On hypothesis testing in ecology and evolution. Am Nat 122(5):602–617
Raebild A, Søndergård Larsen A, Svejgaard Jensen J, Moussa Ouedraogo, De Groote S, Van Damme P, Bayala J, Ousmane Diallo B, Kalinganire A, Dahl Kjaer E (2010) Advances in domestication of indigenous fruit trees in the West African Sahel. New Forests doi:101007/s11056–010–9237–5
Rao MR, Coe R (1992) Evaluating the results of agroforestry research. Agrofor Today 4(1):4–9
Rao MR, Roger JH (1990) Discovering the hard facts. Part 2. Agronomic considerations. Agrofor Today 2(2):11–15
Rao MR, Sharma MM, Ong CK (1991) A tree/crop interface design and its use for evaluating the potential of hedgerow intercropping. Agrofor Syst 13:143–158
Sanchez PA (1987) Soil productivity and sustainability in agroforestry systems. In: Steppler HA, Nair PKR (eds) Agroforestry: a decade of development. ICRAF, Nairobi. pp 205–223
Sanchez PA (1995) Science in agroforestry. Agrofor Syst 30:5–55
Sanchez PA, Palm CA, Davey CB, Szott LT, Russel CE (1985) Trees as soil improvers? In: Cannell MGR, Jackson JE (eds) Attributes of Trees as Crop Plants. Institute of Terrestrial ecology, Huntington. pp 327–358
Simons AJ, Leakey RRB (2004) Tree domestication in tropical agroforestry. Agrofor Syst 61:-167–181
Steel GD, Torrie JH, Dickey DA (1997) Principles and procedures of statistics: a biometrical approach, 3rd edn. The McGraw-Hill Companies, Inc., New York

Chapter 15
On-Farm Agroforestry Research

Abstract Agroforestry research aims to develop new practices or technologies and facilitate their adoption by farmers. Research on farms is done to address adoption issues, by allowing farmers to evaluate or adapt the technology being researched or tested in local conditions. On-farm research, which involves the participation of farmers in the technology generation process, is often used, depending on the objectives of the research, the nature of the questions being investigated, and local conditions, when conducting participatory farming systems, "farmer first", or "augmented designs" research. Participatory farming systems research is the generation of technologies by involving technology users in the planning and evaluation process. *Farmer first* research is when farmers conduct the experimentation and analysis on agricultural innovation themselves, with facilitation and support from scientists. Lastly, *augmented designs* research involves experimental designs that take into account the participation of farmers, thereby allowing the estimation of farmer-augmented defined treatments. Wider dissemination of agroforestry technologies has been largely done through scaling-up approaches, some of which have facilitated the widespread adoption of fodder shrubs among smallholders in the highlands of Kenya, Rwanda, Uganda, and Tanzania. Techniques that have been successfully used include: (i) the collaboration of researchers with large Non-Governmental Organizations (NGO) that promote fodder shrubs; (ii) large scale dissemination of facilitators who train trainers and provide support to extension workers; (iii) farmer-to-farmer dissemination; (iv) private seed vendors; and (v) civil society campaigns bringing together different stakeholders to train farmers by farmers. In the humid tropics of West and Central Africa, grassroots organizations were also effective in disseminating agroforestry innovations.

15.1 Introduction

The development of innovations in agroforestry and the adoption of new technologies by farmers are complex. The participation of farmers is important to ensure the success of the development of new agroforestry technologies. Indeed, farmers have an advantage over research institutions when managing complex experiments specific to a site (Chambers 1989). However, agroforestry technology transfer over the past several years has been a one-way process. Agroforestry research has

A. Atangana et al., *Tropical Agroforestry*, DOI 10.1007/978-94-007-7723-1_15,

borrowed from conventional agricultural research, consisting of technology developed by researchers, taught or demonstrated by extension workers, and adopted by farmers. This approach most often yielded poor farmer adoption rates, especially among poorer farmers. On-farm research (OFR) is a response to the realization that involving farmers in the technology generation process is important. Enabling farmers to partner with researchers in the design, testing, evaluation and modification of new practices helps to:

- Reduce the time needed to introduce new practices to farmers, because some degree of dissemination takes place during testing
- Increase adoption rates

On-farm research involves conducting research or technology testing in the field so that farmers can evaluate or even adapt the technology to suit local conditions (Williams et al. 2001), in order to ultimately facilitate the adoption of the technology by as many farmers as possible. On-farm research uses a combination of methods to increase the potential of technology adoption by farmers. In a book edited by Franzel and Scherr (2002), case studies on the adoption of agroforestry technologies in eastern Zambia and western and central Kenya are examined, and perspectives for future research recommended (Franzel et al. 2002a,b; Swinkels et al. 2002a,b). Methods for assessing the adoption potential of agroforestry technology are also described in the aforementioned book (Franzel et al. 2002c). An important aspect of on-farm research is the development of methods for the spread of efficient agroforestry technologies. One such method is the application of scaling-up approaches in tree domestication (Degrande et al. 2012).

15.2 Rationale and Approaches of On-farm Research

On-farm research is very important in agroforestry, which is a more developmental-driven applied science. On-farm research, which has four key elements (i.e., the farmer, the farmer's land, the farmer's involvement, and the farmer's environment) can take various forms from experimental (i.e., researcher managed, researcher/farmer managed, and farmer managed) to developmental (i.e., exploratory, intermediate, pilot project) (Attah-Krah and Francis 1987), depending on the degree of farmer involvement in the management and technology assessment (Franzel et al. 2002c), and may be undertaken at any stage of research (Scherr 1991). Most often, OFR is implemented after on-station research; but it can also be done simultaneously. The OFR is used in participatory, farming systems (i.e., the generation of technologies is made by involving technology users in the planning and evaluation process; Damhofer et al. 2012), "farmer first", and "augmented designs" research. Participatory research involves scientists and farmers cooperating in the research process. Farming systems research and extension uses OFR as a primary component of its methodology for evaluating alternative technology in the field (Byerlee et al. 1982). "Farmer first" research (i.e., farmers conduct the experimentation and analysis on agricultural innovation themselves, with facilitation and support from scientists) (Chambers 1989; Toulmin and Chambers 1990) and "Augmented Designs for Participatory Research" (i.e., experimental designs that take into account the participation of farmers, thereby allowing the estimation of farmer-augmented defined treatments) (Pinney 1991) are other forms of OFR. The OFR model chosen depends on the objectives of research, the nature of the questions investigated, and local conditions. Scherr (1991) suggested that additional objectives of on-farm research include studying agroforestry systems, learning from farmer's knowledge, accessing real field conditions, or eliciting farmer evaluation of new technology.

Investigations and exploratory surveys to gain an in-depth understanding of farmers' needs are usually the first step of an OFR process. Researchers gather information on the perceptions of farmers on the existing land-use systems practices, as well as diagnose current land-use problems, allowing them to identify the key elements likely to affect the social acceptability of any new technology. This information is integrated into the technology process design. The designed technology is then tested under real field conditions to obtain information about its performance and its acceptability to farmers. In this phase, a strong parallel program of technology-testing with farmers is imperative (Scherr 1991). Once the technology is adopted or modified by farmers, it can then be further disseminated by scaling up. Two types of information can be gained through this process: quantitative data on the biophysical and economic performances of the technology, and qualitative information about the technology, generally related to its reliability and acceptance by farmers. The difference between the two types of information is the level of control over the experiment by stakeholders.

This chapter describes on-farm agroforestry research and presents some methods and techniques that are commonly used.

15.3 Characteristics, Objectives and Types of On-farm Agroforestry Research

The rise of agroforestry coincided with the development of farmer-centered approaches in international agricultural research, facilitating close contact between agroforestry researchers and farmers. Although the knowledge of the scientific community on various agroforestry systems is progressively increasing, the rich experience of farmers, who have developed effective agroforestry systems under various conditions for ages, remains underutilized, calling for increased use of participatory approaches in agroforestry research.

On-farm research is research done on farmer managed units where agroforestry technology would be applicable, such as farms or forests. Scherr (1991) identified characteristics that are unique to agroforestry and that tend to make on-farm research more important in agroforestry than in conventional agricultural research. These characteristics are:

- Poor understanding of farmers' agroforestry strategies
- Lack of empirical information on agroforestry systems
- The complexity and variability in terms of objectives, components, and management and ecological interactions of agroforestry systems
- Lack of locally adapted agroforestry technologies
- Lack of data for research and agroforestry technology development
- The longer technology cycle and period required for farmer and researcher assessment

Altering project objectives to suit farmers' selection criteria and experimental designs to on-farm limitations is an important aspect of agroforestry research (Berr 1991). On-farm research is relevant to agroforestry research, because solving problems at the farm level is already an important goal of agroforestry research.

On-farm experimentation objectives include (Franzel et al. 2002c):

- To permit farmers and researchers to work as partners in the technology development process. Farmers' innovations may serve as a basis for future research or for modifying recommendations;
- Evaluation of the biophysical performance of a practice under a wider range of conditions than is available on station;
- To obtain realistic input-output data for financial analyses, as analyses conducted using on-station data can be unreliable, owing to upward bias of yield response, and unrepresentativeness of estimates of labor use by station laborers on small plots;
- To provide important diagnostic information about farmers' problems, even if diagnostic surveys and appraisals have already been conducted.

On-farm trials are of biophysical, financial and social natures, concerned with the biophysical performance of the technology, profitability of return estimates and feasibility and acceptability regarding farmers' adaptation of the technology (Franzel et al. 2002c). On-farm trials can be classified in the following categories (Franzel et al. 2002c):

- Type 1 trials are designed and managed by researchers. Trials are carried out in farmers' fields, and results are more representative of farmers' biophysical conditions than on-station trials, although the trials are simply on-station trials transferred to farmers' fields.
- Type 2 trials, also known as "researcher-designed farmer-managed trials", are trials designed by researchers, and managed by farmers. Researchers and farmers collaborate in the design and implementation of the trial, and researchers consult farmers on the design of the trial. Each farmer agrees to follow the same prototype, but manages it in his/her own manner. Type 2 trials may provide less reliable biophysical data than type 1 trials, but are more useful for the collection of data on labor and financial returns.
- Type 3 trials are designed and managed by farmers. Farmers are informed about new practices, and can experiment with the new practices as they wish. They are not obliged to plant in plots or include control plots. Researchers simply monitor the farmer's experiments. Farmers' assessments on feasibility, profitability and acceptability of the technology are more accurate in these trials.

Table 15.1 The suitability of types 1, 2 and 3 trials for meeting specific objectives[a] (Franzel et al. 2002c)

Information types	Type 1 trial: researcher-designed and managed	Type 2 trial: researcher-designed, farmer managed	Type 3 trial: farmer-designed and managed
Biophysical response	High	Medium	Low
Profitability	Low	High	Low
Acceptability			
Feasibility	Low	Medium	High
Farmers' assessment of a particular prototype[b]	Low	High	Medium
Farmers' assessment of a particular practice	Low	Medium	High
Other			
Identifying farmers' innovations	None	Low	High
Determining boundary conditions	High	High	High

[a] The suitability involves both the appropriateness of the trial for collecting the information and the ease with which the information can be collected

[b] By particular prototype, we mean a practice that is carefully defined. For example, a prototype of improved fallows would include specific management options such as species, time of planting, spacing, etc.

The appropriateness of the types of trials to use to gather reliable information and the approaches with which information can be collected are illustrated in Table 15.1. The three types of trials presented here constitute points along a continuum, and may not necessarily be carried out sequentially (Franzel et al. 2002c).

15.4 Some Methods Used in On-farm Research

On-farm research consists of testing the solutions to problems that have been identified during the diagnosis. The methodology used to conduct on-farm research is heavily dependent on the study objectives and local conditions. The guidelines for experimental research in agroforestry are developed in Chap. 14.

15.4.1 Stability Analyses

On-farm agroforestry experiments are very complex because they most often have unbalanced designs. Agroforestry experiments commonly have between- or within-site variation in the replicates of the number of different treatments, in farmers' management practices, species' composition and farmers' preferences. Stability analysis addresses such problems, as it assesses the relative performance of a technology under a range of variable environments. Stability analysis is fitted in mixed models, which were developed to handle statistical analysis of unbalanced design experiments where treatments are fixed factors and environment is random (Montgomery 2012). Stability analysis is based on the variance-covariance structures that serve as stability model for treatments within environments. Environment is considered as a repeated measure factor, and statistical analysis tests whether the model obeys the assumptions of compound symmetry. Next, covariance structures are modeled using a variety of random structures. Indices of goodness-of-fit (e.g., Log likelihood, Akaike Information Criterion (AIC), Bayesian Information Criterion) are then used to select the best-fitting variance-covariance structure (the smaller, the better) (Littell et al. 2006).

Stability analysis was used to compare improved resource management, agricultural technology practices, and farmers' practices for the cultivation of several rice and wheat varieties in on-farm trials conducted at 6 locations in the Indo-Gangetic plain of Bangladesh, India and Nepal (Raman et al. 2011). The Shuka's stability variance (i.e., stability model) component and AMMI (Additive Main Effects and Multiplicative Interaction) model provided lowest AIC values for rice and wheat grain yield. Fitting these stability models in the mixed model ANOVA structures, identified reduced-till transplanted rice and reduced-till-drill rice-seeded wheat, and using a power seeder with integrated crops as best for grain yield and stability.

Stability analyses have proven to be efficient in the evaluation of field trials at different locations under various conditions (Russell 1991; Hildebrand et al. 1993; Raman et al. 2011). Modified stability analysis (MSA) allows the evaluation of data from a wide range of environments using farmers' and researchers' criteria. This procedure borrows from the genotype x environment analysis method (Hildebrand 1990), and simultaneously handles biophysical variables and data from changes in management options. Because MSA is based on regression analysis, experimental designs can be restricted to a single block of treatments per field. In that case, however, environmental indices are based on an average crop harvest, not the actual environmental data, and are only one indicator of the differences between environments (Russell 1991). The main drawback of this method is that when a number of interacting factors affect the environment, it is difficult to link these indices to the specific characteristics in a usable manner. The MSA is still effectively used in agroforestry.

15.4.2 Assessment of Agroforestry Adoption Potential

The assessment of agroforestry adoption potential is performed using various methods, such as those used by Franzel et al. (2002 b,c) in Kenya and Zambia. As pointed out by these authors, the assessment of adoption potential "requires an understanding of biophysical performances under farmers' conditions, profitability from the farmers' perspective and its acceptability to farmers (in terms of both their assessment of its value and their willingness and capacity to access the information and resources necessary to manage it well)" (Franzel et al. 2002c). Simply put, "on-farm

Table 15.2 A framework for assessing the adoption potential of an agroforestry practice. (Franzel et al. 2002c)

Factors	Key questions
Biophysical performance	Does the practice result in higher yields, lower variability and provide the anticipated environmental services? Are these biophysically sustainable?
Profitability	Is the practice profitable to the farmer as compared with alternative practices? How variable are returns, and how sensitive are returns to changes in key parameters?
Feasibility and acceptability	Do farmers have the required information and resources, and are they willing and able to establish and manage the practice and cope with problems that occur? Do farmers perceive significant advances using the technology?
Boundary conditions	Under what circumstances (e.g., biophysical, household, community characteristics, market conditions) is the practice likely to be profitable, feasible and acceptable to farmers?
Lessons for effective dissemination: extension policy	What does farmer feedback suggest that will help interest farmers in the practice? What type of extension support do they need most? What types of changes in institutional arrangements, public investments or market conditions would enhance the adoption potential of the practice?
Feedback to research and extension	How do farmers modify the practice? What does farmer experience suggest are research priorities for further modification and development of the practice?

research should determine the biophysical and socioeconomic circumstances under which the practice is likely to be profitable, feasible and acceptable to farmers, and thus adopted by them" (Franzel et al. 2002c). Therefore, the assessment of adoption potential needs to embrace biophysical, socioeconomic and cultural factors (Opio et al. 2001). Mercer (2004) reviewed the theoretical and empirical literature that has developed since the early 1990s on agroforestry adoption from a variety of perspectives, and identified needed future research. Much progress has been made, especially in using binary choice regression models (logistic regression models) to assess influences of farm and household characteristics on adoption and in developing ex-ante participatory, on-farm research methods for analyzing the potential adoptability of agroforestry innovations. Other multivariate techniques have also been used to identify adoption factors in agroforestry such as step-wise discriminant analysis (Opio et al. 2001), multiple regression and logistic model analyses (McGinty et al 2008). Mercer (2004) also identified additional research-needs including developing a better understanding of the role of risk and uncertainty, insights into how and why farmers adapt and modify adopted systems, factors influencing the intensity of adoption, village-level and spatial analyses of adoption, the impacts of infectious diseases such as AIDS and malaria on adoption, and the temporal path of adoption.

The framework for assessing adoption potential in agroforestry is given in Table 15.2.

The assessment of the biophysical performance of an agroforestry technology is done by measuring the products and services of the technology and comparing them against other options (Franzel et al. 2002c). These analyses aim to determine how the differences between options change with the environment (Hildebrand and Russel 1996; Franzel et al. 2002c). Environmental variables should be measured before undertaking an analysis, as the environment is used as a continuous variable during the assessment. The main weakness of this method is the lack of certainty that the comparisons are representative of the choices that farmers would make (Franzel et al. 2002c).

Franzel et al. (2002c) reported three issues that affect the profitability of agroforestry practices. First, the financial net benefits of the new practice must be greater than those of alternative practices, including those that farmers currently use. Second, the variability of benefits across farmers and seasons must be assessed, as well as the sensitivity of results to changes in key parameters. Finally, benefits need to be appraised relative to the total household income in order to assess their potential for contributing to improved household welfare. Financial analysis indicates the profitability of a practice using data from costs and returns (refer to Chap. 16 of this book) from all profitability cases.

The acceptability of a technology should be assessed from the perspective of the farmers (Scherr 1995; Franzel et al. 2002c). Feasibility depends on the availability of resources such as land, labor and capital, the cognizance of the required information and skills, and the ability to cope with any problems that arise. Tools used for assessing feasibility include resource budgets, which are used to compare the availability of resources with the needs of the practice, and the evaluation of the biophysical performance of the technology, as planted and managed by the farmer (Franzel et al. 2002c).

The acceptability of a practice depends on profitability, feasibility, risk, compatibility with farmers' values and farmers' valuation of benefits, cultural and eating habits, and several other factors. Monitoring whether or not farmers continue to use or expand the technology, and whether or not neighboring farmers take it up, is the best way of ascertaining acceptability (Franzel et al. 2002c). However, farmers may expand practices, not because they like it, but because they expect to receive other benefits such as free inputs. Other farmers may wish to expand the practice, but lack access to inputs or critical information (Franzel et al. 2002c). Surveys and farmer workshops may be used to identify the views of farmers on the technology, as well as what farmers perceive as advantages and disadvantages.

15.5 Widespread Dissemination of Agroforestry Technologies

The dissemination of an agroforestry technology follows adoption by farmers in sites where on-farm research has been implemented. Scaling-up approaches are used by ICRAF and its partners to disseminate efficient agroforestry technologies

(Franzel and Wambugu 2007; Wambugu et al. 2011; Degrande et al. 2012). Scaling-up approaches have facilitated the widespread adoption of fodder shrubs including *Calliandra calothyrsus*, *Sesbania sesban* and *Leucaena leucocephala* among smallholders in central and western Kenya, Rwanda, Uganda and Tanzania. Approaches for dissemination of agroforestry innovation in this case included: (i) the collaboration of researchers with large NGOs that promote fodder shrubs, (ii) widespread dissemination of information through facilitators who train trainers and provide support to extension workers, (iii) farmer-to-farmer dissemination, (iv) private seed vendors, and (v) civil society campaigns bringing together different stakeholders to train farmers (Franzel and Wambugu 2007). These practices have facilitated the adoption of studied fodder shrubs by about 200,000 farmers in the highlands of East Africa in 10 years. Other factors associated with the successful widespread adoption of fodder shrubs in East Africa using scaling-up approaches, included the deliberate involvement of fodder technology champions, collective actions in community mobilization and project implementation, pluralistic extension approaches, sustainable germplasm supply systems, broader partnerships, and civil society campaigns (Wambugu et al. 2011). In another study on dissemination pathways for scaling-up agroforestry technologies in western Tanzania, Matata et al. (2013) examined the effectiveness of different dissemination pathways, the government agricultural extension services, farmer trainers and traditional leaders for scaling-up of agroforestry technologies. Seventy-six percent of the farmers interviewed felt that farmer trainers were more effective in providing extension training on improved fallows than other channels (e.g., government extension service and traditional leaders). About 92% of the samples in the western zone of Tanzania were familiar with the concept of improved fallow technology. Modes of communication and effectiveness of agroforestry extension in eastern India has been reviewed by Glendimning et al (2001). Once again in this study, the decision to adopt agroforestry was found to be determined by the "farmers attitude" towards agroforestry, which in turn, was shaped by information received through farmer-to-farmer and farmer-to-extension contact. The mode of communication was important and, to be effective, needs to be customized for each target group.

Scaling-up approaches were also used for the widespread dissemination of tree domestication techniques in the humid tropics of west and central Africa (Degrande et al. 2012). The World Agroforestry Centre has been collaborating with grassroots organizations, local NGOs or community-based relay organizations (ROs) that promote agroforestry techniques in rural resource centers. Rural resource centers are "places where agroforestry techniques are practiced and where farmers can come for information, training and experimentation" (Degrande et al. 2012). A study on the performance of these ROs (Degrande et al. 2012) showed that ROs are efficient in the widespread dissemination of agroforestry innovations, and therefore, should be further involved in the diffusion process of efficient agroforestry techniques.

Bibliography

Attah-Krah AN, Francis PA (1987) The role of on-farm trials in the evaluation of composite technologies: The case of alley farming in southern Nigeria. Agric Syst 23:133–152

Berr J (1991) Implementing on-farm agroforestry research: Lessons learned in Talamanca, Costa Rica. Agrofor Syst 15:229–243

Byerlee DL, Harrington L, Winkelman DL (1982) Farming system research: Issue in research and technology design. Am J Agric Econ 64:897–904

Chambers R (1989) Farmers-first: A practical paradigm for the third-world agriculture. In: Altieri M, Hecht S (eds) Agroecology and Small Farm Development. CRC Press, Boca Rato. Pp. 237–244

Damhofer I, Gibbon D, Dedieu B (2012) Farming systems research into the 12st century: the new paradigm. Springer. p. 490

Degrande A, Franzel S, Siohdjie Yeptiep Y, Asaah E, Tsobeng A, Tchoundjeu Z (2012) Effectiveness of grassroot organisations in the dissemination of agroforestry innovations. In: Kaonga M (ed) Agroforestry for Biodiversity and Ecosystem Services—Science and Practice. InTech, Rijeka, Croatia. Pp. 142–164

Franzel SC, Scherr SJ (2002) Trees on the farm: assessing the adoption potential of agroforestry in Africa. CABI Publishing, Wallingford. p. 207

Franzel S, Wambugu C (2007) The uptake of fodder shrubs among stakeholders in East Africa: Key elements that facilitate widespread adoption. In: Hare MD, Wongpichet K (eds) Forages: A pathway to prosperity for smallholder farmers. Proceedings of an International Symposium, Faculty of Agriculture, Ubon Ratcha thani University, Thailand. Pp. 203–222

Franzel S, Arimi HK, Murithi FM (2002a) *Calliandra calothyrsus*: Assessing the early stages of adoption of a fodder shrub in the highlands of Central kenya. In: Franzel SC (ed) Trees on the farm: Assessing the adoption potential of agroforestry in Africa. CABI Publishing, Wallingford, pp. 125–143

Franzel S, Phiri D, Kwesiga F (2002b) Assessing the adoption potential of improved fallows in Eastern Zambia. In: Franzel SC, Scherr SJ (eds) Trees on the farm: Assessing the adoption potential of agroforestry in Africa. CABI Publishing, Wallingford. Pp. 37–61

Franzel SC, Scherr SJ, Coe R, Cooper PJM, Place F (2002c) Methods for assessing agroforestry adoption potential. In: Franzel SC, Scherr SJ (eds) Trees on the farm: Assessing the adoption potential of agroforestry in Africa. CABI Publishing, Wallingford. Pp. 11–35

Hildebrand PE (1990) Modified stability analysis and on-farm research to breed specific adaptability for ecological diversity. In: Kang MS (ed) Genotype-by-Environment Interaction and Plant Breeding. Louisiana State University, Baton Rouge. Pp. 169–180

Hildebrand PE, Russell JT (1996) Adaptability analysis. A method for the design, analysis and interpretation of on-farm research-extension. Iowa State University Press, Iowa

Hildebrand PE, Singh BK, Bellows BC, Campbell EP, Jama BA (1993) Farming systems research for agroforestry extension. Agrofor Syst 23:219–237

Glendimning A, Mahapatra A, Mitchell CP (2001) Modes of communication and effectiveness of agroforestry extension in eastern India. Human Ecol 29:283–305

Littell RC, Milliken GA, Stroup WW, Wolfinger RD, Schlabenberger O (2006) SAS for mixed models, 2nd edition. SAS Institute, Cary. pp. 814

Matata PZ, Masolwa LW, Ruvuga S, Bagarama FM (2013) Dissemination pathways for scaling-up agroforestry technologies in western Tanzania. J Agric Exten Rural Develop 5:31–36

Mercer DE (2004) Adoption of Agroforestry innovations in the tropics: A review. Agrofor Syst 61–62:311–328

McGinty MM, Swisher ME, Alavalapati J (2008) Agroforestry adoption and maintenance: self-efficacy, attitudes and socio-economic factors. Agrofor Syst 73:99–108

Montgomery DC (2012) Design and analysis of experiments. 8th ed. John Wiley, New York

Pinney A (1991) Farmer-augmented designs for participatory agroforestry research. Agrofor Syst 15:275–289

Raman A, Ladha JK, Kumar V, Sharma S, Piepho HP (2011) Stability analysis of farmers participatory trials for conservation agriculture using mixed models. Field Crop Resear 121:450–459

Opio C, Jacob N, Khasa DP (2001) Factors affecting a sheep vegetation management system in British Columbia, Canada. Agrofor Syst 53:305–312

Russell JT (1991) Yield and yield stability of pure and mixed stands of sorghum [*Sorghum bicolor* (L.) Moench] Varieties in North Cameroon. Ph.D. dissertation, University of Florida, Gainesville

Scherr SJ (1991) On-farm research: The challenges of agroforestry. Agrofor Syst 15:95–110

Scherr SJ (1995) Economic analysis of agroforestry systems: the farmers' perspective. In: Current D, Lutz F, Scherr SJ (eds) Costs, benefits and farmer adoption of agroforestry: Project experience in Central America and the Caribbean. World Bank Environment Paper No 14. The World Bank, Washington, DC, pp 28–44

Swinkels RA, Franzel S, Shepherd KD, Ohlsson E, Ndufa JK (2002a) The adoption potential of short rotation improved tree fallows: Evidence from western Kenya. In: Franzel SC, Scherr SJ (eds) Trees on the farm: Assessing the adoption potential of agroforestry in Africa. CABI Publishing, Wallingford. Pp. 65–86

Swinkels RA, Shepherd KD, Franzel F, Nduka JK, Ohlsson E, Sjogren H (2002b) Assessing the adoption potential of hedgerow intercropping for improving soil fertility, Western Kenya. In: Scherr SJ (eds) Trees on the farm: Assessing the adoption potential of agroforestry in Africa. CABI Publishing, Wallingford. Pp. 89–110

Toulmin C, Chambers R (1990) Farmer-first: achieving sustainable dryland development in Africa. International institute for environment and development. Issue Paper No 19

Wambugu C, Place F, Franzel S (2011) Research, development and scaling-up the adoption of fodder shrub innovations in East Africa. Int J Agric Sustain 9(1):100–109

Williams SE, van Noordwijk M, Penot E, Healey JR, Sinclair FL, Wibawa G (2001) On-farm evaluation of the establishment of clonal rubber in multistrata agroforests in Jambi, Indonesia. Agrofor Syst 53:227–237

Part IV
Economic and Cultural Considerations in Tropical Agroforestry

Chapter 16
Economics in Agroforestry

Abstract Carrying out an economic analysis is a crucial step in the process of any agroforestry project selection. It allows us to determine the profitability of a certain enterprise or technology compared to alternatives. The types of economic analysis used in agroforestry include cost-benefit analysis (CBA), environmental economics, farm budgeting, risk assessment, econometrics and optimization, policy analysis matrix modeling, and regional economic modeling. Cost-benefit analysis is the most commonly used economic analysis in agroforestry. Sensitivity and risk-benefit analyses are methodologies used in agroforestry project analysis to overcome issues related to the variation of prices of inputs and outputs, and the risk from any change or development. Project analysis is based on a long-term analytical approach, whether or not this approach is implemented, and includes discounting, farm budgeting, and evaluation criteria. Criteria used to evaluate agroforestry projects include the net present value (NPV; at a given discount rate), benefit-cost ratio (BC), internal rate of return (IRR), return on labor, return on land, opportunity cost, payback period, land expectation value, farm models, and the equivalent periodic value. The most commonly used criteria in CBA are NPV, IRR and BC. Key elements used in agroforestry production that provide data for economic analysis are capital goods, land, and labor.

16.1 Introduction

In the past, agroforestry research focused more on biophysical and physical aspects affecting productivity. Today, there is a growing interest in the marketing of agroforestry products as well as the socioeconomic factors influencing their adoption (Russell and Franzel 2004). A growing number of studies are being conducted on the costs and benefits of agroforestry systems under real farm conditions (Swinkels and Scherr 1991; Scherr 1992; Price 1995; Swinkels et al. 1997). The economic, ecological, and cultural constraints and opportunities for the expansion of agroforestry practices have also been studied by several authors (e.g., as reviewed by Steffan-Dewenter et al. 2007).

Economic evaluations of several land-use systems are crucially needed, because of the increasing pressure on land, and diversified social perceptions and preferences. There is also a great need for the economic evaluation of the integration of germplasm of species under domestication into existing land-use systems.

A. Atangana et al., *Tropical Agroforestry*, DOI 10.1007/978-94-007-7723-1_16,

Assessing the profitability of agroforestry systems and technologies is a very complex task. Agroforestry simultaneously encompasses economic, social and environmental benefits, and its payback period is usually longer than that of conventional agricultural production systems. Quantifying social and environmental benefits of agroforestry is inherently very difficult. Prior to embarking on a discussion of economics in agroforestry, a distinction must first be made between economic and financial analyses. Financial analysis examines the feasibility of a business from the point of view of private individuals, whereas economic analysis focuses on the desirability of an enterprise from the perspective of society as a whole. This is an important distinction. A proposed project that pays an expected profit to individual farmers could have a negative effect on the economy due to heavy subsidies. More specifically, a financial assessment of profitability of a farm that uses subsidized fertilizer should include in its calculations only the fertilizer costs paid by the farmers. An economic analysis should, however, also include expenses incurred by the government paying subsidies in its calculations of the total cost of fertilizers for the society. In situations where the prices set by the market do not accurately reflect the true social value of an input or an output due to tariffs, price controls or other influences, economic analysis can use shadow prices for a more accurate assessment of the true costs and benefits. Shadow prices can be particularly valuable in the adjustment of price distortions in land and labor, or to assess unmarked environmental effects.

Of the types of analysis used in the agroforestry sector, cost-benefit analysis is the most common (Montambault and Alavalapati 2005). Environmental economics, farm budgeting, risk assessment, and econometrics and optimization are gaining prominence. Several other methodologies of economic analysis are used in agricultural and forest economics (Alavalapati et al. 2004a). The policy analysis matrix model is similar to farm budget models, but also includes market failures and assesses their impact on profitability at a farm or regional level. Regional economic modeling is used to estimate changes in income, employment, and price levels at regional or national scales. The evaluation criteria used in agroforestry project analysis include the Net Present Value (NPV, at a given discount rate), cost-benefit ratio, internal rate of return (IRR), return on labor, return on land, opportunity cost, payback period, land expectation value (Nair 1993; Franzel et al. 2001; Bertomeu 2006; Guo et al. 2006), and farm models. Those are used to assess the increase in annual income (Franzel 2004) and the equivalent periodic value (EPV), which is the periodic and constant value necessary for the payment of an amount equal to the net present value of the current investment option along its useful life (Rezende and Oliveira 1993). With the advent of tree domestication, the market analysis of agroforestry products is becoming more and more important in agroforestry (Russell and Franzel 2004).

16.2 Principles of Economic Analyses

The decision to invest in a business implies the exclusion of alternatives. Economic analysis demonstrates the potential impact and trade-offs that result from an alternative course of action. Economic analyses can reduce the possibility of non-optimal choices by providing a common monetary standard among alternative measures, which, ideally, reflects the true value and scarcity of resources. Economic analysis can help to ensure that the implementation of agroforestry practices will increase crop productivity, income, health, well-being, nutritional and shelter security, social welfare, and diversify, and sustain income and food sources for individual farmers, compared to more traditional land uses. From a macroeconomic perspective, economic analysis can determine the expected economic consequences of an enterprise, and whether net contributions to society justify the expenses (Mercer et al. 2005). Environmental costs and benefits also need to be taken into account (Alavalapati et al. 2004b). Agroforestry not only generates financial gains, but also environmental benefits to the society. Other valuable non-monetary benefits of agroforestry to farmers are more difficult to quantify. Agroforestry contributes to local healthcare through the provision of bark, leaves, roots and liana of medicinal value. When conducting any agroforestry economic analysis, one should keep in mind complementarity and long-term characteristics, illustratable through the use of graphs showing production prospects.

As several different combinations of annual and perennial crops are possible in agroforestry systems, identifying the economic feasibility of specific combinations requires information on, and the evaluation of, relevant agricultural inputs and outputs (Hoekstra 1987; Alavalapati and Mercer 2004). Economic analysis should determine whether and how the realities of a fluctuating market can be integrated with the physical production possibilities in order to result in sustainable and diversified production, as well as the welfare and optimal income of the farmer.

16.3 Analysis of Projects in Agroforestry

Agroforestry project analyses are based on long-term analytical approaches, whether these approaches are implemented or not; approaches such as discount rate, farm budgeting, and other evaluation criteria. Agroforestry project analyses help select appropriate evaluation criteria and reasonable discount rates, identify costs and benefits of a given project and quantify these costs in on-farm budget, implement calculations under selected evaluation criteria, and formulate conclusions on the basis of viability risk.

Discount rate selection in agroforestry, especially for subsistence farmers, needs careful consideration. Applying the same discount rate for all farmers should be avoided if there is clear indication that differences exist due to factors such as wealth, age, risk aversion, or the climate conditions of the area where the farmers live (Hoekstra 1985). Important points to consider in any agroforestry economic analysis, raised by Hoekstra (1990), are:

- Does the system being evaluated make efficient use of its resources?
- In case the project begins, can the project be completed with the funds available?
- Is the system technically feasible under the prevailing employment constraints?

Table 16.1 Financial performance of *Jhum* (shifting cultivation) and agroforestry systems (Rasul and Thapa 2006)

	Jhum	Agroforestry	Relative (*jhum* = 100)
Gross benefits (Tk.[a] ha^{-1})	80,806[b]	100,783[c]	125
Total costs (Tk. ha^{-1})	98,279	78, 647	80
Labor costs (Tk. ha^{-1})[d]	85,363 (87)	56,593 (72)	66
Non-labor costs (Tk. ha^{-1})[e]	12,916 (13)	22,052 (28)	171
Financial performance			
Net financial benefits (NPV) (Tk. ha^{-1})			
With opportunity cost of HH[f] labor	−17,473	22,139	–
Without opportunity cost of HH labor	67,890	78,731	116
Initial establishment costs (Tk. ha^{-1})	0	13,390	–
Return to labor (Tk. person-day^{-1})	72	91	126
BC[g] ratio			
With opportunity cost of HH labor	0.82	1.28	156
Without opportunity cost of HH labor	6.26	4.57	73

[a] 1 US$ ~ Taka (Tk.) 57

[b] In a typical *jhum*, several crops, such as rice, cotton, sesame, chili, and different types of vegetables, banana and root crops such as ginger, turmeric, yam, and cassava, are grown together. The gross benefit was calculated based on average yield of each crop multiplying the average farm-gate price

[c] In a typical agroforestry farm, there are annual crops and tree crops. Trees include both fruit and timber species. Average amount of production and farm gate price of respective produces were considered for calculating the gross benefit

[d] Labor cost includes the value of labor spent on site selection, land preparation, and planting, weeding, fertilization, harvesting and transportation of crops

[e] Non-labor cost includes the cost of seeds, seedlings, fertilizers, pesticides and interest on capital

[f] HH = household

[g] BC = benefit-cost ratio

- Is the system economically viable under given capital constraints?
- What are the risks involved in technology introduction?

16.3.1 Evaluations using "with" and "without" approaches

Agroforestry is concerned with long-term sustainability of production. An analytical long-term approach "with or without implementation" is particularly suitable for economic evaluations of agroforestry systems. A "with" or "without" approach takes into account the positive environmental effects of agroforestry, as well as the costs and benefits of introducing an agroforestry practice. An example of such an analysis compared agroforestry practices and shifting cultivation (*Jhum*) in the Chittagong Hill Tracts, Bangladesh (Rasul and Thapa 2006). The cost of soil erosion was estimated using the replacement cost technique between hedonic pricing and changes in productivity (Magrath and Arens 1989; Enters 1998). The analysis indicated that agroforestry systems provide higher profit than *Jhum* (Table 16.1).

Short-term evaluations of agroforestry usually underestimate total benefits, as most of the benefits of agroforestry are seen after a longer period of time than traditional agriculture.

16.3.2 Discounting

The selection and function of a discount rate is one of the most controversial subjects in economic analysis. Simply put, the discount rate is the interest rate paid by the market, or a bank, on invested capital. It allows the determination of the present value of future cash flows. The discount rate is linked to the time value of money or opportunity cost, and is used in economics to factor time into economic calculations. The discount rate represents the difference in value of the money we receive today and money received in the future. Because costs and benefits in agricultural projects do not occur at the same time, the discount rate is used to compare money made at different periods in time in terms of their worth in the present. Costs and benefits are more easily compared if they are incurred in the same year. The discount rate makes it possible to assess the present value of money earned at different periods of time. However, the economic comparisons between alternatives are only viable if the same discount rates were used for all calculations. Also, the specific choice of a discount rate can lead to the intentional or unintentional manipulation of the results of an analysis. In well-performing economic systems, the discount factor approaches the interest rate that is applied by the banks. The discount factor is very important in operating accounts because it allows the calculation of the profitability of an enterprise over several years. The longer the period of time over which monetary value is being discounted, the more effect the choice of discount rate has on the NPV calculations. This is very important in agroforestry, which involves a sizeable initial investment (tree planting), and benefit flows that come several years later. For poor farmers with no access to credit, or other sources of income, this poses a problem because their discount rate is very high. Even if the profits generated later would be high, the enterprise may not be profitable. This is the reason why it is important to integrate some short-term benefits in agroforestry.

Inflation is the drop in the value of wealth over time. Each unit of currency buys fewer and fewer goods as time passes, due to a rise in prices. Discounting attempts to overcome this issue. The use of a discount rate implies that a unit of currency today does not possess the same intrinsic value several years later. Additionally, the discount rate represents the opportunity cost of foregone alternative investments with a real positive interest rate. Discounting also takes into account the difference in perceived usefulness of one dollar or euro to a poor man compared to a rich man. The decrease of marginal utility of that dollar to the welfare of an individual when financial status changes must be taken into account in monetary comparisons. Discounting is also affected by time preference relative to consumption. Most people have a positive rate of time preference, and would rather spend their money than save it.

The discount rate is the reverse of compound interest. Compound interest is calculated by incorporating the interest earned in the previous period into the inter-

Table 16.2 Mean and variance ($ ha^{-1}) of the simulated probability distributions for the net present values and risk (%) associated with three monocrops and six agroforestry systems in Talamanca, Costa Rica (Ramirez et al. 2001)

Cropping system	Discount rate (*i*)											
	i=4%			*i*=6%			*i*=8%			*i*=12%		
	Mean	Var[a]	Risk	Mean	Var[a]	Risk	Mean	Var[a]	Risk	Mean	Var[a]	Risk
Cacao monocrop[b]	12128	22.4	75.5	10494	18.5	85.9	9194	15.2	91.2	6220	10.4	100.0
Plantain monocrop	11301	0.4	100.0	10391	0.4	100.0	9620	0.3	100.0	8056	0.3	100.0
Laurel monocrop	3096	0.2	100.0	22.78	0.1	100.0	1627	0.1	100.0	225	0.1	100.0
1 cacao: 1 plantain	16670	11.0	12.7	14519	8.4	51.4	12687	6.4	79.8	8766	3.9	100.0
2 cacao: 1 plantain	18455	19.7	4.2	16083	14.7	31.0	14034	11.5	61.3	10091	7.6	93.2
3 cacao: 1 plantain	19267	24.6	3.0	16675	18.0	27.2	14603	13.9	53.6	10598	8.7	89.1
1 cacao: 2 plantain	15480	4.7	19.8	13540	3.7	70.2	11928	2.0	91.3	8464	0.9	100.0
1 cacao: 3 plantain	15132	3.0	21.2	13282	2.4	69.3	11698	1.7	95.1	8128	1.0	100.0
Additive treatment	12450	10.1	78.8	10796	7.6	90.3	9372	5.7	95.9	6543	2.9	100.0

[a] Divided by 10,000.
[b] Cacao (*Theobroma cacao*); plantain (*Musa* AAB); laurel (*Cordial alliodora*)

est earning capital. In essence, the interest previously earned also earns interest. The current value of a certain amount of money in the future is given by the formula:

Current value $V_0 = V_t / (1+i)^t$, for compound interest rate

Where V_t is the amount of money at year *t*, and *i* the compound interest rate.

or

Current value $V_0 = V_t / (1+it)$, for simple interest rate

Where V_t is the amount of money at year *t*, and *i* the simple interest rate.

As Nair (1993) pointed out, the use of a higher discount rate will favor projects that generate significant benefits in the early years, and where the majority of the costs are incurred later. An example is intensive agriculture on fragile tropical soils (Wannawong et al. 1991). At higher discount rates the net present value declines and the risk increases (Tables 16.2 and 16.3). Similarly, if the discount rate increases, the weight attached to long-term effects decreases. The use of a higher discount rate is likely to underestimate the value of long-term environmental costs and benefits, which characterize agroforestry projects.

Interest rates are determined by the markets, and strongly influence the selection of discount rates by companies. In the public sector, social discount rates are established by national authorities, and are used to evaluate projects, especially during periods of high interest rates (Gregory 1987). For instance, in India, the social discount rate for land-use projects is a function of the elasticity of social marginal

Table 16.3 Sensitivity of Eucalyptus-cassava system and a cassava monocrop to different discount rates based on mean *Vigna radiata* yields in Thailand (Wannawong et al. 1991)

System	Net present value (baht rai^{-1})			
	5%	7%	9%	11%
Eucalyptus and cassava	4228.5	4052.0	3887.0	3771.0
Cassava	3009.4	2872.2	2744.0	2623.9

1 Thai baht ~0.0335 USD (in February 2013)
1 rai ~0.16 ha

utility of consumption and the growth rate or per capita real consumption (Sharma et al. 1991). The discount rate should reflect market interest rate and the need for social development. Society's perspective of long-term development is determined in relation to its individual members.

16.3.3 Evaluation Criteria

Among the various tools used to assess investments providing services over several years, Cost-Benefit Analysis (CBA) is the most common. The CBA allows the comparison of the long-term benefits and costs of proposed projects. The most commonly used criteria in CBA are net present value, the benefit-cost ratio, and the internal rate of return, which is the interest rate that needs to be earned in a project to cancel out the effects of discounting and earn a profit that is proportional to the risk of the project. The private sector, the World Bank, FAO, and governments all frequently use net present value (NPV) and internal rate of return (IRR) (Gregory 1987). Usually, private organizations calculate the NPV of an investment for a range of possible interest rates, and then determine the IRR. Governments often use cost-benefit ratios for economic evaluations.

16.3.3.1 Net Present Value

The discounting of all annual costs and benefits of a project or enterprise should be done prior to the calculation of NPV. All costs and profits over the prescribed lifespan of a project or enterprise are first discounted at a prescribed rate. The discounted costs and benefits are then summed as a simple indicator of the long-term value of the project. The NPV is calculated using the following formula (Sang 1988):

$$\mathrm{NPV} = \sum_{t=0}^{n} (B_t - C_t)/(1+r)^t ,$$

where B represents the benefits in year t, C the costs in year t, r the selected discount rate, and n the number of years.

NPV can also be computed using the following formula (Countryman and Murrow 2000):

$$\mathrm{NPV} = \sum_{t=1}^{\infty} \left[R_t + (1+i)^t\right] - \sum_{t=0}^{\infty} \left[C_t + (1+i)^t\right]$$

Where R_t is the revenue (or net cash flow) in period t, C_t the cost in period t, and i the interest rate.

NPV can also be computed using the following formula:

$$NPV = \frac{A_0}{(1+i)^0} + \frac{A_0}{(1+i)^1} + \frac{A_0}{(1+i)^2} + \ldots + \frac{A_n}{(1+i)^n}$$
$$= \sum_{t=0}^{n} \frac{A_t}{(1+i)^t}$$

where A_t is the revenue in period t and *i* the interest rate.

The NPV is the sum of the present values of all individual cash flows. In general, any project with an NPV greater than zero is technically viable, and long-term benefits outweigh the costs. However, the NPV value by itself provides very little information about the scale of capital required. Although a proposed project may have a greater NPV than an alternative, it may also require a greater investment of capital.

The decision rules are the following:

- For independent projects, when NPV > 0, we accept the projects (the present value of future cash inflows is greater than the initial investment cost); when NPV = 0, we are indifferent and when NPV < 0, we reject the projects.
- For mutually exclusive projects, we accept the projects with the highest NPV.

An example of the use of NPV was the assessment of the financial performance of agroforestry practices compared to shifting cultivation or *Jhum* (Rasul and Thapa 2006; ref. Table 16.1). In this study, NPV was used to express return to land. Results indicated that the NPV of *Jhum* is negative, even without taking into account the opportunity costs of household labor. This land-use practice is non-profitable. The NPV of agroforestry was positive both with and without including the opportunity costs of household labor (Rasul and Thapa 2006), indicating that the agroforestry systems were profitable for farmers.

16.3.3.2 Land Expectation Value

Land expectation value (LEV) refers to the present value of the income from an infinite sequence of harvests. The LEV, which indicates the value of bare land in perpetual timber production, was defined by Albuquerque (1993) as the net present value of a reforestation project, given the occurrence of identical forest rotations that are repeated infinitely. The LEV formula depends on whether the land is occupied by crops, trees, or a mixture of trees and crops (Guo et al. 2006). The calculation of LEV can be done using the following Faustmann formula (Faustmann 1849; Klemperer 1996; Guo et al. 2006):

$$\text{LEV}=\left[\sum_{t=0}^{T}(R_t - C_t)^*(1+r)^{T-t} / (1+r)^T - 1\right] - (\text{c}-\text{a})/\text{r}$$

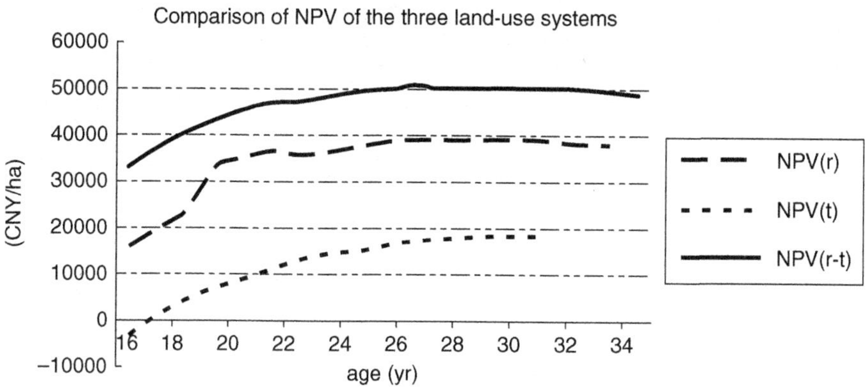

Fig. 16.1 Comparison of net present value (NPV) of the rubber (*r*) monocropping, tea (*t*) monocropping and rubber-tree (*r* - *t*) intercropping in Hainan, China (CNY/ha = Chinese Yuan per ha; 1 US$ = 8.28 CNY, 2001–2002) (Guo et al. 2006)

Where

R_t = revenue in year t
C_t = cost in year t
c = annual cost
a = annual revenue
T = rotation age
r = discount rate

The decision rules are the following:

- For independent projects, when LEV > 0, we accept the projects (the present value of future cash inflows is greater than the initial investment cost); when LEV = 0, we are indifferent and when LEV < 0, we reject the projects.
- For mutually exclusive projects, we accept the projects with the highest LEV.

An illustration of the use of LEV in economic analyses in agroforestry is given by Guo et al. (2006). The authors compared the profitability of rubber and tea monocultures, and rubber-tea intercropping. Results (Fig. 16.1) indicated that rubber-tea intercropping has the highest LEV (NPV_{r-t}), followed by rubber monoculture. The difference between LEV values of rubber-tea intercropping and tea monoculture was over 300,000 Chinese Yuan ha^{-1}, indicating the profitability of agroforestry was far greater than monoculture. At the time of the study, the Chinese Yuan was fixed at a rate of 8.28 Yuan per US dollar. The LEV was also used to estimate the value of bare land if used to grow trees (Bertomeu 2006). The study of Bertomeu (2006) compared the profitability of two maize-agroforestry systems, trees in blocks, and trees in hedgerows, with a maize monocropping in the Philippines. Results indicated that maize monocropping has higher LEV values than maize-tree agroforestry (Tables 16.4 and 16.5). The author speculated that this might be due

Table 16.4 Returns on land and labor of agroforestry with *Gmelina arborea* and maize monocropping over and 8-year tree rotation period (Bertomeu 2006)

System	Maize (t ha^{-1} 9 years^{-1})	Timber (m^3 ha^{-1})	Returns on land: LEV (US$ ha^{-1})		Net returns on labor (US$ work-day^{-1})	
			r=15%	r=20%	r=15%	r=20%
Maize monocropping	70.1	0.0	2,278	1,708	3.8	3.8
Low timber yield:						
Tree hedgerow (1 × 10 m)	12.5	69.1	1,581	1,056	5.7	4.9
Tree block (2 × 2.5 m)	12.9	60.8	1,448	979	4.6	4.0
High timber yield						
Tree hedgerow (1 × 10 m)	12.5	110.6	2,279	1,479	7.4	6.2
Tree block (2 × 2.5 m)	12.9	104.4	2,180	1,422	6.0	5.1

Timber price = PhP4 bf^{-1} or US$ 42.4 m^3; LEV = land expectation value

Table 16.5 Returns on land and labor of agroforestry with *Eucalyptus deglupta* and maize monocropping over a 12-year tree rotation period (Bertomeu 2006)

System	Maize (t ha^{-1} 12 years^{-1})	Timber (m^3 ha^{-1})	Returns on land: LEV (US$ ha^{-1})		Net returns on labor (US$ work-day^{-1})	
			r=15%	r=20%	r=15%	r=20%
Maize monocropping	117.3	0.0	3,245	2,433	4.7	4.7
Low timber yield:						
Tree hedgerow (1 × 10 m)	28.7	146.1	2,204	1,495	5.6	4.9
Tree block (2 × 2.5 m)	14.5	146.1	1,656	1,037	5,9	4.8
High timber yield						
Tree hedgerow (1 × 10 m)	28.7	184.6	2,520	1,662	6.1	5.2
Tree block (2 × 2.5 m)	14.5	184.6	1,972	1,205	6.6	5.3

Timber price = PhP4 bf^{-1} or US$ 42.4 m^3; LEV = land expectation value

to the high competiveness of *Gmelina arborea*, one of the two tree species used in the study, for site resources. This competitiveness reduces maize grain yield, affecting the profitability of agroforestry. The price of *Gmelina* timber is also very low, demonstrating that the choice of species used in an agroforestry system strongly influences the profitability of the said system.

16.3.3.3 Internal Rate of Return

The IRR determines the theoretical maximum interest rate that a project can repay on loans while it recovers all investment and operating costs. Simply put, the IRR determines the purchasing power of money invested in a certain enterprise. The IRR can also be seen as the discount rate that cancels the net value of a series of cash flows. It equals the total costs and benefits expected from a business in real calculation. The IRR can be calculated using the following formula (Randall 1987):

$$\sum_{t=0}^{r}(B_t - C_t)\ /(1+p)^t = 0$$

Where

B = increased profits in year t
C = increased costs in year t
p = IRR

The internal rate of return can also be computed using the following formula:

$$NPV(i^*) = NPV_{cash\ inflows} - NPV_{cash\ outflows} = 0$$
$$= \frac{A_0}{(1+i^*)^0} + \frac{A_1}{(1+i^*)^1} + \ldots + \frac{A_n}{(1+i^*)^n}$$

where i^* is the IRR and A_n the cash flow in period *n*.

The IRR is used as a criterion to identify viable projects. In general, an enterprise with an IRR of 10 % at an 8 % discount rate would be acceptable, whereas a project with 5 % IRR would not. As a rule of thumb, projects with the highest IRR are retained. One of the main advantages of IRR is that it is not necessary to preselect a discount rate for its calculation. However, the IRR is not a consistently useful evaluation criterion when relationships between basic costs and profits change radically during the project, as often happens in agroforestry.

Dube et al. (2002) used IRR to analyze the economic aspects of a eucalyptus-based agroforestry system in the savanna region of Brazil. The study found that the eucalyptus-based agroforestry system had an IRR of 13.49 %, higher than that of eucalyptus monoculture at 12.56 %. Another example is given by Mehta and Leuschner (1997) in a study that compared commercial timber production of cypress and coffee against agroforestry systems (coffee and *Erythrina poeppigiana,* coffee and *Eucalyptus saligna*, coffee and *Leucaena leucocephala*) in Costa Rica. Higher IRR values were found in agroforestry systems, and the highest in the coffee-*Leucaena* agroforestry system with an IRR of 37.2 % (Table 16.6).

Decision rules:

If IRR > MARR (Minimum acceptable rate of return) (NPV > 0): Accept the project

If IRR = MARR (NPV = 0): Indifferent

If IRR < MARR (NPV < 0): Reject the project

16.3.3.4 Benefits and Costs Ratio

Prior to the calculation of the benefit and cost ratio, profits and costs are first identified and discounted at the pre-selected rate. The discounted benefits are then

Table 16.6 Financial analysis of agroforestry and commercial timber production systems in Costa Rica (Mehta and Leuschner 1997)

Criterion	Discount rate	Cypress	Coffee	Coffee/*poro*	Coffee/eucalypt.	Coffee/leucaena
NPV	5%	362.6	1,269.2	1,48620	1,484.0	1,539.0
NPV	10%	100.4	751.6	931.5	925.6	966.4
EAI	5%	33.5	143.2	167.7	167.4	173.6
EAI	10%	12.8	110.3	136.7	135.7	141.8
IRR		13.7%	29.4%	37.0%	36 0%	37.2%
NPVP	5%	669.1	2,864.6	3,353.6	3,348.7	3,472.8
NPVP	10%	128.3	1,103.1	1,367.1	1,358.4	1,418.3

NPV = net present value; EAI = equivalent annual income; IRR = internal rate of return; NPVP = net present value of perpetual rotations; NPV, EAI, and NPVP in thousands of Costa Rican (CR) colónes per hectare. 1 US$ = 180.3 CR colónes in July, 1995

added, and divided by the discounted costs to obtain the BC ratio. In theory, the higher the BC ratio, the more attractive the project is. The BC ratio has its advantages and drawbacks. The BC ratio can be used to compare projects of different sizes. However, this ratio requires preselection of a discount rate. Another inconvenience of BC ratios is that they are very sensitive to the original definition and assessment of the benefits and costs of a project. For example, Jain and Singh (2000) used BC ratios to compare poplar-based agroforestry and traditional agriculture, with and without assistance of a company in Uttar Pradesh, India. The BC ratio (and net present worth) (Table 16.7) indicated that agroforestry is financially viable to farmers with or without the assistance of a WIMCO, the Western Indian Match Company, at discount rates of 12% and 15% over an 8-year rotation (poplar trees are mature after 8 years, but can be harvested after 6 or 7 years). However, using the internal rate of returns, the agroforestry systems were found to be unprofitable over an eight-year period if discount rates were higher than 24% with Wimco assistance, and 14% without.

B/C ratio can be calculated as follows:

$$B/C = \frac{\sum \frac{R_n}{(1+i)^n}}{\sum \frac{C_n}{(1+i)^n}}$$

where R_n and C_n are revenues and costs at time n, respectively, and i is discount rate.

The decision rules are as follows:

- If B/C > 1 (NPV > 0): accept the project
- If B/C = 1 (NPV = 0): indifferent
- If B/C < 1 (NPV < 0): reject the project

Table 16.7 Financial analysis of poplar-based agroforestry with and without Wimco (West Indian Match Company) assistance in the northwestern part of India (Jain and Singh 2000)

Rotational age/year	Discount rate (%)	Without Wimco assistance			With Wimco assistance		
		NPW (in Rupees)	BC ratio	Annuity value (in Rupees)	NPW (in Rupees)	BC ratio	Annuity value (in Rupees)
6	10	40528	1.34	9306	14064	1.10	3229
	12	33982	1.30	8266	8711	1.06	2120
	15	25469	1.25	6549	1824	1.01	469
FIRR (%)			32.40–32			16.02–16	
7	10	46056	1.35	9461	18284	1.12	3756
FIRR (%)	12	37147	1.30	8139	10724	1.07	2349
	15	26364	1.23	6140	1762	1.01	411
			31.30–31			15.76–16	
8	10	46016	1.32	8625	17197	1.10	3224
FIR (%)	12	35410	1.27	7128	8081	1.05	1627
	15	22853	1.19	5093	3483	–	–
			24.45–24			14.29–14	

Net Present Value $(NPW) = \sum_{t-1} t - n(B_t - C_t)/(1+i)^t$

Financial Internal Rate of Return $(FIRR) = \sum_{t-1} t - n\ (B_t - C_t)/(1+i)^t = 0$

Cost-benefit ratio $(BC\ ratio) = \sum_{t-1} t - n\, B_t\ /\ (1+i)^t\ /\ C_t\ /\ (1+i)^t$

Where, B_t=benefit in each year; C_t=cost in each year; n=number of years; t=1, 2,3 ….., n; i=discount rate

16.3.4 Break-Even Relative Additional Cost (BeRAC) and Actual Relative Additional Cost (ARAC)

The BeRAC corresponds to the threshold profitability of additional costs of a new method, relative the value of established new forest stand (Garcia 1996):

$$BeRAC(\%) = (1+i)^{\delta} - 1$$

where i is the discount rate and δ is the time gain for harvesting. Time gain is the parameter to identify the profitability of treatment. In other words, the benefits are correlated to a time gain at the time of harvest.

The ARAC of a new treatment corresponds to the new treatment cost, relative to the value of established new forest stand (Garcia 1996; Opio et al. 2009):

$$ARAC\ (\%) = \frac{C - C}{L + C} \times 100$$

where *C* is the new total cost ($/ha), C is the establishment cost ($/ha) and L is the LEV ($/ha).

The decision rules for independent projects are the following:

If ARAC < BeRAC, the new treatment is profitable

ARAC = BeRAC, we are indifferent about the new treatment

ARAC > BeRAC, the new treatment is unprofitable

16.3.5 Payback Period

In order to calculate the payback period formula of the given investment in agroforestry projects, we require the cash inflows of each year of the agroforestry business. There is no mathematical formula of calculating payback period for a series of uneven cash flows. However for even cash flow, we can calculate the payback period as follows:

16.3.5.1 Payback Period (PP) = Cost of Investment/Annual Cash Inflows of the Investment

Discounted payback period (DPP) takes the real value of the cash flows. The cash flows are measured with respect of their market value and are discounted at a particular interest rate that is called discounted interest rate. Discounted payback period is the period that is required to recover the initial costs of the investment while considering the real values of the cash flows of the business over the years. It can be calculated as follows:

Discounted PP = Year Before Recovery of the Investment + Uncovered Cost at the Start of Year/Discounted Cash Flow Over the Year

There are two main reasons why the DPP is preferred to the sample PP analysis. Firstly the financial managers' actual payback period is that which will be required to recover their initial costs at the market value of the cash flows. Secondly, discounted payback period gives a more accurate and finely tuned estimate of the time required to recover initial costs. DPP helps the managers select their projects in a precise and careful manner. In DPP the decision rule is that if the DPP is less than target period that the management should accept the project, otherwise they should reject it.

16.3.5.2 Farm Budgeting and Partial Budgets

The farm budget is the most common basic unit in the economic analysis of agroforestry systems. It allows the identification of the costs and benefits of agroforestry projects at a household level. Davis (1989) reported two common approaches in farm budgeting. The first approach consists of selecting and modeling several representative farm projects. The overall impact is determined by multiplying the results

Table 16.8 Partial budget: Extra costs and benefits of using calliandra as a substitute for dairy meal in milk production, central Kenya ($US year^{-1}, 2001) (Franzel 2004)

Year	Extra costs		Extra benefits		Net benefits
	Item	$US	Item	$US	$US
1	Tree seedlings	3.5		0	
	Planting labor	3.3			
	Subtotal	7.14			−7.14
2	Cutting; feeding labor	10.03	Saved dairy meal cost	129.72	
			Saved dairy meal transport	4.02	
			Interest on capital	1.11	
	Subtotal	10.03		134.85	124.82

Years 3–10 same as year 2; Net present value at 20% discount rate = $US 413.36; Net benefit per year after year 1 = $US 124.82; Annualized net benefit treating establishment costs as deprecation = $US 122.44; Note: base farm model same as in Table 16.7; coefficients are from Appendix A of Franzel (2004)

of the individual models by the number of similar farms, and then adding the results together. This method is time-consuming when a large number of different types of farms are involved in the project. The second approach consists of building a simple but wide-scale model to simultaneously simulate all projects, regardless of farm type or scale of operation. When the costs and benefits of the model farm are determined, they are multiplied by the total number of farms to assess the economic feasibility. This approach is advantageous because it only requires the design of a single model, but can become very complex if dealing with a large and heterogeneous project.

Partial budgeting is a financial technique in agricultural economics that takes into account only those changes in costs and returns that result from the change in practice (Upton 1987, 1996). Partial budgeting assesses the benefits and costs of a practice relative to not using the practice. For instance, Franzel (2004) used partial budgeting to assess the profitability of calliandra as a substitute for dairy meal in Kenya (Table 16.8).

16.3.6 Quantification and Valorization

The valorization of a commodity refers to assigning a value to that commodity. The accuracy of any economic evaluation depends on the accuracy of the data used. Consequently, the designation of sustainable agroforestry interventions depends on the exactness of the estimates of costs and revenues (Price 1995; Alavalapati and Mercer 2004). Relationships between farm inputs and outputs can be described by a production function, which helps to identify the key elements requiring examination:

$$Y = g(K, L, R_0)$$

Where Y is the income (or output) of the farm, g the technology production used, K the capital goods, L the mental and physical labor, and R_0 the natural resources used (e.g., land).

16.3.6.1 Capital Goods

Capital goods are quantifiable items manufactured or purchased and used to produce goods and services. They are usually valued at the market price for the end user (Hoekstra 1990). Agroforestry capital goods consist of i) inputs/consumables such as fertilizers, feed, seeds, herbicides, which are "consumed" throughout the project and therefore will need to be purchased more than once in the lifespan of the project, and ii) capital items/equipment, which are tools that are usually bought at the start of the project. Capital items can be taken into consideration using two different methods, either through depreciation (i.e., by dividing the value of the equipment over the number of years that the equipment will be used) or as a cost at the first year of the project. Examples of capital items in agroforestry are seedlings, marcotts, watering cans, non-mist propagators, nursery shelter and fences. If capital items have a longer life than the project, the terminal or recovery value of the input is commonly included as a benefit in the final year of analysis.

In subsistence agriculture practiced by most farmers in the Congo Basin, Latin America and Southeast Asia, capital goods are rare compared to other production factors. Poor farmers in the tropics do not have enough cash to both fulfill their needs and purchase agricultural inputs. Many sub-Saharan countries subsidized agriculture in rural areas until the end of the 1980s. Farmers, who had grown accustomed to receiving inputs from the government, found it difficult to purchase inputs, especially with the price drop of coffee, cocoa and cotton, the main cash crops of the region. In the early 1990s, the governments of sub-Saharan countries have been obliged by Bretton Woods institutions (i.e., the World Bank and the International Monetary Fund (IMF)) to stop subsidizing agriculture because of the economic crisis affecting the region at that time. Poverty increased in rural areas of sub-Saharan countries because of this decision to stop subsidizing agricultural inputs, and farmers now have to count on their own resources to invest in any agricultural project. For that reason, agroforestry projects that are supposed to be implemented in the region should require only inexpensive and easy to make capital items such non-mist propagators that are used in tree domestication programs (for more details on non-mist propagators, please see Chapter 6).

16.3.6.2 Labor

Given the limited resources in capital and land (often considered capital in agriculture) of smallholder and subsistence farmers, labor is typically the most important input in small-scale or subsistence farming. For instance, Steven and Jabara (1988) reported that labor accounts for 80–85 % of the total value of all farm resources used in traditional farming systems.

In economic analysis, labor usually refers to the physical and mental contribution of men and women to the production of output. Labor is generally expressed in days or hours, and often categorized by the age or gender of the contributor. Hired labor is most often valued based on the current market salary, while family labor is valued by its opportunity cost, the wage that could be earned elsewhere if the person was not engaged in the enterprise under evaluation (Hoekstra 1990).

Most agroforestry interventions require a degree of change in the distribution or total requirement of labor. Under conditions of under-employment or unemployment, agroforestry will greatly improve labor efficiency, while a shortage of labor will present serious constraints to the adoption of practices such as alley cropping, such as in the humid tropics of Cameroon (Degrande and Duguma 2000; Degrande et al. 2007). When the family labor available is insufficient, it is possible to hire outside labor for particularly lucrative agroforestry practices on the farm.

In any case, estimating the real value of agricultural labor remains a challenge. In labor-intensive agroforestry systems, the use of virtual low wages is often advocated, especially under conditions of widespread under- or unemployment (Prinsley 1990). Real wage rates in the market are a more accurate measure of value when the demand for labor is high and there is competition with other agricultural or non-agricultural enterprises.

16.3.6.3 Land

In economics, land refers to the natural resources such as soil and water that contribute to production. For practical purposes, only the resources for which there is a recognized monetary value are included in financial assessments. Any economic analysis of a particular enterprise should take into account the evaluation of the natural resources in terms of what their contribution could be to alternative projects.

Land is most often quantified in terms of the physical area and categorized by tenure status, production capacity, or use. Land valuation requires the establishment of land prices in a market setting. If these prices are not establishable, opportunity costs are used as an approximate value. If land resources are abundant, opportunity costs would be close to zero. However, the allocation of land to agroforestry in densely populated areas will surely require an exclusion of other activities. The appropriate assessment of land in these circumstances would be the monetary contribution of land to output of a known farm (Prinsley 1990). Where land is rented, the appropriate cost would be the amount of rent paid.

16.3.6.4 Benefit Valuation of Agroforestry Products

Agriculture most often aims to increase production, while agroforestry seeks to increase harvest through sustainability or the reduction of required inputs. Another objective of agroforestry is the diversification of income sources, improvement of the quality and quantity of food produced by farmers, the temporal diversification of agricultural and tree products, so that the products of different species are harvested at different times during the year to avoid "welding periods" where food and money are in short supply. Benefits can be financially quantified by converting the physical output to a monetary value. An important issue is off-season production, which provides what are commonly called "off season products". The development of varieties of trees that bear fruits during periods other than the usual period of production adds value to the product. For instance, a safou fruit from *Dacryodes edulis* typically costs 10 Central African Francs (XAF) in Cameroon in production period (i.e., in August, from year 2005 to 2008), about 2 cents US, but can bring as much as 30 XAF francs in a non-production period (i.e., in October-November) in the food markets of Yaoundé, Cameroon.

16.3.6.5 Direct Production

Currently, the trade of agroforestry products, such as indigenous fruit, leaves, kernels, nuts, and bark, lacks established commercial channels in the tropics, which negatively affects the valuation of these products. Some agroforestry products from West and Central Africa, such as *Irvingia gabonensis*/*I. wombolu* kernels, *Dacryodes edulis* fruits, *Gnetum africanum* leaves and *Ricinodendron heudelotii* kernels, are sold in Europe and North America (Tabuna 1999; Russell and Franzel 2004). Proper trade channels for these goods need to be organized by linking farmers to markets and facilitating the flow of information on markets of agroforestry products (Russell and Franzel 2004). In local markets, fruit are sold in piles, and leaves in bundles, and that makes their valuation difficult. Valuation is also more difficult when dealing with subsistence level agriculture, where almost all production is either consumed or sold on the farm. When commercial channels exist, the analytical market price of agroforestry products is the price at the point of first sale. Most often, on-farm consumption of agroforestry products is not included in economic analyses, which can lead to the underestimation of the real returns of the investment in agroforestry related to market-oriented systems (Prinsley 1990; Mercer et al. 2005). There are two accepted methods of pricing such assets, (i) the labor used in their production and (ii) the cost consumers would be willing to pay for sold substitutes.

The valuation of agroforestry timber products is also difficult: pricing depends heavily on market utilization. Logs are sold in cubic meters and poles by length. Fodder leaves are usually sold in fresh or dry mass, making their valuation easier. In on-farm consumption of fodder, the benefits will be reflected in the increased production of livestock. Similarly, the value of green manure and litter from foliage used by the household to fertilize their field crops will be included in increased harvest value.

16.3.6.6 Environmental Benefits

The indirect effects of agroforestry practices should be taken into account in economic assessment of agroforestry. The effects of watershed maintenance, erosion control, land rehabilitation, biodiversity maintenance, pest management and greenhouse gas mitigation (through carbon sequestration) on the overall welfare and social development of people should be included in economic evaluation of agroforestry practices. With the advent of the REDD Program (Reducing Emissions from Deforestation and Forest Degradation; see chapter 20 for detailed information), which is supported by the World Bank, carbon sequestration in agroforestry systems can be easily valued through carbon payments/compensation to farmers. Economic analysis takes into account the desirability of an enterprise from the perspective of the entire society. Agroforestry project analysis should therefore consider these benefits as key factors in the decision to promote a certain agroforestry enterprise (Mercer 2004). These benefits can heavily influence adoption of agroforestry practices, but these effects are not easily quantified, especially in the short term.

Soil conservation benefits in agroforestry may be provided by the market value of increased or sustained agricultural production. "With" or "without" analysis that takes into account the positive environmental effects of agroforestry is useful to highlight the environmental impacts associated with the introduction of a particular agroforestry enterprise.

16.3.6.7 Risk Assessment in Agroforestry

One of the objectives of agroforestry is to reduce financial risks to farmers through diversification and making products available year-round and fostering the sustainability of production over years. The decision to adopt any agricultural technology, which may negatively affect revenues, such as a delay between planting and harvesting, unfavorable climate, or fluctuating markets, is of critical importance to farmers. It is unrealistic to assume near-perfect knowledge and a stability of prices when performing economic assessments, especially in agroforestry, where projects have a longer lifespan than in agriculture.

Risks in agroforestry and agriculture can be defined as the probability that a required minimum family income is not attained (Ramirez et al. 2001). For instance, in a study aimed at assessing the financial returns, stability and risk of cocoa-plantain-timber agroforestry systems, Ramirez et al. (2001) found that risk is lower for technologies with larger mean NPV. They also found that even under a discount rate of 4 %, monocrops are very risky, showing risk levels in excess of 75 %, whereas agroforestry systems presented low risks, below 22 % (see Table 16.2). Results from Ramirez et al. (2001) indicated that the diversification of products seems to lower risk in agroforestry.

16.3.6.8 Sensitivity Analysis

Due to inflation and market fluctuations caused by supply and demand interaction, the future price of inputs will always be uncertain, influencing the selection of the discount rate, the amount of expected harvest, and market prices. In such situations, sensitivity analysis can help determine how an economic evaluation will be affected if changes occur in key variables and assumptions. This methodology evaluates the effects of changing circumstances by varying the price of inputs, outputs and other important variables in an evaluation by a percentage or a fixed amount, and recalculating key indicators such as NPV, IRR and CB ratios. The data are then presented as a range of possible outcomes and associated probabilities, with the best estimate in the middle of the row. Sensitivity analysis can thus help determine how risky a project is.

Franzel (2004) used sensitivity analysis to determine how changes in key parameters (maize yield and price, wood yield and price, wage rate, and discount rate) would affect the returns on land and labor in rotational woodlots and maize monocropping in Tabora District, Tanzania. Changes in maize yield and price induced changes in returns on land and labor, whether using rotational woodlots or in maize monocropping (Table 16.9). Changes in wage rate and discount rate affected returns on land, but did not affect returns on labor in maize monocropping, whereas changes in wood yield and price did not affect returns on land or labor in maize monocropping (Franzel 2004), indicating the importance of sensitivity analysis for the determination of the profitability of agroforestry systems.

16.3.6.9 Risk-Benefit Analysis

Risk-benefit analysis is based on the concept that any change or development will involve some degree of risk. Risk-benefit analysis compares the risks of an enterprise with its related benefits. The decision to invest in a certain enterprise will depend on the trade-off between risks and increased productivity (Randall 1987).

An example of risk-benefit analysis in agroforestry is given by Blandon (2004). The study reported the expected net present values (ENPV) for two crop systems (*Fraxinus excelsior* L. (ash), and rye-grass pasture supporting sheep) in temperate agroforestry in North Wales, United Kingdom, using the portfolio approach (Blandon 1985). Though the context of the study is different from that of tropical countries where access to markets is not perfect, the author demonstrated that the portfolio theory is applicable in the tropics (Blandon 2004). The study also compared risks and benefits associated with the combination of ash production and rye-grass pasture in coarse-level and in an agroforestry system. The ENPV for sheep production was higher (£ 7,665; 1 £≈1.514 US$ on February 27, 2013) than that of ash production (£ 6,119) (Blandon 2004). Sheep husbandry was also less risky than ash production (risk = 1,306 and 6,647 for forestry and sheep production, respectively) (Blandon 2004). Using 3.7 % of the farm area for ash production and the remaining 96.3 % for sheep husbandry minimized the risk (1,282) at coarse level, and yielded slightly similar ENPV than rye-grass system (£ 7,607) (Blandon 2004). The author reported that agroforestry increases the ENPV and risk because of the interaction

Table 16.9 Sensitivity analysis of the results of the financial analysis of rotational woodlots to changes in key parameters, Tabora District, Tanzania (Franzel 2004)

Parameter	Rotation woodlots		Maize without trees	
	Return on land (Net present value, $US ha^{-1})	Returns on labor ($US workday^{-1})	Return on land (Net present value, $US ha^{-1})	Returns on labor ($US workday^{-1})
Base analysis*	389	2.67	61	1.31
50 % decrease in maize yield	272	2.1	−56	−0.12
50 % increase in maize yield	476	3.49	179	2.56
50 % decrease in maize price	298	2.19	−60	−0.11
50 % increase in maize price	479	3.15	182	2.72
50 % decrease in wood yield	155	1.42	61	1.31
50 % increase in wood yield	622	3.92	61	1.31
50 % decrease in wood price	155	1.42	61	1.31
50 % increase in wood price	622	3.92	61	1.31
50 % decrease in wage rate	443	2.67	86	1.31
50 % increase in wage rate	334	2.67	36	1.31
30 % discount rate	302	2.51	55	1.31
10 % discount rate	510	2.84	70	1.31

*From data on the financial analysis of rotational woodlot as compared to a maize fallow system in Tabora District, Tanzania (For more details, please refer to Franzel et al. 2004)

between system components (animal-trees, tree-crops, etc.), but it is better to mix sheep and trees in a true agroforestry system than a "coarse-level" (Price 1995) mixing, or growing crops without any spatial arrangement (Blandon 2004). Similar studies are needed in the tropics, where subsistence farmers produce crops from their own consumption rather than for the markets, so as to determine the risks and benefits associated with agroforestry, and with monocropping systems.

16.3.7 Econometrics in Agroforestry

Econometrics refers to the application of mathematical models and statistical methods to economic data. According to Montambault and Alavalapati (2005), econometrics explores the relationships between economic variables, and is an emerging field in agroforestry. For instance, Adesina et al. (2000) quantified the factors determining the adoption of alley cropping in Cameroon using an econometric model. The model used was a Logit model, and the explanatory variables related to adoption were land tenure rights, the socio-economic characteristics of the farmers, and village-specific characteristics. The most important factors influencing alley cropping adoption in Cameroon were, respectively, membership in the farmers' association, contact with agroforestry extension agents, village fuelwood scarcity index, village land pressure, and sex (gender of plot owner) (Adesina et al. 2000). The Logit model estimated that

31 % of the farmers surveyed in the study adopted alley cropping, and that adoption was lower in areas with very high population pressure (Adesina et al. 2000).

In Cameroon, Molua (2005) also used econometrics to model the profit of agroforestry, which was used to identify the socio-economic factors influencing the profitability of agroforestry farms. The outputs were related to tree products, staple food crops and livestock products, and inputs included labor and credit (Molua 2005). Market prices, farm operating costs, and contact with extension workers were important positive covariates of agroforestry production in the region. The study also revealed that farm holdings are profitable in the study area, and that agroforestry farms are amongst the most profitable ones (Molua 2005). For instance, 97 % of agrisilvicultural farms, 57 % of silvopastoral farms, and 56 % of agrosilvopastoral farms were profitable, whereas agroforestry generated on average a mean profit of 116,000 XAF (1 USD = 500.634 XAF on February 28, 2013) per farmer (Molua 2005). In Burkina Faso, a Logit model integrating technology profitability as an explanatory variable was used to study farmers' decision-making processes for live hedges adoption (Ayuk 1997). The results indicate that water availability and the profitability of the technology itself enhance the probability of adopting live hedges.

16.3.8 Optimization in Agroforestry

The decision to invest in a project is most often based on the selection of the best alternative from a set of available options. This is referred to as optimization, the identification of the best available values of a function. Optimization involves the minimization of costs and maximization of profits. In agroforestry, optimization compares and selects the best land use design for a particular site and the farmer's situation (e.g., labor and land constraints and wealth) and objectives. In other words, optimization in agroforestry aims to choose the land use design that provides the highest and sustainable yields over time, and least amount of risk while considering the farmer's objectives and environment characteristics (i.e., soil fertility, demographic pressure, sloping land etc.). For that reason, optimization requires a good understanding of tree-crop (or animal-tree-crop) interactions and economic evaluations of all alternatives. This is possible when large data sets on all possible alternatives are available, so as to model optimization function to allow the selection of best alternatives.

Two optimization functions are highly used in agroforestry, i.e., the land equivalent ratio (LER) and the cost equivalent ratio (CER) (Wojtowski 2002). The LER is based on a combination of design variables (Table 16.10), whereas the CER is derived from the costs associated with the variables.

Optimization functions are used to draw possibility curves or boundaries, which are standard analytical tools for the selection of the best alternative (i.e., the one predicting the maximum production level, or the marginal rate of transformation also known as *opportunity cost*). Two possibility curves are used for this purpose, i.e., the production possibility curve (PPC) and the cost possibility curve (CPC). The PPC compares the rates of production of several commodities that use the same

Table 16.10 An example of design variables that could be used to develop optimization functions in agroforestry (modified from Wojtowski 2002)

Design variable
Species
Primary design species
Secondary species (one or more)
Spatial patterns
Pattern of spatial arrangements
Dimensions
Canopy patterns
Root patterns
Temporal patterns
Rotational
Seasonal
Intra-seasonal
Production cycle (i.e., the products of different species in the system are harvested at different times during the year)
Management options/inputs
Labor inputs
Weeding
Mulching
Tree establishment
Seed
Seedling
Cutting (leafy stem cutting, marcot)
Germplasm
Tree maintenance
Pruning
Coppicing
Lopping
Pollarding
Watering (especially for plants in the nursery)
Thinning
Root pruning
Supplemental inputs
Fertilizers
Insecticides
Herbicides
Irrigation
Root barriers
Terraces (for sloping areas)
Stem/branch grafting
Planting or tillage method
Use of fire

factors of production, whereas the CPC is used in cost-based studies where it is derived using raw data.

Because agroforestry is complex, full optimization (i.e., involving all the design variables) in agroforestry is scarce and represents an ideal situation. For this reason, bioeconomic models have been used in agroforestry modeling (Wojtowski and Cubbage 1991; Wojtowski et al. 1991; Lawson et al. 1995; Menz and Grist 1996; Graves et al. 2006). For example, Wojtowski and Cubbage (1991) developed an optimization tool called the "bordered matrix approach" that was used to model agroforestry production systems in order to identify the optimal planting density by meeting specific economic criteria. The tool successfully modeled a banana-cassava mixture, indicating its effectiveness for use in mixed cropping systems (Wojtowski and Cubbage 1991).

Land Equivalent Ratio (LER) (Mead and Willey 1980) is an indicator of productivity of mixtures. It is a measure of how much land would be required to achieve intercrop yields with crops grown as pure stands. It can be calculated as follows:

$LER = Sum(Relative Yields) = RY_{tree} + RY_{crop} = (Mixed Yield 1/Pure Yield 1) + (Mixed Yield 2/Pure Yield 2)$

The decomposition of LER requires a model that provides details of light, water, carbon dynamics in the system and has the possibility of annual increment data (annual) or of cumulative increment data (integrated). LER = 1 when the agroforestry system (AF) does not have an advantage over a monocroping system, meaning that the inter- and intraspecific interactions are equivalent. The LER value > 1 indicates mixed systems are advantageous and when the LER is less than 1, no over yielding is occurring and the sole crops are more productive than the intercrop.

16.4 Economic Studies in Agroforestry

There are very few economic assessments in agroforestry, as compared to biophysical investigations. Most previous economic evaluations in agroforestry were performed before project implementation (*ex ante*). However, evaluations are increasingly being carried out after the project implementation (*ex post*).

16.4.1 General Studies

Overviews of the issues in the evaluation of agroforestry projects were provided by Betters (1988), Prinsley (1990), Hoekstra (1990), Sullivan et al. (1992), Cacho (2001), Knapp and Sadorsky (2000) and Molua (2005), among others. For instance, Cacho (2001) developed a model to assess the externalities of forests and marginal costs of soil degradation, on the basis of the general economic analysis of land subjected to degradation in the presence of the positive externalities of forests. This technique can be used to provide cost estimates as a basis for negotiations among the stakeholders, to develop an approach to the common ownership of land management in a watershed.

In Cameroon, Molua (2005) modeled a profit equation of agroforestry farms. The model was used to assess the socio-economic factors influencing the profitability of agroforestry farms in Cameroon. The analysis focuses on the importance of market prices, cost of farming, and contact with extension workers as positive covariates for production in the studied area. Knapp and Sadorsky (2000) have developed a more robust and dynamic model for agroforestry management in saline soils. The model predicted soil salinity changes in response to water application, and revealed that the harvest (of wood biomass) decision is robust to changes in most of the economic and physical parameters (Knapp and Sadorsky 2000).

In the Chittagong Hill Tracts, Bangladesh, Rasul and Thapa (2006) used the cost-benefit ratio, NPV, and labor output methods to assess the financial and economic benefits of agroforestry and shifting cultivation. Results from the study indicated that the economic returns of agroforestry were better than those of shifting cultivation (Rasul and Thapa 2006).

16.4.2 Agroforestry System Economic Studies

More and more economic assessments are being done in order to assess the profitability of agroforestry practices and technologies. The adoptability of agroforestry depends in part on its profitability.

In India, Jain and Singh (2000) reported the economic analysis of poplar-based agroforestry systems in terms of income, employment and environmental impact from the farmer's perspective. The study revealed that poplar-based agroforestry systems are not only viable, but also more profitable than many rotation crops in the study area. Sensitivity analysis indicated that this system is not high risk. Another example of economic analysis of agroforestry practices is provided by Ramirez et al. (1992), who analyzed the impact of improved agroforestry practices based on the management of the natural regeneration of commercially valuable timber trees in coffee farms associated with herbaceous legumes in Ecuador. The proposed agroforestry practices improved the long-term productivity of both land and labor while sparing the need for external inputs and hired labor, making these practices more adoptable and sustainable. The study also indicated that the proposed agroforestry practices are economically feasible and technically possible.

In Central America, Ramirez et al. (2001) compared the financial returns, stability, and risk of six agroforestry systems based on cocoa, *Cordia alliodora,* and plantain (*Musa* AAB), and their corresponding monocultures. This *ex post* analysis was used to model and simulate the probability distribution function over time of the three commodity prices. Net revenues expected from agroforestry systems were higher than those of monocultures (Ramirez et al. 2001). In addition, agroforestry presented lower risks, as agroforestry systems with more cocoa than plantain were less risky but also less sTable (Ramirez et al. 2001). The tree component (*C. alliodo-*

ra) was a key factor in reducing the financial risk (Ramirez et al. 2001). Similar results were found in a previous study in Costa Rica by Reeves and Lilieholm (1993), who assessed the financial risk associated with agroforestry with the fluctuations in net income of alternative farming systems. Coffee production in monoculture provided the highest expected net income, but was also the most economically risky (Reeves and Lilieholm 1993). Also in Costa Rica, Mehta and Leuschner (1997) carried out an economic and financial analysis of coffee/tree agroforestry systems, coffee plantations and tree plantations in a forest reserve. Coffee/tree systems showed an IRR of over 30 %, higher than that of coffee without trees. In addition, the coffee/tree combination reduced the risks from fluctuating coffee prices, and reduced the need for chemical fertilizers. The income of a block planting of cypress trees was a small fraction of that of a coffee/tree combination or coffee monocropping, but required less initial investment.

The profitability of agroforestry to farmers was also studied by Current et al. (1995), who investigated the cost of agroforestry to farmers. The study was carried out in Central America and the Caribbean, and revealed that many agroforestry practices are profitable under a wide range of conditions, and probably applicable to a wider scale. Agroforestry projects offered a broad basket of species and systems (Current et al. 1995), hereby diversifying income sources.

Reforestation through agroforestry is also subject to economic evaluation. In Thailand, Niskanen (1998) evaluated the financial and economic profitability of reforestation. Three systems were compared, namely plantations of *Eucalyptus camaldulensis*, *Tectona grandis* (teak), and agroforestry systems where *Manihot esculenta* (cassava) was used as an intercrop for three years. It was more profitable to invest in reforestation from the societal point of view, as well as from that of a private investor (Niskanen 1998). Teak planting was more profitable than eucalyptus, and cassava cultivation between rows decreased the financial and economic profitability of reforestation (Niskanen 1998).

16.4.3 Alley Cropping and Improved Fallows

Economic studies of alley cropping systems were carried out during the 1990s, when this tree-crop system was popular in the tropics. Several studies showed that alley cropping is profitable under certain conditions. In Nigeria, Ruhigwa et al. (1994) conducted an economic analysis of mulching in a "cut-and-carry" system with *Pennisetum purpureum*, compared to mulching with several agroforestry species (*Alchornea cordifolia* Müll.Arg., *Dactyladenia barteri* Prance and White, *Gmelina arborea* Roxb., *Senna siamea* (Lam.) Irwin et Barneby syn. *Cassia siamea* Lam.) in alley cropping systems. The "cut-and-carry" with *Pennisetum purpureum*, used as mulch for plantain, showed the highest production of plantain in terms of bunches, but incomes were similar to that in *Dactyladenia* systems. Other agroforestry species produced lower incomes. The increase in the amount of land required to produce *Pennisetum* mulch lowered net income (which was negative) per ha over three years of cultivation. Agroforestry systems exhibited higher returns to labor

Table 16.11 Labor requirements, maize production and return on land and labor of cajanus fallow compared to natural fallow over a six-year period, Cameroon (Degrande 2001)

	Cajanus fallow	Natural fallow
Present value of		
Returns on land (CFA ha^{-1})	912,000	307,000
Returns on labor (CFA day^{-1})	2,180	1,860
Total maize produced (tons ha^{-1})	15.18	4.98
Total groundnut produced (tons ha^{-1})	1.83	0.40
Total number of workdays	1,435	483

1 CFA=0.00198867 USD (on February 16, 2012)

than *Pennisetum* treatment, and *Dactyladenia* systems were found to be the most profitable among those studied (Ruhigwa et al. 1994).

Jabbar et al. (1994) conducted a study in southwestern Nigeria, and found that continuous alley cropping is more profitable than cropping without corridors or alley cropping with fallow. The study also reported that the incorporation of small ruminants increases the profitability of alley cropping, and that alley cropping remains profitable even when compensation costs are internalized in the final cycle of a current project. However, a study conducted on acidic and densely populated uplands of Burundi showed that the economic benefits of alley cropping with *Leucaena diversifolia* (Schltdl.) Benth. were comparable to the control after 5 years (Akyeampong and Hitimana 1996).

Swinkels et al. (1997) conducted an evaluation of a short rotation improved fallow with *Sesbania sesban* intercropped with maize in the densely populated highlands of Kenya. This study found that improved fallow requires less labor than continuous cropping, indicating that as the opportunity cost of labor increases, so does the profitability of improved fallows (Swinkels et al. 1997). Degrande (2001) compared *Cajanus cajan* fallows with natural fallows using economic tools over 6 years in Cameroon. The net present value per ha of shrub fallow is three times higher than that of natural fallow (Tables 16.11 and 16.12). Changing key parameters also showed that *Cajanus* fallow is more profitable than natural fallows in the study area (Table 16.13). Franzel (2004) found that a continuously fertilized maize system is more financially profitable than an improved fallow with *Sesbania sesban* (Table 16.14). The benefits of improved fallows were reduced labor requirement, firewood production, increased maize yield between years 3 to 5, and reduced land preparation and weeding costs.

16.4.4 Economic Assessment and Commercialization of Other Agroforestry Practices

Economic studies have been conducted on small-scale woodlot, homestead tree and shrub growing and boundary tree and shrub growing in Ethiopia (Duguma 2013); on farming systems consisting of intercrops with legumes for fodder production on

Table 16.12 Returns on land and on labor of cajanus fallows compared to natural fallows for seven individual farms, Cameroon (Degrande 2001)

Farm	Returns on land (CFA ha^{-1})		Returns on labor (CFA $workday^{-1}$)	
	Cajanus fallow	Natural fallow	Cajanus fallow	Natural fallow
1	1,150,000	474,000	2,500	2,300
2	1,513,000	726,000	3,000	3,100
3	292,000	130,000	1,400	1,400
4	1,036,000	202,000	2,300	1,600
5	1,455,000	295,000	2,900	1,800
6	466,000	50,000	1,600	900
7	568,000	115,000	1,700	1,300

100 CFA=0.198867 USD (on February 16, 2012)

Table 16.13 Sensitivity analysis showing the effects of changes in key parameters, Cameroon (Degrande 2001)

	Returns on land (CFA ha^{-1})		Returns on labor (CFA $workday^{-1}$)	
	Cajanus fallow	Natural fallow	Cajanus fallow	Natural fallow
Base analysis	912,000	307,000	2,200	1,900
Maize price –50%	364,000	56,000	1,500	1,200
Maize price+50%	1,459,000	558,000	2,900	2,600
Shrub seed price+50%	903,000	307,000	2,200	1,900
Discount rate 10% instead of 20%*	1,194,000	339,000	2,200	1,800
Discount rate 30% instead of 20%*	724,000	281,000	2,200	1,900

100 CFA=0.198867 USD (on February 16, 2012); *Instead of 20% because the base analysis that is used is 20% as discount rate and since this is a sensitivity where the parameters were changed, this was changed to 10% and 30% respectively to indicate different levels of discounting

terraces of the middle hills of Nepal (Neupane and Thapa 2001); on cocoa agroforests of southern Cameroon (Gockowski et al. 2010), multi-layered cropping (von Platen 1992), and contour tree buffer strips (Countryman and Murrow 2000). It was shown that agroforestry is profitable, and contour strips are economically competitive with or without subsidies, indicating that contour strips are an economically feasible form of erosion control.

Since the advent of domestication of tree species providing non-timber tree products, commercialization studies have gained importance in agroforestry. The expansion of market opportunities for smallholders has been overlooked by agroforestry research in the past, but is now recognized as critical to the success of agroforestry innovations (Russell and Franzel 2004). A method of evaluating non-timber forest products has been developed by Godoy et al. (1993). According to these authors, the median value of a non-timber forest product is $50. In the humid tropics of West Africa, an approach has been developed by ICRAF and its local partners to assist smallholder farmers in developing marketing skills and knowledge (Facheux et al.

Table 16.14 Labor requirements, maize production, and returns on land and labor of *Sesbania sesban* improved fallows and continuously cropped maize over a 5-year period, using an average farm budget[a], in eastern Zambia (Franzel 2004)

Option	Workdays ha^{-1}	Tons of maize ha^{-1}	Returns on land: net present value ($US ha^{-1})	Returns on labor: discounted net returns ($US $workday^{-1}$)
	Over a 5-year period[b]		1996	1996
Continuous fertilized maize	499	4.8	5	0.42
Improved 2-year Sesbania fallow	433	7.3	115	0.85
Continuous fertilized maize	649	21.9	203	0.93

[a]The means of values from individual budgets of the twelve trials were used. Details on budgets and coefficients are provided in Appendix C of Franzel (2004)

[b]A 5-year period is used because that is the period needed to complete a cycle of the improved fallow practice: two years of fallow and three years of cropping

2007). Drawing heavily on *Ricinodendron heudelotii* kernel (njansang) commercialization, Facheux et al. (2007) found that farmers involved in njansang production realized an average of 31 % increase in their selling price and more than 80 % increase in their revenue derived from njansang by using the sub-sector approach in combination with the development of post-harvest technologies. Ndoye et al. (1997) highlighted the economic importance of non-timber forest products, reporting that the commercialization of *Irvingia gabonensis* kernels from Cameroon, Equatorial Guinea, Gabon, Nigeria and Central African Republic generates US$ 260,000 annually. The development of marketing strategies for non-timber forest products is now part of the ICRAF's tree domestication program.

Bibliography

Adesina AA, Mbila D, Nkamleu GB, Endamana D (2000) Econometric analysis of the determinants of adoption of alley cropping by farmers in the forest zone of southwest Cameroon. Agric Ecosyst Environ 80:255–265

Ayuk ET (1997) Adoption of agroforestry technology: the case of live hedges in the central plateau of Burkina Faso. Agric Syst 54:189–206

Akyeampong E, Hitimana L (1996) Agronomic and economic appraisal of *Leucaena diversifolia* on an acid soil in the highlands of Burundi. Agrofor Syst 33:1–11

Alavalapati JRR, Mercer DE (2004) Valuing agroforestry systems: Methods and applications. Advances in Agroforestry, Springer

Alavalapati JRR, Mercer DE, Montambault JR (2004a) Agroforestry systems and valuation methodologies. In: Alavalapati JRR, Mercer DE (eds) Valuing agroforestry systems. Kluwer, The Netherlands, pp 1–8

Alavalapati JRR, Shrestha RK, Stainback GA, Matta JR (2004b) Agroforesty development: an environmental perspective. Agrofor Syst 61:299–310

Albuquerque JL (1993) Avaliação econômica de alternativas de financiamento da produção florestal no estado de Minas Gerais. MSc Dissertation, Universidade Federal de Viçosa, Viçosa

Bertomeu M (2006) Financial evaluation of smallholder timber-based agroforestry systems in Calveria, Northern Mindanao, the Philippines. Small-Scale For Econ Manag Policy 5(1):57–82
Betters DR (1988) Planning optimal economic strategies for for agroforestry systems. Agrofor Syst 7:17–31
Blandon P (1985) Agroforestry and portfolio theory. Agrofor Syst 3:239–249
Blandon P (2004) Analyzing risk in agroforestry systems using a portfolio approach. In: Alavalapati JRR, Mercer DE (eds) Valuing agroforestry systems: methods and applications. Advances in agroforestry, Kluwer, pp 95–122
Cacho O (2001) An analysis of externalities in agroforestry systems in the presence of land degradation. Ecol Econ 39:131–143
Countryman DW, Murrow JS (2000) Economic analysis of contour tree buffer strips using present net value. J Soil Water Conserv 55(2):152–160
Current D, Lutz E, Scherr SJ (1995) The costs and benefits of agroforestry to farmers. World Bank Res Obs 10(2):151–180
Davis L (1989) Analysis of agroforestry systems. Winrock International, Morrilton
Degrande A (2001) Farmer assessment and economic evaluation of shrub fallows in the humid lowlands of Cameroon. Agrofor Syst 53:11–19
Degrande A, Duguma B (2000) Adoption potential of rotational hedgerow intercropping in the humid lowlands of Cameroon.ODI Network paper No 103, DFID, UK
Degrande A, Asaah E, Tchoundjeu Z, Kanmegne J, Duguma B, Franzel S (2007) Opportunities for and constraints to adoption of improved fallows: ICRAF's experience in the humid tropics of Cameroon. In: Bationo A, Waswa B, Kihara J, Kimetu J (eds) Advances in integrated soil fertility management in sub-saharan Africa: challenges and opprtunities. Springer, Dordrecht, pp 901–910
Dube F, Couto L, Silva ML, Leite HG, Garcia R, Araujo GAA (2002) A simulation model for evaluating technical and economic aspects of an industrial eucalyptus-based agroforestry system in Minas Gerais, Brazil. Agrofor Syst 55:73–80
Duguma LA (2013) Financial analysis of agroforestry land uses and its implications for smallholder farmers livelihood improvement in Ethiopia. Agrofor Syst 87:217–231
Enters T (1998) Methods for the economic assessment of the on-and-off site impacts of soil erosion. International Board for Soil Research and Management, Bangkok
Facheux C, Tchoundjeu Z, Foundjem-Tita D, Degrande A, Mbosso C (2007) Optimizing the production and marketing of NTFPs. African Crop Science Conference Proceedings 8:1249–1254
Faustmann M (1849) On the determination of the value which forest land and immature stands possess for forestry. J For Econ (Reprinted) 1(1):7–44
Franzel S (2004) Financial analysis of agroforestry practices. In: Alavalapati JRR, Mercer DE (eds) Valuing agroforestry systems: methods and applications. Kluwer, The Netherlands, pp 9–37
Franzel S, Scherr SJ (2002) Trees on the farm: assessing the adoption potential of agroforestry practices in Africa. CABI Publishing in association with ICRAF, Wallingford
Franzel S, Coe R, Cooper P, Place F, Scherr SJ (2001) Assessing the adoption potential of agroforestry practices in sub-Saharan Africa. Agric Syst 69:37–62
Garcia O (1996) Easy evaluation of establishment treatments in Forest management planning and managerial economics. Proceedings of the S4.04 meetings sponsored by International Union of Forestry Research Organizations (IUFRO) XX World Congress. August 6–12, 1995, Tampere, Finland. Gockoswki J, Tchatat M, Dondjang J-P, Hietet G, Fouda T (2010) An empirical analysis of the biodiversity and economic returns to cocoa agroforests in southern Cameroon. J Sustain For 29:638–670
Gockowski J, Tchatat M, Dondjang J-P, Hietet G, Fouda T (2010) An empirical analysis of the biodiversity and economic returns to cocoa agroforests in southern Cameroon. Journal of Sustainable Forestry 29(6-8): 638–670
Godoy R, Lubowski R, Markandya A (1993) A method for the economic valuation of non-timber tropical forest products. Econ Bot 47(3):220–233
Graves AR, Burgess PJ, Liagre F, Terreaux J-P, Dupraz C (2006) Development and use of a framework for characterizing computer models of silvoarable economics. Agrofor Syst 65:53–65
Gregory GR (1987) Resource economics for foresters. Wiley, New York

Guo Z, Zhang Y, Deegen P, Uibrig H (2006) Economic analyses of rubber and tea plantations and rubber-tea intercropping in Hainan, China. Agrofor Syst 66:117–127

Hoekstra DA (1985) Choosing the discount rate for analysing agroforestry systems/technologies from a private economic viewpoint. For Ecol Manag 10(1–2):177–183

Hoekstra DA (1987) Economics of agroforestry. Agrofor Syst 5:293–300

Hoekstra D (1990) Economics of agroforestry. In: MacDicken K, Vergara N (eds) Agroforestry classification and management. Wiley, New York, pp 310–331

Hoekstra DA (1994) Factors influencing the economic viability of three agroforestry systems in Eastern and Southern Africa. For Ecol Manag 64:265–276

Jabbar MA, Larbi A, Reynolds L (1994) Profitability of alley farming with and without fallow in southwest Nigeria. Exp Agricu 30:319–327

Jain SK, Singh P (2000) Economic analysis of induatrial agroforestry: poplar (*Populus deltoides*) in Uttar Pradesh, India. Agrofor Syst 49:255–273

Kang BT, Reynolds L, Attah-Krah AA (1990) Alley farming. Adv Agron 43:315–359

Klemperer WD (1996) Forest resource economics and finance. McGraw-Hill, New York, pp 551

Knapp KC, Sardosky PA (2000) Economics of agroforestry production in irrigated agriculture. J Agric Resour Econ 25(1):285–306

Lawson GJ, Crout NMJ, Levy PE, Mobbs DC, Wallace JS, Cannell MGR, Bradley RG (1995) The tree-crop interface: representation by coupling of forests and crop-process models. Agrofor Syst 30:199–21

Magrath W, Arens P (1989) The costs of soil erosion on Java: a natural resource accounting approach, the world bank, policy planning and research staff. Environment department working paper No. 18

Matthews RB, Lungu S, Volk J, Holden ST, Solberg K (1992) The potential of alley cropping in improvement of cultivation systems in the high rainfall areas of Zambia II. Maize production. Agrofor Syst 17:241–261

Mead R, Willey RW (1980) The concept land equivalent ratio and advantages in yield from intercropping. Exp Agric 16:217–228

Mehta NG, Leuschner WA (1997) Financial and economic analyses of agroforestry systems and a commercial timber plantation in the La Amistad Biosphere reserve, Costa Rica. Agrofor Syst 37:175–185

Menz KM, Grist P (1996) Increasing rubber planting density to shade *Imperata*: a bioeconomic modeling approach. Agrofor Syst 34:291–303

Mercer DE (2004) Adoption of agroforestry innovations in the tropics: a review. Agrofor Syst 61-62(1):311–328

Mercer DE, Haggar J, Snook A, Sosa M (2005) Agroforestry adoption in the Calakmul biosphere reserve, Campeche, Mexico. Small-Scale For Econ 4(2):163–184

Molua EL (2005) The economics of tropical agroforestry systems: the case of agroforestry farms in Cameroon. For Policy Econ 7:199–211

Montambault JR, Alavalapati JRR (2005) Socioeconomic research in agroforestry: a decade in review. Agrofor Syst 65:151–161

Nair PKR (1993) An introduction to agroforestry. Kluwer, Dordrecht

Ndoye O, Perez MR, Eyebe A (1997) The markets of non-timber forest products in the humid forest zone of Cameroon. Rural Development Forestry Network Paper 22 c, ODI, London

Niskanen A (1998) Financial and economic profitability of reforestation in Thailand. For Ecol Manag 104:57–68

Neupane RP, Thapa GB (2001) Impact of agroforestry intervention on farm income under the subsistence farming system of the middle hills, Nepal. Agrofor Syst 53:31–37

Palada MC, Kang BT, Claassen SL (1992) Effect of alley cropping with *Leucaena leucocephala* and fertilizer application on yield of vegetable crops. Agrofor Syst 19:139–147

Opio C, Fredj K, Wang B (2009) Profitability of manual brushing in young lodgepole pine plantations. BC J Ecosyst Manag 10(1):15–21

Price C (1995) Economic evaluation of financial and non-financial costs and benefits in agroforestry development and the value of sustainability. Agrofor Syst 30:75–86

Prinsley R (1990) Agroforestry for sustainable production: economic implications. Commonwealth Science Council, London

Ramirez A, Seré C, Uquillas J (1992) An economic analysis of improved agroforestry practices in the Amazon lowlands of Ecuador. Agrofor Syst 17:65–86
Ramirez OA, Somarriba E, Ludewigs T, Ferreira P (2001) Financial returns, stability and risk of cacao-plantain-timber agroforestry systems in central America. Agrofor Syst 51:141–154
Randall A (1987) Resource economics. Wiley, New York
Rasul G, Thapa GB (2006) Financial and economic suitability of agroforestry as an alternative to shifting cultivation: the case of the Chittagong hill tracts, Bangladesh. Agric Syst 91:29–50
Reeves LH, Lilieholm RJ (1993) Reducing financial risk in agroforestry planning: a case study in Costa Rica. Agrofor Syst 21:169–175
Rezende JLP, Oliveira AD (1993) Avaliação de projetos florestais. UFV Imprensa Universitaria, Viçosa, pp 47
Ruhigwa BA, Gichuru MP, Spencer DSC, Swennen R (1994) Economic analysis of cut-and-curry, and alley cropping systems of mulch production for plantains in south-eastern Nigeria. Agrofor Syst 26:131–138
Russell D, Franzel S (2004) Trees of prosperity: agroforestry, markets and the African smallholder. Agrofor Syst 61:345–355
Sang H (1988) Project evaluation: techniques and practices for developing countries. Wilson, New York
Scherr SJ (1992) The role of extension in agroforestry development: evidence from western Kenya. Agroforest Syst 18: 47–68
Sharma RA, McGregor MJ, Blyth JF (1991) The social discount rate for land use projects in India. J Agric Econ 42(1):86–92
Steffan-Dewenter I, Kessler M, Barkmann J, Bos MM, Buchori D, Erasmi S, Faust H et al (2007) Tradeoffs between income, biodiversity, and ecosystem functioning during tropical rainforest conversion and agroforestry intensification. PNAS 104(12):4973–4978
Steven RD, Jabara CC (1988) Agricultural development principles: Economic theory and empirical evidence. John Hopkins University Press, Baltimore
Swinkels RA, Scherr SJ (1991) Economic analysis of agroforestry technologies: an annotated bibliography. ICRAF, Nairobi
Swinkels RA, Franzel S, Shepherd KD, Ohlsson E, Ndufa JK (1997) The economics of short rotation improved fallows: evidence from areas of high population density in western Kenya. Agric Syst 55(1):99–121
Tabuna H (1999) The market for Central African non-wood forest products in Europe. In: Sunderland TCH, Clark LE, Vantomme P (eds) Current research issues and prospects for conservation and development. FAO, Rome, pp 251–263
Upton M (1987) African farm management. Cambridge University Press
Upton M (1996) The economics of tropical farming systems. Cambridge University Press, Oakleigh
Verinumbe I, Knipsheer H, Enabor EE (1984) The economic potential of leguminous tree crops in zero-tillage cropping in Nigeria: a linear programming model. Agrofor Syst 2:129–138
Von Platen HH (1992) Economic evaluation of agroforestry systems of cacao (*Theobroma cacao*) with laurel (*Cordia alliodora*) and poro (*Erythrina poeppigiana*) in Costa Rica. In: Sullivan et al. (eds) Financial and economic analysis of agroforestry systems: proceedings of a workshop held in Honolulu, Hawaii. July 1991. Nitrogen Fixing Tree Association, Paia, USA, pp 174–189
Wannawong S, Belt GH, McKetta CW (1991) Benefit-cost analysis of selected agroforestry systems in Northeastern Thailand. Agrofor Syst 16:83–94
Wojtowski PA (2002) Agroecological perspectives in agronomy, forestry and agroforestry. Science Publishers, Enfield
Wojtowski PA, Cubbage F (1991) A bordered matrix approach to the bioeconomic modeling of agroforestry systems. Agrofor Syst 14:81–97
Wojtowski PA, Jordan CF, Cubbage FW (1991) Bioeconomic modeling in agroforestry: a rubber-cacao example. Agrofor Syst 14:163–177

Chapter 17
Socio-Cultural Aspects of Agroforestry and Adoption

Abstract Agroforestry has to consider both biophysical and socio-cultural aspects of the practice, the latter being the consideration of the influences that agroforestry has on society and culture, which can, in part, be determined by its social acceptability at the farmer's level, as farmers are considered to be the primary beneficiaries of agroforestry practices. The social acceptability of agroforestry is influenced by heterogeneity in village structure, land and tree tenure arrangements, division of gender roles, and local perceptions and attitudes towards trees. Important socio-cultural factors to be considered in agroforestry include land tenure, labor requirement, marketing of products, local knowledge, local organization, cultural and eating habits, gender, and well-being and age of landowners. Agroforestry technologies should be simple but robust and initially designed to satisfy the needs of poor farmers so as to facilitate social acceptability. Less-risky agroforestry practices are more likely to be accepted in rural areas. Studies on social benefits and costs, land and tree tenure and adoption, structure, functioning, and evolution of social institutions in communities, as well as identification of factors affecting the adjustment and the response by the community to different types of innovation can help agroforestry researchers plan and prepare strategies and actions for the dissemination of agroforestry technologies. The success of any agroforestry project is influenced by public policies and regulations that provide incentives to integrate trees on farms and promote the use of products from these trees.

17.1 Introduction

Socio-cultural factors important for agroforestry development include heterogeneity in social organization of the village, land and tree tenure arrangements, division of gender roles, and local perceptions and attitudes towards trees (Wiersum 1987). Understanding local communities by identifying livelihood strategies and different manners of resource use is crucial for the implementation of any agroforestry project. Therefore, agroforestry involves both biophysical and social dimensions. Poor farmers are usually the primary beneficiaries of agroforestry, and are increasingly becoming the drivers of agroforestry development (Simons and Leakey 2004). Agroforestry technologies must therefore be relevant and applicable to small-scale land users with limited capital, should require few resources, and fulfill primarily

A. Atangana et al., *Tropical Agroforestry*, DOI 10.1007/978-94-007-7723-1_17,

basic household needs. Technologies must also be market-oriented, to increase, sustain and diversify farmers' income sources. Consequently, social acceptability is an important factor in the success of agroforestry technologies, which should be simple and robust, and initially designed to satisfy the needs of poor farmers.

17.1.1 The Social Aspects of Agroforestry Research

Agroforestry as a science in the early days was focused on biophysical and ecological aspects. However, increasing concern about the adoption and acceptability of agroforestry practices raised the importance of the social aspects of agroforestry. This social dimension is highly recognized, and agroforestry has been referred to as "a very social science" (Pawlick 1989). Montambault and Alavalapati (2005) analyzed over 500 publications dealing with socio-economic issues in agroforestry from 1992 to 2002, highlighting the importance of socio-economic aspects in agroforestry adoption. Agroforestry is not unique in terms of the recognition and principles of the social sciences. Recognition of the social factors that play an important role in all applications of biological sciences in developing countries has been widely established. However, it is rightly argued that socio-cultural considerations are particularly important in agroforestry. Burch (1991) reviewed the theoretical perspectives of the contribution that the social sciences can make to agroforestry. He postulated that five types of studies illustrate the range of empirical studies that can help researchers in designing agroforestry interventions:

- Evaluation studies: social benefits and costs are measured and assigned to the government, project, village and households respectively.
- Tenure studies: theories and methods to examine the influence of tree and land tenure on the practice of agroforestry.
- Institutional studies: the use of knowledge about the structure, functioning, and evolution of social institutions in communities and their impact on agroforestry practices.
- Community studies: factors affecting the adjustment and the response by the community to different types of innovation.
- Adoption studies: the elements that influence the adoption process of new or improved technology by different client groups.

Mercer and Miller (1998) identified the priority research areas of agroforestry socio-economics, including the empirical and theoretical analysis of agroforestry adoption decisions, whether it is at the local, national, or regional level. Adoption decisions in agroforestry depend on several socio-cultural factors, most of which are discussed in this chapter.

17.1.2 *Important Socio-Cultural Factors for Adoption of Agroforestry*

17.1.2.1 Land Tenure

Land tenure is a crucial factor for the implementation of long-term land-use practices. A land tenure system that does not guarantee continuous ownership and control of the land is not likely to be favorable to the adoption of long-term agroforestry investments. This also applies to short-term practices, some of whose benefits would be achieved in the long run. Land rights are very important in determining if the benefits of agroforestry will reach the intended beneficiaries (Bruce and Fortmann 1988). Traditional reservations of poor farmers have always included concerns about the loss of control of land reclaimed through the planting of trees (Gregersen and McGaughey 1985). The total area of land owned or cultivated by farmers, and the total amount of land belonging firmly to farmers, had a significant positive influence on the establishment of tree plantations in Sumatra (Zhang and Owiredu 2007). Otsuka et al. (2001) demonstrated that the granting of individual secure properties to farmers spurred earlier tree planting. A forecasted increase in tenure security led to early planting of trees (Otsuka et al. 2001). Also, in regard to soil conservation in agroforestry, securing land tenure was associated with long-term investments in terraces in Ethiopia (Gebremedhin and Swinton 2003). However, tenure security for women, tenants and pastoralists in the rural areas in West Africa, especially in Niger and Benin, is limited by both customary land rights systems and state law (Neef 2001).

Planting trees is seen as a sign of ownership in certain regions of Africa, such as Benin, Niger (Neef 2001), and Cameroon (Degrande et al. 2006), and customary tenure does not allow non-landowners to plant trees. The decision to plant trees is largely dependent on security of land tenure (Degrande et al. 2006). Also, the motivation to invest in soil fertility improvement for future use of the land is low, unless the benefits will accumulate for the tree planter (Francis 1989).

Land tenure is sometimes different from tree tenure, and tree rights are often distinct from land rights. The tenure of the tree is not always the same as that of the land, and rights to tree products are treated differently than tree removal rights (Fortmann 1988). Tree tenure is the right to have or inherit trees, plant trees, use trees and tree products as well as exclude others from these uses, and the right to dispose of tree products (Fortmann 1988; Fortmann and Riddell 1985). These various rights differ widely across cultures, and may even be different depending on the species of tree. This invariably has a major influence on the social acceptability of any new initiative in agroforestry.

17.1.2.2 Labor

The practice of agroforestry most often requires more work and changes in farm activities from the norms. Work requirements are reviewed by farmers before they decide to adopt a new agroforestry practice (Hoskins 1987). Families traditionally use strategies that employ different family members at various times of the year for different tasks. For example, there is a division of labor on farms at the household level on a gender basis in the humid tropics of Cameroon. The division of labor on the farm typically involves men who clear the land and women who sow seeds and weed the farm when necessary for food crops. Men are responsible for cash crops such as cocoa or coffee. Labor input on farms peaks in December (land clearing), and March-April (seeding), and during the rainy season (weeding). Additional work for people already fully occupied in peak working seasons is considered more expensive than additional demands during the off-peak season. For example, alley cropping is labor-intensive, with most of the labor demand occurring during the rainy season, the busiest period of the year. The production cost will dramatically increase if additional labor is required at that time. Although these additional costs are compensated for by additional benefits later, the immediate need for additional work could sometimes be a deterrent for the adoption of this practice (Kang et al. 1990). This can be seen in the fact that the additional work requirements in alley cropping systems have been a major cause for non-adoption in the humid lowland forest areas of Cameroon (Degrande and Duguma 2000; Degrande et al. 2007).

17.1.2.3 Marketing of NTFPs and AFTPs and Adoption of Agroforestry

The income that can be obtained from a land use system is an important criterion in judging social acceptability. The processing and sale of agroforestry products is an important potential source of income for people involved in the market chain. Farmers involved in njansang (*Ricinodendron heudelotii*) production can realize a more than 80 % increase in their revenue derived from njansang with proper marketing (Facheux et al. 2007). The commercialization of *Irvingia gabonensis* kernels from Cameroon, Equatorial Guinea, Gabon, Nigeria and Central African Republic annually generates US $ 260,000 (Ndoye et al. 1997). The kernels of *I. gabonensis* are also exported to Europe and North America, where West African natives use these kernels for soup thickening. Boosting the market opportunities for agroforestry products is critical to the adoption of agroforestry practices (Russell and Franzel 2004). The increase in demand for these products will require the development of provisions to meet such demands. It should be noted that certain agroforestry products, such as the seeds of *Irvingia gabonensis*, *Irvingia wombolu*, *R. heudelotii*, *Garcinia kola*, and *Cola acuminata*, *Dacryodes edulis* fruits, *Gnetum africanum* leaves, and the bark of *Annickia chlorantha*, *Pausinystalia johimbe* (yohimbe), and *Prunus africana* already have established commercial value. The demand for these products is not only local, but also international (Tabuna 1999). To sustainably supply the local and international market of non-timber forest products (NTFPs), it is necessary to domesticate the species that produce these NTFPs and integrate them into agricultural systems. Market information systems are necessary for local producers to be able to meet the demand. A functioning market system would allow exporters to know what quantities are available and where, and what offer is available at what price. The increased interest in non-timber forest products, such as those from *Allanblackia floribunda*, yohimbe, and *A. chlorantha*, increases the need for improvements in their marketing, and on the benefits that farmers could obtain by investing in the integration of these species into their land-use systems.

Along with difficulties in gaining access to markets, a lack of skills in management and organization are amongst the major constraints to the growth of small-scale agricultural enterprises (FAO 1987). In the humid tropics of West and Central Africa, ICRAF and its local partners are training grassroots farmers' organizations and poor farmers on entrepreneurship building and business management (Tchoundjeu et al. 2006; Facheux et al. 2007; Awono et al. 2010).

The processing of non-wood products adds value and allows long-term storage of those non-wood products. Because most products targeted by tree domestication are perishable, processing allows the products to be more easily transported and traded, both locally and internationally. Farmers have developed traditional methods to preserve some NTFPs (Tchoundjeu et al. 2005). Research could build upon those techniques to develop processing methods that allow long-term preservation without altering the nutritional and sensory qualities of these products. Some forest products are used as raw materials by agro-industrial companies, necessitating quality standards for the marketing of novel food products from agroforestry (Leakey 1999). The development of methods for pest control during storage is also important to reduce the loss of stocks during transportation and trade.

17.1.2.4 Other Social Factors Affecting the Acceptability of Agroforestry

The economic feasibility of any project or enterprise has a major influence on its social acceptability. Projects that are less risky and have lower initial costs are more likely to attract people. A survey on the acceptance of agroforestry with 300 farmers in Bendel State, Nigeria, found that the social acceptability of a given project was heavily based on cost-sharing arrangements between the government and farmers (Osemebo 1987), with which prospects were higher for the introduction of trees into traditional land-use systems. Additionally, the acceptance of agroforestry was affected by the availability of a viable extension service, and the potential for a direct economic return from the tree component in the agroforestry system. During the survey, farmers indicated their willingness to plant trees under the following conditions (Osemebo 1987):

- Seedlings are available free of charge.
- The possibility to grow crops between trees without adverse effects from the trees on food crops.
- The possibility to earn an income from these trees.

Many other social factors are extremely important in the introduction, development and scaling-up of agroforestry technologies. Local knowledge, local organization and participation in tree management, cultural and eating habits, land tenure, external and internal on-farm income, food security, and demographic factors such as the health, well-being, gender and age of farmers are all critical issues to the successful introduction and development of agroforestry. Because social and cultural contexts vary from country to country, it would be difficult to select and describe universally critical socio-cultural variables for agroforestry. Social and cultural situations can vary depending on the locality, environment, and main production activities. Degrande et al. (2006), in a survey in lowland humid forest zone of Cameroon and Nigeria, identified market access, land tenure, land-use and access to forest resources as key elements determining farmers' tree-growing strategies.

17.1.3 Farmers' Perceptions of Planting Trees

The success of almost any activity involving trees and people is strongly influenced by government regulations, perceptions, preconceptions and preferences. Government officials do not always understand the needs of farmers. The misunderstanding of farmers' needs by officials and experts may be due to (1) the biased or inadequate knowledge of experts about farmers, and (2) the way that most project assessments about the success of tree planting are carried out. A study by Dove (1992) on beliefs about Pakistan's farmers illustrates the first point, and that of Campbell (1992), on the preferences of farmers regarding tree planting and regeneration in Haiti, illustrates the second.

In the humid zone of West and Central Africa, fruit trees are preferred over timber trees for planting on farms by farmers. The most important factors influencing the strategies of tree planting appear to be market access, the use of land, and access to forest resources (Degrande et al. 2006). Farmers have a preference for trees they know and are familiar with and the primary reason for planting fruit trees over others is consumption (Table 17.1). The study found that 52 % of the 9,202 fruit trees inventoried in the various study sites in Nigeria and Cameroon were native to the area. In addition, all of the farmers interviewed in Nigeria, as opposed to only 73 % in Cameroon, expressed a preference for local fruit tree species (Degrande et al. 2006). These results indicate that any decision about the woody species introduced into agroforestry projects should be made after a careful study of the preferences and perceptions of farmers.

17.1.4 Public Policies and Implementation of Agroforestry

The influence of policy and legislation on smallholder's decisions to plant trees was reported by Foundjem-Tita et al. (2012) who used Cameroon as a case study. The authors found that though the Government of Cameroon is committed to include NTFPs and agroforestry tree products (i.e., products that are gathered from trees on farms, AFTPs) in the poverty reduction strategies, legislation contradicts the poverty reduction goals. Further, legislation did not distinguish between products from trees in the wild (i.e., NTFPs) and products that are gathered from trees on farms

Table 17.1 The proportion (%) of farmers in communities in Cameroon and Nigeria that gave specific reasons for planting or retaining fruit trees in a case study provided in Degrande et al. (2006)

Community (and market access)	Chop farm (dirt road)	Elig-Nkouma (seasonal dirt road)	Nko'ovos II (tarmac road)	Makènènè Est (in community)	Ilile (tarmac road)	Uguaji (tarmac road)
Number of fruit trees on farm	365	1845	1694	3184	951	888
Consumption	56	62	49	19	68	47
Sale	14	12	27	69	27	50
Shade	0	4	4	5	0	0
Border marking	2	3	2	1	1	0
Other or unknown	28	19	18	6	4	3

(i.e., agroforestry tree products), an oversight that would not facilitate the shift from the collection of NFTPs to AFTPs. The explanation for this is that informal taxes are applied on NTFPs in Cameroon, and there is no regulation allowing AFTPs not to be under the application of these taxes. For example, 530 USD are requested from a driver of a car transporting 1.5 t of *Gnetum* spp. in a distance of about 450 km in Cameroon (Ndoye and Awono 2010). Obviously, such taxes would negatively affect the market chain of AFTPs, and there would be no incentive to promote their cultivation and collection on farms. Systems market constraints and inconsistencies in institutional and regulatory frameworks in agroforestry were previously reported in a study that analyzed the policy terrain affecting agroforestry around protected areas in Cameroon, Mali and Uganda (Ashley et al. 2006). It is necessary that governments of tropical countries develop and apply policies and regulations that provide incentives to plant trees on farms.

17.1.5 Social Acceptability

The poor benefit from agroforestry only when the system is designed to meet the social needs of the poor (Chowdhry 1985). The best measure of the social success of an innovation is the speed with which it propagates. The factors that explain technology adoption within an economic framework are preferences, resource endowments, market incentives, biophysical factors, risks, and uncertainties (Pattanayak et al. 2003). An analysis of 32 studies on agroforestry and related investments showed that preferences and resource endowments were the factors most studied related to adoption (Pattanayak et al. 2003). However, adoption behavior is probably most influenced by risk, resource and biophysical factors. The full realization of agroforestry's benefits requires a fundamental understanding of how and why farmers make long-term decisions, and apply this knowledge to the design, development, and commercialization of innovations (Mercer 2004). Areas of identified research in agroforestry technology adoption include the following: developing a better understanding of the role of risk and uncertainty, ideas on why and how farmers adapt and modify innovations, factors influencing the intensity of adoption, and a spatial analysis of adoption at the village level (Mercer 2004).

An example of changes to agroforestry technology made by farmers is given by Mbile et al. (2004). In this example, farmers slightly modified the design of a non-mist poly propagator to accommodate the resources available to them. This modification facilitated the adoption of the technology by local farmers who, in the case of participatory domestication of agroforestry species, popularized their own technology adopted through the rural resources centers (Tchoundjeu et al. 2006).

The adoption of a technology by farmers is the result of a complex process related to a matrix of factors, including household characteristics, community factors, socio-economic incentives, access to information, local institutional arrangements, and agriculture macro-policies (Ajayi et al. 2003). The adoption of improved fallows is an on-going process that must be facilitated by appropriate and conducive policies and institutional incentives (Ajayi et al. 2003). Relationships between the factors influencing this adoption, as well as the relative importance of the various factors themselves, are important. For example, Bannister and Nair (2003) noted the importance of household and farm characteristics in the adoption of agroforestry technologies in Haiti in a study that aimed at understanding the circumstances that have led to an attitude change in adopters of agroforestry. The study revealed that households that had more total years of school or secure land tenure planted more trees. Also, male-headed households planted more tree seedlings than female-headed ones (Bannister and Nair 2003).

Bibliography

Ajayi OC, Franzel S, Kuntashula E, Kwesiga F (2003) Adoption of improved fallow technology for soil fertility management in Zambia: empirical studies and emerging issues. Agrofor Syst 59:317–326

Ashley R, Russell D, Swallow B (2006) The policy terrain in protected landscapes: challenges for agroforestry in integrated landscape conservation. Biodivers Conserv 15:663–689

Awono A, Ndoye O, Preece L (2010) Empowering women's capacity for improved livelihoods in non-timber forest product trade in Cameroon. Int J Soc For 3(2):151–163

Bannister ME, Nair PKR (2003) Agroforestry adoption in Haiti: the importance of household and farm characteristics. Agrofor Syst 57:149–157

Bruce A, Fortmann L (1988) Whose trees? Westview Press, Boulder

Burch RB Jr (1991) Thinking social scientifically about agroforestry. In: Burch RB Jr, Parker JK (eds) Social science application in Asian agroforestry. Winrock International, Arlington

Campbell EP (1992) Do farmers in a deforested environment need help to grow trees? MSc dissertation, University of Florida, Gainesville

Chowdhry K (1985) Social forestry: who benefits? Community forestry: socioeconomic aspects. FAO/East-West Center, Bangkok

Dagang ABK, Nair PKR (2003) Silvopastoral research and adoption in Central America: recent findings and recommendations for future directions. Agrofor Syst 59:149–155

Degrande A, Duguma B (2000) Adoption potential of rotational hedgerow intercropping in the humid lowlands of Cameroon.ODI Network paper No 103, DFID, UK

Degrande A, Schreckenberg K, Mbosso C, Anegbeh P, Okafor V, Kanmegne J (2006) Farmer's fruit growing strategies in the humid forest zone of Cameroon and Nigeria. Agrofor Syst 67:159–175

Degrande A, Asaah E, Tchoundjeu Z, Kanmegne J, Duguma B, Franzel S (2007) Opportunities for and constraints to adoption of improved fallows: ICRAF's experience in the humid tropics of Cameroon. In: Bationo A, Waswa B, Kihara J, Kimetu J (eds) Advances in integrated soil fertility management in sub-saharan Africa: challenges and opportunities. Springer, Dordrecht, pp 901–910

Dove MR (1992) Foresters' beliefs about farmers: a priority for social science research in social forestry. Agrofor Syst 17:13–41

Ehrlich M, Conway F, Adrien N, LeBeau F, Lewis L, Lauwerysen H, Lowenthal I, Mayda Y, Paryski P, Smucker G, Talbot J, Wilcox E (1987) Haiti: country environmental profile. USAID, Washington DC

Facheux C, Tchoundjeu Z, Foundjem-Tita D, Degrande A, Mbosso C (2007) Optimizing the production and marketing of NTFPs. In: African Crop Science Conference Proceedings 8:1249–1254

FAO (1987) Small-scale forestry-based processing enterprises. FAO, Rome

Fortmann L (1988) Tree tenure factors in agroforestry with particular reference to Africa. In: Fortmann L, Bruce JW (eds) Whose trees—proprietary dimensions of forestry. Westview Press, Boulder, pp 16–33

Fortmann L, Riddell J (1985) Trees and tenure: an annotated bibliography for agroforesters and others. Land Tenure Center, Univ. of Wisconsin, USA/ICRAF, Nairobi

Foundjem-Tita D, Tchoundjeu Z, Speelman S, D'Haese M, Degrande A, Asaah A, van Huylenbroeck G, van Damme P, Ndoye O (2012) Policy and legal frameworks governing trees: incentives or disincentives for smallholder tree planting decisions in Cameroon? Small-Scale For. doi:10.1007/s11842-012-9225-z

Francis PA (1989) Land tenure systems and the adoption of alley farming. In: Kang BT, Reynolds I (eds) Alley farming in the humid and subhumid tropics. IDRC, Ottawa, pp 182–195

Franzel S, Jaenicke H, Janssen W (1996) Choosing the right trees: setting priorities for multipurpose tree improvement, research report, 8. ISNAR, The Hague, p 87

Gebremedhin B, Swinton SM (2003) Investment in soil conservation in northern Ethiopia: the role of land tenure security and public programs. Agric Econ 29:69–84

Gregersen H, McGauhey SE (1985) Improving policies and financing mechanisms for forestry development. Economic and social development department, Inter-American development bank, Washington DC

Hoskins MW (1987) Agroforestry and the social milieu. In: Steppler HA, Nair PKR (eds) Agroforestry: a decade of development. ICRAF, Nairobi, pp 191–203

Kang BT, Reynolds L, Atta-Krah AN (1990) Alley farming. Adv Agron 43:315–359

Leakey RRB (1999) Potential for novel food products from agroforestry trees. Food Chem 66:1–14

Mbile P, Tchoundjeu Z, Degrande A, Avana M-L, Tsobeng A (2004) Non-mist propagation by resource poor, rural farmers of the forest zone of Cameroon: some technology adaptations to enhance practice. For Trees Livelihoods 14:43–52

Mercer DE (2004) Adoption of agroforestry innovations in the tropics: a review. Agrofor Syst 61–62 (1–3):311–328

Mercer DE, Miller RP (1998) Socioeconomic research in agroforestry: progress, prospects, priorities. Agroforest Syst 38: 177–193

Montambault JR, Alavalapati JRR (2005) Socio-economic research in agroforestry: a decade in review. Agrofor Syst 65:151–161

Ndoye O, Perez MR, Eyebe A (1997) The markets of non-timber forest products in the humid forest zone of Cameroon. Rural development forestry network paper 22c, ODI, London

Ndoye O, Awono A (2010) Policies for *Gnetum* spp. Trade in Cameroon: overcoming constraints that reduce benefits and discourage sustainability. In: Laird SA, McLain RJ, Wynberg RP (eds) Wild product governance: finding policies that work for non timber forests products. People and plants International conservation series. Earthscan, London

Neef A (2001) Land tenure and soil conservation practices—evidence from West Africa and Southeast Asia. In: Stott DE, Mohtar RH, Steinhardt GC (eds) Sustaining the global farm. Selected papers from the 10th International soil conservation organization meeting held May 24–29, 1999, at Purdue University and the USDA-ARS National Soil Erosion Research Laboratory, pp 125–130

Osemebo GJ (1987) Smallholder farmers and forestry development: a study of rural land-use in Bendel, Nigeria. Agricultural Syst 24:31–51

Otsuka K, Suyanto S, Sonobe T, Tomich TP (2001) Evolution of land tenure institutions and development of agroforestry: evidence from customary land areas of Sumatra. Agric Econ 25:85–101

Pattanayak SK, Mercer DE, Sills E, Yang J-C (2003) Taking stock of agroforestry adoption studies. Agrofor Syst 57:173–186

Pawlick T (1989) Agroforestry: a very social science. Agroforestry Today (ICRAF) 1(2):2–5

Raintree JB, Warner K (1986) Agroforestry pathways for the intensification of shifting cultivation. Agrofor Syst 4:39–54

Russell D, Franzel S (2004) Trees of prosperity: agroforestry, market and the African smallholder. Agroforest Syst 61: 345–355

Simons AJ, Leakey RRB (2004) Tree domestication in tropical agroforestry. Agrofor Syst 61:167–181

Tabuna H (1999) The market for Central African non-wood forest products in Europe. In: Sunderland TCH, Clark LE, Vantomme P (eds) Current research issues and prospects for conservation and development. FAO, Rome, pp 251–263

Tchoundjeu Z, Atangana AR, Degrande A (2005) Indigenous methods in preserving bush mango kernels in Cameroon. Am J Appl Sci 2(9):1337–1342

Tchoundjeu Z, Asaah EK, Anegbeh PO, Degrande A, Mbile P, Facheux C, Tsobeng A, Atangana AR, Ngo Mpeck ML, Simons AJ (2006). Putting participatory domestication into practice in West and Central Africa. For Trees and Livelihoods 16:53–69

Wiersum KF (1987) Significance of social organization and cultural attitudes for agroforestry development. Adv Agrofor Res 23–33

Zhang D, Owiredu EA (2007) Land tenure, market, and the establishment of forest plantations in Ghana. For Policy Econ 9:602–610

Part V
Outlook on Tropical Agroforestry

Chapter 18
Tropical Agroforestry for Biofuels Production

Abstract Biofuels are a promising alternative to fossil fuels whose combustion is a major source of greenhouse gas emissions, which is linked to global warming. Biofuel production using agroforestry practices is emerging as a valuable co-benefit of agroforestry. Feedstock for biofuel includes wood biomass, starch, sugars, and vegetable oils. Biofuel crops are cultivated in the tropics and some of them have potential for exploitation within agroforestry systems. Agroforestry tropical woody species valued for biodiesel production include *Azadirachta indica*, *Balanites aegyptica*, *Calophyllum inophyllum*, *Jatropha curcas*, *Elaeis guineensis* Jacq., *Nephelium lappaceum*, *Mesua ferrea*, *Pongamia glabra* and *Pongamia pinnata*. Agricultural crops such as sugarcane (*Saccharum officinarum* L.) and corn (*Zea mays* L.) are being promoted for bioethanol production whereas soybean (*Glycine max* L. Merr.) is for biodiesel production. Biofuel production in the tropics using agroforestry practices is environmentally friendly, as it addresses climate change and food security concerns, thereby reducing pressure exerted by local farmers on forests. However, to promote biofuel production using agroforestry practices in the tropical context, it is imperative to design viable tree-crop systems that include biofuel crop species. The use of agricultural crops from agroforestry systems that are in the human food chain for biofuel production is not a socially or economically viable option in the tropics because it can exacerbate the problems of food insecurity. The agroforestry systems designed to improve food security while producing feedstock for biofuel are socially and economically acceptable while providing environmental benefits.

18.1 Introduction

The burning of fossil fuels is one of the major causes of the increase in atmospheric greenhouse gas concentrations. To address this issue, several initiatives have been taken, the use of biofuels as substitutes for fossil fuels being one example. Several products have been investigated for their efficiency as raw materials for biofuels, including bio-waste, cellulosic materials, oilseed crops, sugar- and starch-rich crops. Bio-waste includes forestry (bark, sawdust, waste lumber), agricultural, and animal waste (e.g., cattle dung). Vegetable oil crops are used mostly for biodiesel production, whereas starchy- and sugar-rich feedstocks are used to produce bioethanol

A. Atangana et al., *Tropical Agroforestry*, DOI 10.1007/978-94-007-7723-1_18,

fuels. Cellulosic feedstocks are used to produce bioethanol fuels. Starchy feedstocks used around the world for ethanol production include cereal grains (e.g., maize and wheat), potato, sweet potato, and cassava. Sugar feedstocks include sugarcane, sugar beet, sweet sorghum, and various fruits. For instance, Brazil has developed one of the most successful fuel ethanol programs from sugarcane in the world. Cellulosic feedstocks being abundant and outside the human food chain makes them relatively inexpensive feedstocks for ethanol production. To date, biofuels for industrial use include oilseeds, sugarcane and timber. The majority of biofuel crops worldwide are oilseed species, including oil palm (*Elaeis guineensis*), rapeseed (*Brassica napus* L.), soybean (*Glycine max* (L.) Merr.) and sunflower (*Helianthus annuus* L.), and a starch crop (maize (*Zea mays*)). Other biodiesel sources include *Prunus dulcis* Batsch (almond), *Arachis hypogaea*, *Camelina sativa* L. Crantz, *Carapa guianensis* Aublet, *Caryocar* spp., *Cocos nucifera* L. (coconut), *Cynara cardunculus* L., *Dipterix odorata* (Aubl.) Willd., *Jatropha curcas* L., *Lesquerella fendleri* (Gray) S.Wats., *Linus usitatissimum* Linnaeus (linseed), *Madhuca indica* (J.Konig) J.F.Macbr., *Azadirachta indica* A.Juss. (neem), *Orbignia* spp., *Milletia pinnata* (L.) Panigrahi syn. *Pongamia glabra* Vent., *Shorea robusta* Roth, *Ricinus communis* L. (castor bean), *Hevea brasiliensis* Müll.Arg. (rubber), *Sesamum indicum* L. (sesame), *Sorghum bicolor* (L.) Moench, *Nicotiana tabacum* L., and *Triticum* spp. (Demirbas 2009; Singh and Singh 2010). Even though sugarcane, oil palm, maize, and soybean are found in the tropical agricultural landscape, they are rarely used for biodiesel production in the tropics. Instead, cattle dung and wood sticks are commonly used as energy sources in rural areas in the tropics. However, some agroforestry species have proven to be efficient biofuel crops for industrial use in the tropics. These are mostly multipurpose species that provide additional benefits to farmers, including food, fodder, medicine, shelter or atmospheric nitrogen fixation.

18.2 Tropical Agroforestry Species with Potential for Biofuel Production

Even though most of the work on biofuel production using agroforestry is recent, some agroforestry species have been proven to produce effective biofuels. *Jatropha curcas* (Euphorbiaceae), a multipurpose drought-resistant perennial plant used to protect fields from erosion, to reclaim land, and to use as live fence, produces seeds containing inedible oil easily convertible into biodiesel (Achten et al. 2007; Heller 1996; Kumar and Sharma 2008). Indeed, the cetane number (52.3) of *J. curcas* seed oil makes it suitable for biodiesel production (Azam et al. 2005). Seed cake and glycerin obtained from Jatropha biodiesel production are also valued as biofertilizer and for the production of cosmetics, respectively. Some countries, such as India, are promoting the cultivation of *Jatropha* in wastelands to increase biodiesel production (Leduc et al. 2009). Achten et al. (2010a) reported that *Jatropha* cultivation on such lands in India improved the structural ecosystem quality, but reduced the

Table 18.1 Analysis of engine test results of diesel blends containing various percentages of Desert date biodiesel (Chapagain et al. 2009)

Biodiesel (%)	Rpm	Fuel consumption (mL min^{-1})	Moment (Nm)	Co (%)	CO_2 (%)	O_2 (%)	HC (ppm)	NO (ppm)
0%	1,200	59.45	107	0.036	ND	ND	28	390
	1,500	85.61	125	0.043	ND	ND	38	490
	1,800	109.39	132	0.057	ND	ND	34	500
	2,000	128.42	113	0.035	ND	ND	32	480
	2,200	121.28	104	0.028	ND	ND	23	460
5%	1,200	59.29	103	0.03	ND	ND	50	340
	1,500	83.01	119	0.034	ND	ND	51	425
	1,800	109.1	126	0.038	ND	ND	48	435
	2,000	111.47	115	0.025	ND	ND	36	425
	2,200	113.84	102	0.013	ND	ND	20	395
100%	1,200	66.26	105	0.036	ND	ND	44	384
	1,500	93.68	122	0.037	ND	ND	43	487
	1,800	116.53	125	0.034	ND	ND	37	486
	2,000	116.53	112	0.03	ND	ND	31	470
	2,200	118.82	100	0.011	ND	ND	7	433

HC hydrocarbons, *NO* nitric oxide, *ppm* parts per million, *rpm* revolutions per minute of the engine, *ml* milliliter, *Nm* Newton meter, *ND* non-determined

functional ecosystem quality, probably due to N-fertilizer application. The comparison of environmental impacts of the production and use of *Jatropha* biodiesel with that of the life cycle of fossil fuel in Allahabad, India, revealed that *Jatropha* biodiesel induces an 82% decrease in non-renewable energy requirements, and a 55% reduction in global warming potential (Achten et al. 2010a). *Jatropha* is the target of domestication projects (Achten et al. 2010b). However, to advance domestication efforts of *Jatropha,* a semi-wild plant, information is needed on the regeneration ecology and genetic diversity within the species. Also needed are data on *Jatropha*'s breeding system, and inbreeding and outbreeding effects (Achten et al. 2010b).

The suitability of the desert date (*Balanites aegyptiaca*) as an oil crop for sustainable, large-scale biodiesel production was assessed by Chapagain et al. (2009). The study found that *B. aegyptiaca* kernels may contain up to 47% oil, mostly palmitic, stearic, oleic and linoleic acids. The authors successfully developed biodiesel production from oil-enriched powder, and the quality of the biodiesel produced from desert date easily met international biodiesel standards. Desert date is an effective bioresource for biofuel production (Chapagain et al. 2009), confirmed by engine tests (Table 18.1). The domestication of desert date should be focused on the selection and development of oil-rich varieties for biodiesel, and fruit varieties whose fruit flesh is rich in sugar. Another tree with high potential for large-scale vegetable oil production in the tropics is *Pongamia pinnata* (Scott et al. 2008). *Pongamia pinnata* is used for the bioamelioration of degraded lands, and nodulates effectively with three strains of rhizobia (*Bradyrhizobium japonicum* strain CB1809, *Bradyrhizobium* spp. strain CB564 and *Rhizobia* spp. strain NGR234) (Scott et al. 2008).

Important characteristics of biodiesels were reported by Azam et al. (2005) and Demirbas (2009). These characteristics include fatty acid methyl esters (FAMEs), iodine value (IV), cetane number (CN), saponification number, viscosity, density,

Table 18.2 Production of non-edible oilseeds and bioresidues in India (Adapted from Singh and Singh 2010)

Species	Oil faction (%)	Estimated seed production (10^6 tones year^{-1})	Oil production (Tons ha^{-1} year^{-1})
Ricinus communis	45–50	0.25	0.5–1.0
Jatropha	50–60	0.20	2.0–3.0
Madhuca indica	35–40	0.20	1.0–4.0
Shorea robusta	10–12	0.20	1.0–2.0
Linseed	35–45	0.15	0.5–1.0
Azadirachta indica	20–30	0.10	2.0–3.0
Pongamia spp.	30–40	0.06	2.0–4.0

cloud and pour points, distillation range, flash point and high heating value. Cetane number (i.e., combustion quality of diesels after being injected) and viscosity were listed as the most important characteristics for biodiesels. For that reason, good biodiesels should have low viscosity and high CN values. The viscosity of *P. pinnata* seed oil at 40 °C ranges between 3.8–4.8 mm^2 s^{-1} (Demirbas 2009), which fits within the International Standard (EN 14214) requirement for biodiesels (3.5–5.0 mm^2 s^{-1} at 40 °C). Also, the flash point (135–150 °C) and the cetane number (55.84) of *P. pinnata* seed oil fit within the EN 14214 standard for biodiesel (Azam et al. 2005; Demirbas 2009). The pour point (lowest temperature at which the oil flows) and cloud point of *P. pinnata* seed oil are 2.1 and 8.3 °C, respectively, indicating that *P. pinnata* seed oil has good potential for use as biodiesel.

Numerous tropical fruit, crop and tree species have also been found to produce vegetable oil with good properties as raw material for biodiesels, including *A. indica*, *Calophyllum inophyllum* (takamaka), *Carica papaya* (papaya), *Mesua ferrea* (nahor) and *Nephelium lappaceum* (rambutan). Indeed, CN values of neem oil, takamaka oil, papaya seed oil, nahor oil and rambutan oil were 57.83, 57.3, 56.27, 54.6–55.10 and 61.17, respectively (Azam et al. 2005; De and Bhattacharyya 1999; Winayanuwattikun et al. 2008).

Large quantities of non-edible oilseeds are used as biofuel feedstock in the tropics. Castor beans are among the most abundantly produced oilseeds in India for biodiesel production (Table 18.2). They are also produced in Brazil, China and other countries, though castor oil contributes to less than 0.15 % in the international oilseed trade market (Scholz and da Silva 2008). As biodiesel for internal combustion engines, castor oil is a safety risk owing to its high viscosity and water content (Scholz and da Silva 2008).

In rural areas in the tropics, poor farmers use firewood as the main energy source. One of the major benefits of agroforestry, especially in savanna areas, is the provision of fuelwood for domestic consumption and sale. Large amounts of fuelwood are consumed in rural areas for heating and cooking. In Kenya, more than 15 million tons of firewood was consumed in 1997 and 17.1 million tons converted to charcoal (Kituyi et al. 2001). The same study also reported the consumption of 1.4 million tons of crop residues as domestic fuel. Fuelwood is also converted to charcoal for domestic use in the tropics. Agroforestry species are used to produce charcoal with valuable properties for biofuel (Fuwape 1993). Indeed, heat combustion from charcoal obtained from *Leucaena leucocephala* and *Tectona grandis* was higher than that of wood (Fuwape 1993).

18.3 Studies on Biofuels in Tropical Agroforestry

Studies on biofuels in agroforestry mostly investigated the suitability of agroforestry products as biofuels and the risks involved in establishing large-scale monocrop plantations of biofuel crops. Also investigated is the increase in the use of edible crops as alternatives to fossil fuels, as it could increase food insecurity in developing countries.

The suitability of vegetable oils as sources of biodiesel is determined by their chemical and physical properties, which should be similar to those of diesel. Therefore, the suitability may be dependent on the climatic zone in which the vegetable oil is produced. The kinematic viscosity (ratio of absolute viscosity, i.e., the resistance of a fluid to flow, to density) at 38 °C, CN ratio, heating value, pour point and flash point are 3.06 $mm^2 s^{-1}$, 50, 43.8 MJ kg^{-1}, 16 and 150 °C, respectively for vegetable oils (Barnwal and Sharma 2005), indicating that biodiesels obtained from these oils may be used under a wide range of temperatures. The pour point of vegetable oils from tropical species is more often higher than 0 °C (for example, the pour point of *Pongamia* vegetable oil is 2.1 °C), indicating that these oils are not suitable as sources of biodiesel for temperate countries in the winter. However, biodiesels from tropical seed oils are suitable in the tropics, as their CN value is sometimes higher than that of diesel. Indeed, the CN values of neem oil, *P. glabra* oil, *P. pinnata* oil, *M. ferrea* oil and *J. curcas* oil are 57.83, 56.2, 55.84, 554.6, 54.6 and 52.31, respectively (De and Bhattacharyya 1999; Azam et al. 2005; Barnwal and Sharma 2005). However, the oil production from these tropical tree and crop species should be assessed in an agroforestry context, so as to avoid large monocrop plantations, which could have negative effects on biodiversity in the tropics.

In the tropical context, biofuel production should take into account food security and environmental concerns. Indeed, biofuel produced from food crops would cause inflation in food prices, thereby impacting local and global food security. Also, producing biofuels from crops growing on arable lands, especially in large monocrop plantations, would affect not only food security, but also negatively impact biodiversity in the tropics. Danielsen et al. (2008) discussed oil palm plantations on forested lands in Indonesia for biofuel production. They concluded that reductions in the rate of deforestation more effectively mitigated climate change than conversion of forests for biofuel production. Similar findings have been reported by Phalan (2009). Also, Carlson et al. (2012) reported that oil palm plantation caused direct loss of 27 % of total and 40 % of peatland forests during the 2007–2008 period in West Kalimantan, Indonesia.

Agricultural expansion for biofuel feedstock production would also increase the demand for irrigation, especially in the semiarid tropics. Also, biorefineries producing bioethanol are high water demanding and would likely have an impact on water availability (Pate et al. 2007; Phillips et al. 2007).

Biodiesel crops with agroforestry potential should be promoted in the tropics. Agroforestry tree/shrub crops such as *Jatropha* and pongamia may be promoted for cultivation on marginal lands in the tropics for biofuel production. Tree crop species with potential for biodiesel production could also be used in tree-based agroforestry systems in the tropics, such as shelterbelts, live fences, silvopastures, soil reclamation systems, woody hedgerows and multipurpose tree systems.

Next steps in biofuels research in the tropics should focus on developing land use systems for biofuel crop production using agroforestry practices. These studies should investigate best management practices for biodiesel production, tree-crop interactions, and the environmental benefits and social acceptability of these practices.

Bibliography

Achten WMJ, Verchot L, Singh VP, Aerts R, Muys B (2007) *Jatropha* biodiesel fueling sustainability? Biofuels, Bioprod. Bioref 1:283–291. doi:10.1002/bbb

Achten WJM, Almeida J, Fobelets V, Bolle E, Mathijs E, Singh VP, Tewari DN, Verchot LV, Muys B (2010a) Life cycle assessment of *Jatropha* biodiesel as transportation fuel in rural India. Appl Energy 87:3652–3660

Achten WMJ, Nielsen LR, Aerts R, Lengkeek AG, Kjaer ED, Trabucco A, Hansen JK, Maes WH, Graudal L, Akinnifesi FK, Muys B (2010b) Towards domestication of *Jatropha curcas*. Biofuels 1(1):91–107

Azam MM, Waris A, Nahar NM (2005) Prospects and potential of fatty acid methyl esters of some non-traditional seed oils for use as biodiesels in India. Biomass Energy 29:293–302

Barnwal BK, Sharma MP (2005) Prospects of biodiesel production from vegetable oils in India. Renew Sustain Energy Rev 9:363–375

Carlson KM, Curran LM, Ratnasari D, Pittman AM, Soares-Filho BS, Asner GP, Trigg SN, Gaveau DA, Lawrence D, Rodrigues HO (2012) Committed carbon emissions, deforestation and community land conversion from oil palm plantation expansion in West Kalimantan, Indonesia. PNAS early edition www.pnas.org/cgi/doi/10.1073/pnas.120045109

Chapagain BP, Yehoshua Y, Wiesman Z (2009) Desert date (*Balanites aegyptiaca*) as an arid lands sustainable bioresource for biodiesel. Bioresour Technol 100:1221–1226

Danielsen F, Beukema H, Burgess ND, Parish F, Brühl CA, Donald PF, Murdiyarso D, Phalan B, Reijnders L, Struebig M, Fitzherbert EB (2008) Biofuel plantations on forested lands: double jeopardy for biodiversity and climate. Conserv Biol 23(2):348–358

De BK, Bhattacharyya DK (1999) Biodiesel from minor vegetable oils like karanja oil and nahor. Fett/Lipid 101(10):404–406

Demirbas A (2009) Progress and recent trends in biodiesel fuels. Energy Conserv Manag 50:14–34

Fuwape JA (1993) Charcoal and fuel value of agroforestry tree crops. Agrofor Syst 22:175–179

Heller J (1996) Physic Nut. *Jatropha curcas L.* Promoting the Conservation and use of Underutilized and Neglected Crops. International Plant Genetic Resources Institute, Rome

Kituyi E, Marufu L, Wanfiga SO, Jumba IO, Andreae MO, Helas G (2001) Biofuel availability and domestic use patterns in Kenya. Biomass Bioenergy 20:71–82

Kumar A, Sharma S (2008) An evaluation of multipurpose oil seed crop for industrial uses (*Jatrophas curcas* L.): a review. Ind Crops Prod 28(1):1–10

Leduc S, Nataraja K, Dotzauer E, McCallum I, Obersteiner M (2009) Optimizing biodiesel production in India. Appl Energy 86:125–31

Pate R, Hightower M, Cameron C, Einfeld W (2007) Overview of Energy-Water Interdependencies and the Emerging Energy Demands on Water Resources. Report SAND 2007-1349C. Sandia National Laboratories, Los Alamos, New Mexico, USA

Phalan B (2009) The social and environmental impacts of biofuels in Asia: an overview. Appl Energy 86:S21–S29

Phillips S, Aden A, Jechura J, Dayton D, Eggeman T (2007) Thermochemical ethanol via indirect gasification and mixed alcohol synthesis of lignocellulosic biomass. Technical Report NREL/TP-510-41168. National Renewable Energy Laboratory, Golden, Colorado, USA. http://www.nrel.gov/ docs/fy07osti/41168.pdf

Scholz V, Silva JN da (2008) Prospects and risks of castor oil as a fuel. Biomass Bioenergy 32(2):95–100

Scott PT, Pregelij L, Chen N, Hadler JS, Djordjevic MA, Gresshoff PM (2008) *Pongamia pinnata*: an untapped resource for the biofuels industry of the future. Bioenergy Res 1:2–11

Singh SP, Singh D (2010) Biodiesel production through the use of different sources and characterization of oils and their esthers as the substitute of diesels: a review. Renew Sustain Energy Rev 14:200–216

Winayanuwattikun P, Kaewpiboon C, Piriyakananon K, Tantong S, Thakernkarnkit W, Chulalaksananukul W, Yongvanich T (2008) Potential plant oil feedstock for lipase-catalyzed biodiesel production in Thailand. Biomass Bioenergy 32:1279–1286

Chapter 19
Phytoremediation in Tropical Agroforestry

Abstract Agroforestry has great potentials for providing numerous environmental services, including, among others, soil conservation, land rehabilitation, ground water-table stabilization, erosion control, and phytoremediation of soils contaminated with heavy metals and other pollutants. Phytoremediation, one of those environmental services provided by agroforestry, is the use of plants to decontaminate polluted soils. Phytoremediation involves extraction of soil pollutants by roots and accumulation or transformation by plants, e.g., hyperaccumulators. Hyperaccumulators are plants that can tolerate metals and organic pollutants and extract them from contaminated soils and accumulate them at concentrations far exceeding what normally would be found in plant tissues. Agroforestry systems restore contaminated soils through the decontaminating effects of legumes, hyperaccumulators, and hydraulic lift (a major mechanism behind soil water redistribution between soil layers in agroforestry systems). Legume species that are used in tropical agroforestry for nitrogen fixation in soils most often have the ability to decontaminate polluted soils, as symbiotic associations between legumes and symbionts (mycorrhiza and *Rhizobium* spp.) and actinomycorrhizal plants (mycorrhiza and *Frankia* spp.) enhance phytobial remediation. Hyperaccumulators are used for soil decontamination in several agroforestry systems, including riparian buffer systems, tree-crop combinations, and short woody rotation crops. Research to increase understanding of the functions of agroforestry with regard to environmental services, and of the impact of these benefits to landscape health, is emerging. Society has not yet exploited the full potential of agroforestry. This chapter will focus on an under-exploited agroforestry benefit in the form of phytoremediation.

19.1 Introduction

Agroforestry has potential for environmental cleanup through atmospheric carbon sequestration, rehabilitation of waste lands, phytoremediation of soil and groundwater contaminants, and redistribution of groundwater from wet zones to dry zones in drought-affected land through hydraulic lift (Burken and Ma 2006; Liste and White 2008; Armas et al. 2010). Phytoremediation may occur in the following five forms to decontaminate polluted soils, including phytostabilization, phytoextraction,

A. Atangana et al., *Tropical Agroforestry*, DOI 10.1007/978-94-007-7723-1_19,

phytovolatilization, phytofiltration and rhyzodegradation. Phytoremediation also uses hyperaccumulators, which are plants that tolerate metals and extract heavy metals from contaminated soils to then be accumulated in the plant. Some legume species that are used in tropical agroforestry may also act as hyperaccumulators, and phytoremediation success most often depends on the success of association between the host plant and symbionts (Meier et al. 2012). Indeed, mycorrhizobial (arbuscular mycorrhizae + *Rhizobium spp.* + legumes) and actinomycorrhizal (*Frankia spp.* + mycorrhizae + non-legumes) plants used in agroforestry systems have potential to extract organic and inorganic pollutants, including hydrocarbons, chlorinated compounds, salts and heavy metals, from contaminated soils (Dommergues 1997; Eweis et al. 1998; Bento et al. 2012). This is possible as arbuscular mycorrhizal fungi and *Frankia* or *Rhizobium* provides a direct link between soils and roots. Actinomycorrhizal and mycorrhizobial plants have been shown to enhance phytoremediation (Dommergues 1997; Zimmer et al. 2009; Meier et al. 2012; Ho et al. 2012).

The potential for agroforestry systems to capture and store atmospheric carbon has been discussed in detail in Chapter 10. Also, soil rehabilitation and soil water redistribution mechanisms in agroforestry have been discussed in Chapters 7, 8 and 9. This chapter will focus on the phytoextraction of soil contaminants using plant-symbiont systems in agroforestry systems, and agroforestry practices with good potential to decontaminate polluted soils.

19.2 Tropical Agroforestry Species and Plant-Symbiont Associations with Potential for Phytoremediation

Few studies have been carried out to identify hyperaccumulators (or metallophytes) that provide additional benefits in the tropics, though several plants have been found to grow on iron (Jacobi et al. 2007), cobalt and copper contaminated soils (Brooks et al. 1986; Faucon et al. 2009). Several metal hyperaccumulators have been reported in the tropics (as reviewed in Reeves 2003), but few are valued in agroforestry. However, a legume, *Pearsonia metallifera* (Fabaceae) accumulated high quantities of nickel (Ni) in serpentine soils in Zimbabwe (as reviewed by Reeves 2003). However, some species used in tropical agroforestry in Central Africa have been reported as metallophytes, including *Annona senegalensis* and *Albizia adianthifolia* (Brooks et al. 1986). Also, a few agroforestry species such as *Leucaena leucocephala*, *Bidens pilosa* and *Crotalaria micans*, effectively extracted Ni in contaminated soils (Ho et al. 2012). *Leucaena leucocephala* also exhibited potential for phytoextraction of cadmium and zinc (Saraswat and Rai 2011), whereas *Sesbania rostrata* was efficient for hyperaccumulation of cadmium and copper in contaminated soils (Chen et al. 2009).

Other agroforestry tree and shrub species have potential for the extraction and accumulation of soil pollutants. Hybrid poplar clones have been found to extract and accumulate boron (B) and selenium (Se) from saline effluents. In central California, one *Populus trichocarpa* x *Populus deltoides* clone (clone 49177) accumulated the greatest amounts of B and Se (Bañuelos et al. 1999) in a sand-culture study investigating the phytoaccumulation of B and Se in eight poplar clones irrigated with synthetic effluent. The study also found that *Populus* species effectively removed and accumulated B and Se at salinity levels less than 7 dS m^{-1} (Bañuelos et al. 1999). Phytoaccumulation of Se by agroforestry species grown on clay loam soil was also investigated in a greenhouse in Punjab, India (Dhillon et al. 2008). The selenium content of the leaves of *Eucalyptus* hybrids, *Acacia tortillas*, *Dalbergia sissoo*, *Morus alba* L., *Terminalia arjuna* (Roxb.) Wight & Arn., *Syzygium cumini* (L.) Skeels. and *Melia azedarach* trees planted on seleniferous soils was found to increase with increasing levels of applied Se, with *T. arjuna* being the most effective in Se removal from soils (Dhillon et al. 2008). In India, *Albizia amara* Boivin, *Casuarina equisetifolia*, *Tectona grandis* and *L. leucocephala* seedlings were found to accumulate chromium in their roots, with *A. amara* being the best Cr accumulator among the species studied in a pot experiment conducted in the greenhouse (Shanker et al. 2005).

The decontamination success of several nitrogen-fixing agroforestry species depends on the success of association between plants and symbionts (Leyval et al. 1997; Zimmer et al. 2009; Bento et al. 2012). Associated microorganisms of nitrogen-fixing plants contribute to phytoremediation through biosorption, bioaccumulation, and transformation of metals (Meier et al. 2012). These associated microorganisms are involved in heavy-metal tolerance by plants (Leyval et al. 1997). They also protect roots against stress induced by fungicides (Campagnac et al. 2010) and increase phytoextraction of metals in contaminated soils (Zimmer et al. 2009). For instance, *Acacia angustissima* (Mill.) Kuntze, *Mimosa caesalpiniaefolia* Benth and *Albizia saman* F. Muell. potential to remediate petroleum-contaminated soils was enhanced by association with microorganisms (Bento et al. 2012). Mycorrhizal associated with non-legume agroforestry tree species also enhance phytoremediation of heavy-metal polluted soils. For instance, ectomycorrhizal fungi enhanced cadmium phytoextraction by poplars (Sell et al. 2005).

Despite the potential for agroforestry species to restore contaminated soils, limited efforts have been made to include phytoremediation potential in the criteria of tree selection for introduction in tropical agricultural landscapes. To expand the benefits of tropical agroforestry to the environment, it is imperative to understand the metal-extracting properties of multipurpose species used by farmers for their everyday needs for food, cash, medicine, and shelter, especially in areas that suffer soil-contamination. More work should be done on the phytoremediation properties of the association between agroforestry species and associated microorganisms in metal and fungi contaminated or saline soils in the tropics.

19.3 Agroforestry Systems with Potential for Phytoremediation in the Tropics

Hydraulic lift, which is one of the first benefits of agroforestry, as it redistributes water between agroforestry components, plays an important role in phytoremediation. Riparian buffers and short woody rotation crops have more phytoremediation potential than other agroforestry systems. Sequential rapeseed (*Brassica napus* L.) -based agroforestry systems (i.e., rapeseed followed by *Cajanus cajan*, *Crotalaria juncea* L., or *Gossypium arboreum* L.) significantly reduced Se amounts in contaminated soils in India (Dhillon and Dhillon 2009). In the same country, soil losses have been controlled using several silvopastoral systems including combinations of *Acacia catechu*-forage grasses, *Leucaena-Pennisetum purpureum*, *Tectona grandis-Leucaena- Eulaliopsis binata* (Retz.) C.E. Hubb. (bhabar), *Eucalyptus-Leucaena-Curcuma longa* L. (turmeric) and poplars-*Leucaena*-bhabar (Prasad 2007). Also, silvopastoral systems combining *Prosopis juliflora* and *Leptochloa fusca* (L.) Kunth phytoremediated sodic and saline soils, whereas several combinations of agroforestry trees (*Eucalyptus tereticornis*, *Populus deltoides*, *T. arjuna*, *Acacia auriculiformis*, *S. cumini*, *Albizia lebbek*, *D. sissoo*, and *Pongamia pinnata*) and grasses (*Urochloa mutica* (Forssk.) T.Q. Nguyen, *Spartina* (cordgrass), *Cymbopogon*, and Setaria grass) controlled physical soil deterioration due to water flooding in India (Prasad 2007).

19.3.1 Short-Rotation Woody Crop Systems and Phytoremediation

Fast-growing trees with metal-extraction potential including poplars and eucalypts have been used to remediate contaminated soils. Poplars are cultivated on rotations of less than 15 years for timber production. Poplar-based agroforestry systems are common in India (Das and Chatuverdi 2005). Eucalypts are used as fast-growing shade trees in coffee plantations in the tropics (Schaller et al. 2003). However, short-rotation woody species can concentrate only low levels of contaminants. *Eucalyptus* spp. and *Populus* spp. that are cultivated at close spacing for rotations of 10 years or less can be used in phytoremediation (Rockwood et al. 2004). Also, growing short-rotation woody crops to ameliorate degraded soils could cause large soil water deficit in drought-prone areas (Sudmeyer and Goodreid 2007). In a study aimed at assessing the effects of short rotation woody crops on soil water storage and crop and pasture growth in Australia, Sudmeyer and Goodreid (2007) found that *Eucalyptus polybractea* dried out the soil down to 10 m, creating a soil water deficit of 1,350 mm within 6 years of planting.

Despite the fact that fast-growing nitrogen-fixing species including *Acacia angustissima*, *Acacia mangium*, *Inga edulis* and *Albizia* sp., that are used in alley cropping in the tropics, have potential for heavy-metal extraction and accumulation, less is known on the phytoremediation potential of alley cropping. More work needs to be done to assess the use of fast-growing agroforestry species in numerous agroforestry systems, so as to identify systems that have a large potential for phytoremediation, while providing local people with food and other benefits.

19.3.2 Riparian Buffers and Phytoremediation in Tropical Agroforestry

Riparian forest buffers are multi-species vegetation established at the interface between croplands and surface-water to remove sediments and chemical pollutants in run-off and shallow water from agricultural land. Therefore, riparian buffers act as a filtration zone for streams and lakes, and are effective in controlling run-off erosion.

The success of riparian buffers for phytoremediation depends on the species used. Poplars and forage grasses in multi-species buffer zones were found to phytoremediate atrazine (Chang et al. 2005; Lin et al. 2008) and remove phosphorus from soil solution (Kovar and Claassen 2009). Indeed, *Panicum virgatum* L., a grass that is grown on riparian buffer systems, has good potential to uptake, degrade and detoxify atrazine in the rhizosphere (Lin et al. 2008). The phytoremediation property of vegetative grasses involves many processes, including soil microbial activities for herbicide degradation (Lin et al. 2005), denitrification of groundwater, channel stabilization, and nutrient uptake by fast-growing species (Dosskey et al. 2010).

Few studies on phytoremediation using riparian buffer zones in tropical zones have been reported, despite the benefits these agroforestry systems provide to local people. Indeed, tree species grown in riparian buffer zones produced large amounts of woody biomass and substantially removed sediments in channels leading into headwater streams in Peninsular Malaysia (Gomi et al. 2006). Riparian buffers also provide food, firewood, forage, and bark (for medicinal use) to local people, as non-timber forest product species can be collected in riparian zones. In North America, fruit from chokecherries (*Aronia melanocarpa*) and serviceberries (*Amelanchier*), used to produce wine, are collected from riparian buffer zones. Fruits from mango trees (*Mangifera indica* L.), imli (*Tamarindus indica*) and jamun (*Syzygium jambos* L. (Alston) are collected from riparian zones in India (Vyas et al. 2012). On the other hand, bamboo (*Bambusa vulgaris*), a non-timber forest product valued in handicraft, is grown in tropical riparian zones. Riparian buffer zones also provide other environmental services. For example, riparian buffer zones preserve biodiversity as they provide habitats and corridors for wildlife, while aquatic riparian buffer zones also harbor species that are part of the food web.

19.3.3 Hydraulic Lift and Phytoremediation in Tropical Agroforestry

Plant roots can redistribute water from wet zones deeper in the soil, to dry soil zones close to the surface in dry regions, a process known as hydraulic lift (Richards and Caldwell 1987). Indeed, some plants have root systems that absorb water at moister

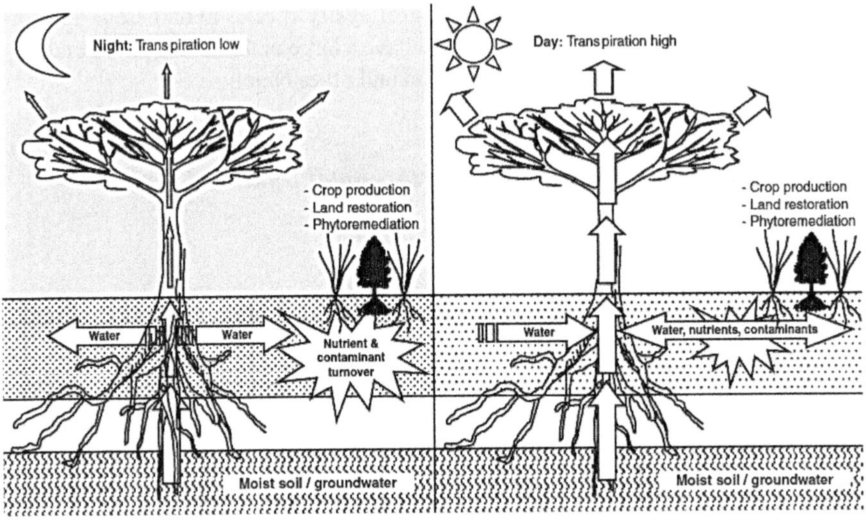

Fig. 19.1 Hydraulic lift and potential benefits. (Liste and White 2008)

layers (usually at depth) when transpiration is high during the day, and re-distribute the absorbed water to the rhizosphere in drier layers when plant transpiration is low (at night); this phenomenon is used in agroforestry for water redistribution between woody species and crops, and to assist in phytoremediation (Fig. 19.1).

Tree and shrub species that perform hydraulic lift are often used for phytoremediation of contaminated soils. Hydraulic lift enhances phytoremediation via rewetting the rhizosphere in the shallow soil layers, keeping microbes active for biodegradation of organic chemicals in soils. Rewetting soil also enhances the release of chemicals, thereby facilitating their acquisition by metallophytes (Liste and White 2008). Therefore, hydraulic lift makes pollutants more available for uptake by roots. Hydraulic lift also keeps fine roots wet, which facilitates the absorption of pollutants by roots.

Hydraulic lift is a major process behind water redistribution between agroforestry system components. In addition, the ability to extract soil pollutants by several agroforestry tree and shrub species has been found to increase by hydraulic lift. Numerous agroforestry tree and shrub species have been found to phytoremediate metal- or chemical-contaminated soils in the tropics. For example, *Acacia tortilis* (Ludwig et al. 2003; Dhillon et al. 2008), Eucalyptus (Bafeel 2008; Caldwell et al. 1998; Hamada et al. 2003), *Dalbergia sissoo*, *Melia azedarach*, *Morus alba*, poplars, *Szygium cumini* and *Terminalia arjuna* (Dhillon et al. 2008) have been found to phytoremediate contaminated soils that benefit from hydraulic lift. Dhillon et al. (2008) reported that an agroforestry farming system composed of poplar and *Mentha viridis* or wheat can remove up to 4207 g Se ha^{-1}, reducing Se content in the surface soil layer by 43–65 %, and in the whole profile by 13–20 %.

19.4 Tropical Agroforestry and Phytoremediation: Next Steps

Despite the great potential for tropical agroforestry to phytoremediate contaminated soils, few have investigated soil decontamination using agroforestry systems in the tropics. For example, several mining sites are operational in sub-Saharan Africa, where local people rely on agroforestry and agriculture for food and other agricultural products. Alley cropping, rotational woodlots, rotational tree fallows, improved fallows, silvopastures and agroforestry parklands most often include legumes or hydraulifting plants such as *Acacia*, *D. sissoo*, *Albizia* spp. *Eucalyptus*, *P. juliflora* and *L. leucocephala* and *Annona senegalensis* that are known to have phytoremediation properties. However, few have investigated the use of such species in agroforestry systems for reclamation of contaminated mining sites in sub-Saharan Africa, so as to provide local people with food and forage, while decontaminating polluted soils. Brooks et al. (1986) and Reeves (2003) reported some metallophytes found in mining sites of Central and Southern Africa. However, no one has investigated the use of these species in an agroforestry system for the decontamination of polluted soils in sub-Saharan Africa.

Similarly, there is a lack of research on this topic in Latin America and Southeast Asia, though Prasad (2007) reported silvopastoral systems with good potential to phytoremediate and restore lands in India. Several agroforestry species have potential to phytoremediate polluted soils in Southeast Asia and Latin America (Bañuelos et al. 1999; Chen et al. 2009; Dhillon et al. 2008; Prasad et al. 2007; Saraswat and Rai 2011; Shanker et al. 2005). Next steps in phytoremediation studies using agroforestry practices in the tropics should focus on the assessment of decontamination potential of agroforestry tree and shrub species for polluted soils. Other topics that should be investigated are the phytoremediation potential of tropical agroforestry systems and the toxicity level of food products from these systems for human consumption.

Although hydraulic lift facilitates the absorption of pollutants by plant roots by increasing water availability in the rooting zone, its beneficial effect for understory crops is limited due to competition for water between agroforestry components (Ludwig et al. 2003). Therefore, much more research is needed to understand the phytoremediation potential of agroforestry systems as it is affected by hydraulic lift and to design agroforestry systems that minimize competition for water while favoring hydraulic lift for phytoremediation of polluted soils in the tropics.

Bibliography

Armas C, Padilla FM, Pugnaire FI, Jackson RB (2010) Hydraulic lift and tolerance to salinity of semiarid species: consequences for species interaction. Oecologia 162:11–21

Bafeel SO (2008) Contribution of mycorrhiza in phytoremediation of lead contaminated soils by *Eucalyptus rostrata* plants. World Appl Sci J 5(4):490–498

Bañuelos GS, Shannon MC, Ajwa H, Draper JH, Jordahl J, Licht J (1999) Phytoextraction and accumulation of Boron and Selenium by poplar (*Populus*) hybrid clones. Int J of Phytoremediation 1(1):81–96

Bento RA, Saggin-Júnior OJ, Pitard RM, Straliotto R, da Silva EMR, de Lucena TSR, de Landa FHTG, Martins LF, Volpon AGT (2012) Selection of leguminous trees associated with symbiont microorganisms for phytoremediation of petroleum-contaminated soil. Water Air Soil Poll. doi 10.1007/s11270-012-1305–3

Brooks RR, Naidu SM, Malaisse F, Lee J (1986) The element content of metallophytes from the copper/cobalt deposits of Central Africa. Bulletin de la Société Royale de Botanique de Belgique 2:179–191

Burken JG, Ma X (2006) Phytoremediation of volatile organic compounds. In Mackova M, Dowling DN, Macek T (eds) Phytoremediation rhizoremediation. Springer, pp 199–216

Caldwell MM, Dawson TE, Richards JH (1998) Hydraulic lift:consequences of water efflux from the roots of plants. Oecologia 113:151–161

Campagnac E, Sahraoui ALH, Debiane D, Fontaine J, Laruelle F, Garçon G, Verdin A, Durand R, Shirali P, Grandmougin-Ferjani A (2010) Arbuscular mycorrhiza partially protect chicory roots against oxidative stress induced by two fungicides, fenpropimorph and fenhexamid. Mycorrhiza 20:167–178

Campagnac E, Yonli H, Sanon K, Vézina MM, Baudrand J, Khasa DP (2011) Les symbiotes racinaires: Présence et utilisation en biorestauration des rejets miniers aurifères. Symposium sur les mines et l'environnement, du 6 au 9 novembre 2011 à Rouyn Noranda, Canada

Chang SW, Lee SJ, Je CH (2005) Phytoremediation of atrazine by poplar trees: toxicity, uptake, and transformation. J Environ Sci Health Part B 40:801–811

Chen F, Yang Z, Yuan J (2009) Cadmium uptake and accumulation by *Sesbania rostrata*, a N-fixing plant: implications for the mechanism of heavy metal tolerance. Front Biol China 4(2):200–206

Das DK, Chatuverdi OP (2005) Structure and function of *Populus deltoids* agroforestry systems in eastern India: 2. Nutrient dynamics. Agrofor Syst 65:223–230

Dhillon SK, Dhillon KS (2009) Phytoremediation of selenium-contaminated soils: the efficiency of different cropping systems. Soil Use Manage 25:441–453

Dhillon KS, Dhillon SK, Thind HS (2008) Evaluation of different agroforestry tree species for their suitability in the phytoremediation of seleniferous soils. Soil Use Manage 24:208–216

Dommergues YR (1997) Contribution of actinorhizal plants to tropical soil productivity and rehabilitation. Soil Biol BioChem 29(5/6):931–941

Eweis JB, Schroeder ED, Chang DPY, Scow KM (1998) Biodegradation of MTBE in a pilot-scale biofilter. In Wickramanayake GB, Hinchee RE (eds) Natural attenuation: chlorinated and recalcitrant compounds. Battelle Press, Columbus, Ohio, pp 341–346

Faucon MP, Parmentier I, Colinet G, Mahy G, Luhembwe MN, Meerts P (2009) May rare metallophytes benefit from disturbed soils following mining activities? The case of the *Crepidorhophalon tenuis* in Katanga (D. R. Congo). Restor Ecol 19(3):333–343

Dosskey MG, Vdon P, Gurwick NP, Allan CJ, Duval TP, Lowrance R (2010) The role of riparian vegetation in protecting and improving chemical water quality in streams. J Am Water Source Assoc 46(2):261–277

Ho CP, Hseu ZY, Chen NC, Tsai CC (2012) Evaluating heavy metal concentration of plants on a serpentine site. Environ Earth Sci. doi 10.1007/s12665–012-2115-z

Gomi T, Sidle RC, Noguchi S, Negishi JN, Nik AR, Sasaki (2006) Sediments and wood accumulations in humid tropical water streams: effects of logging and riparian buffers. For Ecol Manag 224:166–175

Jacobi CM, do Carmo FF, Vincent RC, Stehmann JR (2007) Plant communities on ironstone outcrops: a diverse and endangered Brazilian ecosystem. Biodivers Conserv 16:2185–2200

Hamada H, Tomi R, Asada Y, Furuya T (2002) Phytoremediation of bisphenol A by cultured cells of *Eucalyptus perriniana* regioselective hydroxylation and glycosylation. Tetrahedron Lett 43(22):4087–4089

Kovar JL, Claassen N (2009) Growth and phosphorus uptake of three riparian grass species. Agron J 101:1060–1067

Liste H-H, White JC (2008) Plant hydraulic lift of soil water — implications for crop production and land restoration. Plant Soil 313:1–17

Leyval C, Turnau K, Haselwandter K (1997) Effect of heavy metal pollution on mycorrhizal colonization and function: physiological, ecological and applied aspects. Mycorrhiza 7:139–153

Lin CH, Lerch RN, Kremer RJ, Garrett HE, Udawatta RP, George MF (2005) Soil microbiological activities in vegetative buff er strips and their association with herbicides degradation. In Brooks et al. (eds) Moving agroforestry into the mainstream. Proc. Of the 9th Conf. on Agroforestry in North America, 12–15 June 2005. Dep. of Forest Resources, Univ. of Minnesota, St. Paul, pp 1–10.

Lin CH, Lerch RN, Garrett HE, George MF (2008) Bioremediation of atrazine-contaminated soils by foresge grasses: transformation, uptake and detoxification. J Enviiron Qual 37:196–206

Liste HH, White JC (2008) Plant hydraulic lift of soil water—implications for crop production and land restoration. Plant Soil 313:1–17

Ludwig F, Dawson TE, Kroon H, Berendse F, Prins HHT (2003) Hydraulic lift in *Acacia tortilis* trees on an East African savanna. Oecologia 134:293–300

Meier S, Borie F, Bolan N, Cornejo P (2012) Phytoremediation of metal-polluted soils by arbuscular mycorrhizal fungi. Crit Rev Environ Sci Technol 42:741–775

Prasad MNV (2007) Phytoremediation in India. Method Biotechnol 23:435–454

Reeves RD (2003) Tropical hyperaccumulators of metals and their potential for phytoextraction. Plant Soil 249:57–65

Richards JH, Caldwell MM (1987) Hydraulic lift: substantial nocturnal water transport between soil layers by *Artemisia tridentata* roots. Oecologia 73:486–489

Rockwood DL, Naidu CV, Carter DR, Rahmani M, Spriggs TA, Lin C, Alker GR, Isebrands JG, Segrest SA (2004) Short-rotation woody crops and phytoremediation: an opportunity for agroforestry? Agrofor Syst 61:51–63

Saraswat S, Rai JPN (2011) Prospective application of *Leucaena leucocephala* for phytoextraction of Cd and Zn and nitrogen fixation in metal polluted soils. Int J Phytoremediation 13(3):271–288

Schaller M, Schroth G, Beer J, Jiménez F (2003) Species and site characteristics that permit the association of fast-growing trees with crops: the case of *Eucalyptus deglupta* as coffee shade in Costa Rica. For Ecol Manag 175(1-3):205–215

Sell J, Kayser A, Schulin R, Brunner I (2005) Contribution of ectomycorrhizal fungi to cadmium uptake of poplars and willow from a heavily polluted soil. Plant Soil 277:245–253

Shanker AK, Ravichandran V, Pathmanabhan G (2005) Phytoaccumulation of chromium by some multipurpose-tree seedlings. Agrofor Syst 64:83–87

Shively GE, Zelek CA, Midmore DJ, Nissen TM (2004) Carbon sequestration in a tropical landscape: an economic model to measure its incremental cost. Agrofor Syst 60:189–197

Sudmeyer RA, Goodreid A (2007) Short-rotation woody crops: a prospective method for phytoremediation of agricultural land at risk of salinisation in southern Australia? Ecol Eng 29:350–361

Vyas V, Kumar A, Wani SG, Parashar V (2012) Status of riparian buffer zone and floodplain areas of River Narmada, India. Int J Environ Sci 3(1):659–674

Zimmer D, Baum C, Leinweber P, Hrynkiewicz K, Meissner R (2009) Associated bacteria increase the phytoextraction of cadmium and zinc from a metal-contaminated soil by mycorrhizal willow. Int J Phytoremediation 11:200–213

Chapter 20
Agroforestry and the Carbon Market in the Tropics

Abstract To slow the increase of the atmospheric concentration of greenhouse gases responsible for climate change, initiatives such as the United Nations REDD (Reducing Emissions from Deforestation and Forest Degradation) Programme have been taken. The UN-REDD programme supports REDD+ (i.e., conservation and sustainable management of forests, and enhancement of C stocks, on top of REDD) readiness efforts in the design and implementation of national programs and in national REDD+ action through common approaches and interventions. The REDD+ policies propose to financially compensate countries that improve forest conservation and management to reduce greenhouse gases (GHG; i.e., carbon dioxide (CO_2), methane (CH_4) and nitrous oxide (N_2O)) emissions and mitigate against climate change. The REDD+ initiative has recently evolved to REDD++ (i.e., low carbon (C) emission or low C footprint land use systems through eco-agricultural practices on top of REDD+). Eco-agricultural practices, which aim at producing more food while conserving wild biodiversity, include agroforestry systems such as perennial tree-crop systems, windbreaks, and live fences. Agroforestry systems also store C and may qualify as an afforestation practice as is defined in the Kyoto Protocol, and could be included in the C market under the REDD+ scheme. The Kyoto protocol that deals with environmental issues, especially climate change, is heavily based on clean development mechanism (CDM) as a strategy to mitigate atmospheric greenhouse gas concentrations. The inclusion of agroforestry in CDM is hampered by the lack of standardized methods to estimate C stocks, as well as land tenure issues in the tropics, especially in Africa. Another challenge for the inclusion of agroforestry to CDM is the payment for environmental (or ecosystem) services (PES) option that should be implemented in C contracts. This chapter discusses the opportunities for including agroforestry in C markets, as well as the difficulties and PES options linked to it.

20.1 Introduction

The increased burning of fossil fuels since the beginning of industrialization has caused a rapid increase in the atmospheric concentration of greenhouse gases (GHG), including carbon dioxide (CO_2), methane (CH_4) and nitrous oxide (N_2O). The rate of increase of carbon dioxide emissions was 1.3 % in the 1990s but was

A. Atangana et al., *Tropical Agroforestry*, DOI 10.1007/978-94-007-7723-1_20,

3.3 % between 2000 and 2006 (Canadell et al. 2007). In 2013, the average concentration of CO_2 in the atmosphere is approximately 400 ppm, reaching prehistoric levels (more than 2.5 million years ago, an era known as the Pliocene, when the Canadian Arctic boasted forests instead of icy wastes), from a pre-industrial average of around 280 ppm. The international community has agreed that 450 ppm, linked to a rise of 2 °C in global average temperatures, should not be exceeded.

The United Nations Framework Convention on Climate Change (UNFCCC), signed in 1992 and the Kyoto Protocol (1997–2012) adopted in 1997 by 34 countries are landmark international treaties to address climate change. Carbon offsets (also commonly referred to as "carbon credits" or "certified emission reductions") are generated by C sequestration or emissions reduction activities that are quantified, reported, verified, validated, and certified via the regulatory or voluntary market. Offsets are a form of climate change mitigation that provide assurance that a given amount of greenhouse gas emissions will be avoided, reduced, or actively removed from the atmosphere, thereby balancing or compensating for unavoidable emissions generated through activities such as industrial production or deforestation. The Kyoto protocol that deals with environmental issues, especially climate change, is heavily based on clean development mechanism (CDM) as a strategy to mitigate atmospheric greenhouse gas concentrations. For that reason, initiatives have been taken for the identification, development and diffusion of methods and practices that have been proven to be efficient in the sequestration of atmospheric CO_2. One of the strategies to reduce atmospheric CO_2 concentrations is C sequestration, which consists of C uptake and storage in long-lived pools, including terrestrial, oceanic, or freshwater aquatic ecosystems through abiotic and biotic technologies (Lal 2008). Abiotic techniques such as CO_2 injection in deep ocean, geological strata, old coal mines and oil wells, and saline aquifers along with mineral carbonation of CO_2, have great potential and may be available for routine use by 2025 and beyond, but are expensive and have high risks of leakage. In comparison, biotic techniques are natural and cost-effective processes, have numerous ancillary benefits, and are immediately applicable but have finite sink capacity. In biotic approaches, atmospheric CO_2 is captured via plant photosynthesis and stored in plant tissues (terrestrial C sequestration) and phytoplanktons (freshwater aquatic and oceanic C sequestration). Using a process-based biogeochemistry model, the Terrestrial Ecosystem Model (TEM), Galford et al. (2010) estimated that deforestation is the largest source of greenhouse gas emissions, but land use following clearing accounts for a substantial portion (24–49 %) of the greenhouse gas emissions from deforestation. Emissions of N_2O are amplified by chemical fertilizers' application. Consequently, land use systems with a large potential as atmospheric C sinks were included in programs to reduce greenhouse gas emissions under the CDM, as one of the mechanisms defined in the Kyoto Protocol (IPCC 2007). The Protocol came into effect on 16 February 2005 in accordance with Art. 23. Currently, there are 192 Parties (191 States and 1 regional economic integration organization) to the Kyoto Protocol and to the UNFCCC. At the UNFCCC's 17th Conference of the Parties (COP 17) in Durban (December 2011), the European Union (EU) agreed to a second commitment period for the Kyoto Protocol (KP2), but this was not finalized.

In Doha (COP 18, December 2012), the second commitment period was agreed upon, allowing it to move forward for another eight-year period (1 January 2013 to 31 December 2020), ensuring the future of the CDM mechanism.

Besides the CDM mechanism, the REDD (Reducing Emissions from Deforestation and Forest Degradation in developing countries, a United Nations-based program) mechanism, established in 2008, implements a C-payment system, operating under the principle that countries that improve forest conservation and management to reduce GHG emissions and mitigate against climate change should be financially compensated. Simply put, farmers from developing countries (mostly in the tropics), who implement practices that sequester C (afforestation, reforestation, forest conservation), may receive payment from polluters (industries) under the C market, as approved under the Kyoto Protocol. The UN-REDD Programme supports REDD+ (i.e., conservation and sustainable management of forests, and enhancement of C stocks, on top of REDD) readiness efforts directly in the design and implementation of UN-REDD national programs. Complementarily this program supports national REDD+ action through common approaches and interventions developed under the UN-REDD Global Programme. The "REDD+" concept goes beyond deforestation and forest degradation and includes the role of conservation, sustainable management of forests (community forestry) to explicitly promote the enhancement of forest C stocks. Carbon stocks enhancement in the REDD+ scheme is achieved through afforestation and reforestation within the framework of *land use, land-use change and forestry* (LULUCF), *agriculture, forestry and other land use* (AFOLU), or *reduced emissions for all land use* (REALU). Integral to successful REDD+ program implementation are, among others, the full inclusion of social, environmental and governance safeguards, such as free prior informed consent, stakeholder engagement, active participation in project design, and respect for the rights of indigenous people and other forest-dependent communities, and transparent benefit distribution mechanisms. In Doha (Conference of the Parties or COP 18), two objectives for the REDD+ programme were discussed. The first was to reach an agreement on the basic principles of REDD+, such as guidelines on monitoring, review and verification (MRV) so that a concrete REDD+ based new market mechanism (NMM) could be developed; the second was to ensure that REDD+ remained on the COP agenda.

The REDD+ scheme proposes several approaches to reducing deforestation and forest degradation including establishment of protected areas and agricultural intensification to motivate land sparing. Indeed, the REDD+ policy on agricultural intensification is based on the principle that increasing agricultural inputs to improve per-hectare yields reduces agriculture encroachment into forest. The REDD+ initiative has evolved to REDD++ (i.e., low carbon (C) emission or low C footprint land use systems through eco-agricultural practices on top of REDD+). Ecoagriculture refers to the management of landscapes for both agricultural production and the conservation of ecosystem services, in particular wild biodiversity (Scherr and McNeely 2007). Examples of landscapes supporting ecoagriculture are multi-strata agroforests (Scherr and McNeely 2007), windbreaks, and live fences. Therefore, agroforestry systems could be included in both the REDD+ and REDD++ schemes.

Agroforestry is gaining momentum as a strategy for terrestrial C sequestration that deserves to be included in C markets (Albrecht and Kandji 2003; Montagnini and Nair 2004; Schoeneberger 2009; Nair et al. 2009a, 2010). In addition, agroforestry provides many environmental services as co-benefits. First, agroforestry systems store C in biomass (Schmitt-Harsch et al. 2012) and in the soil (Nair et al. 2009b, 2010). Second, management practices and plant species diversity enhance C sequestration in agroforestry systems (Häger 2012). Third, agroforestry contributes to reducing methane emissions, as it is a good alternative to slash-and-burn agriculture that is widespread in the tropics (CH_4 is released when vegetation is burned). Fourth, agroforestry most often does not need the application of chemical fertilizers that release N_2O. Last but not least, agroforestry helps reduce the combustion of fossil fuels, as it produces valuable biofuels (Achten et al. 2007, 2010a, b; Azam et al. 2005; Chapagain et al. 2009). As agroforestry is widely practiced in the tropics by farmers who rely on this practice for their everyday needs of food and cash, it is a wonderful win-win land use strategy (Leakey 2001) to clean-up the environment, and to enhance and sustain food and nutritional security in developing countries.

Three hundred and thirty-eight REDD+ projects and other C projects were recorded in 52 countries by the Global database of REDD+ and other Forest Carbon Projects as of June 13, 2013, with the highest number of REDD+ projects and other C projects recorded in Brazil (56) and Indonesia (44), respectively (www.forestclimatechange.org/redd-map/). The REDD+ projects consist of REDD+ readiness projects and REDD+ demonstration projects. This chapter will present and discuss the various aspects and challenges for the inclusion of agroforestry in C markets, as well as the different ways of implementing payments for PES for this land use system.

20.1.1 Carbon Payment Mechanisms and Tropical Agroforestry

The C market consists of industries in the developed countries buying C credits from farmers and other land managers to offset their C budget within the CDM. The regulatory C market accepts only afforestation and reforestation practices, excluding soil C sequestration projects. This is disadvantageous to agroforestry, as agroforestry systems can store 30–300 Mg C ha^{-1} in the soil, but only 0.29–15.21 Mg C ha^{-1} $year^{-1}$ aboveground (Nair et al. 2009a, b, 2010). Dhakal (2009) pointed out that the revenues from the sale of C credits are likely to be meager for forest dependent people, and the C market is volatile. Taking into account soil C stocks in C markets would substantially increase net revenues from C sales to poor farmers in developing countries. This is possible if C projects are properly managed. Tree-based cropping systems that were poorly managed responded positively to C projects, as discovered by Nelson and de Jong (2003) when they investigated the response of local communities to C mitigation projects in the State of Chiapas, Mexico. According to their study, the number of C contracts increased ten-fold in 5 years, growing from 43 in 1997 to 454 in 2001 (Nelson and de Jong 2003), indicating that farmers are interested in C mitigation projects, which contradicts the argument that the 'REDD programme is highly inappropriate to developing countries, where people have limited access to alternative income and employment opportunities' (Dhakal 2009).

Agroforestry diversifies and sustains farmers' sources of income, and ensures food security for farmers in the tropics. Revenues from the C market would constitute a supplementary agroforestry benefit for farmers. For this reason, there is a need to establish C projects in various parts of the tropics, and monitor and assess the response of farmers to these projects. In addition, the soil C sequestration effect of agroforestry needs to be taken into consideration in the regulatory context, as soil C represents the largest C pool of terrestrial ecosystems. Although C revenues would constitute new agroforestry benefits to farmers, C prices still need to be appropriate to justify the opportunity costs of land owners (Flugge and Abadi 2006). Indeed, the income from selling C credits should provide returns that are higher than alternative uses of land to make REDD+ attractive. In a study carried out in Australia, the C price needed to be approximately \$ 25–\$ 46 t^{-1} of CO_2 equivalent (CO_2-e) higher than what is expected from other uses in order to make growing trees a worthwhile investment. The price to pay for C credits in the tropics can be extrapolated from a model based on the opportunity cost of land diverted from annual crop production (Shively et al. 2004). Applying this model in the Philippines identified a C storage cost between US \$ 3.30 t^{-1} on fallowed lands and US \$ 62.50 t^{-1} on land that otherwise supports high value crops (Shively et al. 2004). The study also found that C costs for conversion of low- and high-input farms to agroforests range between US \$ 24.20 and US \$ 25.30 t^{-1}, and between US \$ 46.70 and \$ 48.00 t^{-1}, respectively (Shively et al. 2004). Economic analyses of C markets were also done in the tropics of Africa. For example, Takimoto et al. (2008) estimated the net present values for standard live fence and fodder bank with and without C credit sale in the African Sahel; with C credit sale they found an increase of NPV of \$ 13.9 for live fence and \$ 20.5 for fodder bank. In the Miombo woodlands, Bond et al. (2010) suggested that C price varying between US \$ 2.49 and US \$ 3.71 t^{-1} of CO_2-e would make REDD+ attractive to farmers.

The payment for C emission reductions is rewarded in the form of C credits. Therefore, it is expected that C credits will reflect C offset prices. Unfortunately, temporal increases in the price of C offset induce a decrease in C credits, as reported by Diaz et al. (2011). For instance, the price for C offset from primary forests rose from US \$ 3.8 t^{-1} of CO_2-e in 2008, to US \$ 4.5 t^{-1} CO_2-e in 2009, and up to US \$ 5.5 t^{-1} of CO_2-e in 2010, whereas C credits fell from US \$ 4.7 t^{-1} of CO_2-e in 2009 to US \$ 4.5 t^{-1} of CO_2-e in 2010 (Diaz et al. 2011). This reduction in the price of C credits over time is likely to negatively influence the adoption of C sink strategies by poor farmers, who would rather deal with concerns that are immediate such as feeding their family and achieving a balanced household budget. Agroforestry has the advantage of providing food, cash and shelter to farmers while sequestering C and should therefore be promoted in poor rural areas. However, the potential of agroforestry to reduce GHG emissions has been overlooked by decisions-makers in the global strategy on climate change adaptation (Montagnini and Nair 2004; Udawatta and Jose 2011). Nonetheless, more and more efforts are made to highlight the contribution of some agroforestry systems, such as shelterbelt systems, to terrestrial C sequestration (Schoeneberger 2009; Nair et al. 2009a; reviewed by Nair et al. 2010), so as to promote the use of the co-benefits of this agroforestry system. Indeed, C-payments in agroforestry constitute additional benefits to farmers, understanding that agroforestry C projects are not profitable *per se* on their own, but rather as an add-on to other revenue generating activities.

The implementation of C projects in agroforestry is challenging, as the methods used to quantify C stocks are often erroneous (Kim 2012; Nair 2011), and vary widely (Nair 2011, 2012). A possible explanation to the variation in C stock measurement is that soil CO_2 emissions occur in agroforestry systems harboring nitrogen-fixing plants that reduce the amount of C stored (Kim 2012). Therefore, estimates of C stocks in agroforestry systems that did not take these emissions into account are erroneous (Kim 2012). Also, aboveground C stock estimates are derived as a percentage of standing aboveground biomass or calculated using equations determined by optimizing the accuracy of woody biomass estimates associated with labor costs (Zhou et al. 2007), or allometric equations developed from regression analysis used to determine most important aboveground biomass (Hager 2012; Kort and Turnock 1999; Schmitt-Harsh et al. 2012; Tamang et al. 2012). However, these allometric equations have not yet been developed for all species (Nair 2012), and each agroforestry plot is unique. For these reasons, methods and tools for C stock estimation in agroforestry systems need to be standardized.

The standardization of methods would help set criteria for estimating C stock in agroforestry systems and would provide the basis for the comparison of C stocks in agroforestry systems worldwide, especially in the tropics. Also, models for estimating C stocks could be obtained in the process, and these models could integrate the variables known to influence C stock variation in agroforestry systems, including tree species, planting density, rotation length, and age of the trees of studied systems. Soil characteristics should also be taken into account in the models, and trials should be conducted in different eco-regions to validate these models. These models would substantially contribute to the UN-REDD Programme entitled "Measurement, Reporting and Verification and Monitoring".

Another challenge for the inclusion of agroforestry in the C market is the PES option: should farmers be compensated according to the amount of C stored per unit area, or standing biomass? Indeed, temporal variation and the amount of C sequestered per unit surface area are two important characteristics that would influence the inclusion of agroforestry in C payment mechanisms as implemented by REDD+. For example, is the C contract based on the amount of C stored at a given time (i.e., based on the total C per unit area), or on the year-to-year variation of C accumulation (i.e., pay someone as they go)? These questions still need to be answered.

20.1.1.1 Carbon Contract Options in Agroforestry

Most often, C contracts in agroforestry are based on the number of trees retained/grown on agricultural lands. For example, in Costa Rica, farmers were paid

US $ 1.30 per tree for a minimum of 350 trees and a maximum of 3500 on their farms in 2007 (Cole 2010). Obviously, PES for agroforestry vary with the number of trees in agricultural lands and, consequently, with the type of agroforestry system (i.e., homegardens, alley cropping, cocoa farms), as was found in Indonesia (Seeberg-Elverfeldt et al. 2009). For that reason, the PES option based on the number of trees on farms can be beneficial or not to the farmer depending on the agroforestry system that is used (Seeberg-Elverfeldt et al. 2009). To overcome this issue, PES for agroforestry can learn from what is being done for other land use systems. In Costa Rica, for example, PES per ha was higher for reforestation (US $ 537) than for sustainable forest management (US $ 327) and forest conservation easements ($ 210) for a five-year period (http://www2.gsu.edu/~wwwcec/special/lr_ortiz_kellenberg_ext.pdf). However, implementing this C payment option for agroforestry would not be adequate, as C stocks vary widely with agroforestry systems and plots, and methods and tools to estimate C stocks in agroforestry systems are not yet standardized. Also, there is a temporal change in soil C following afforestation (Paul et al. 2002).

The rate of aboveground C accumulation in agroforestry largely depends on the trees grown. Indeed, fast-growing tree species accumulate biomass faster than slow-growing ones. An explanation is that fast-growing plant species tend to have greater photosynthetic capacity than slow-growing plants (Aerst and Chapin 2000). Carbon storage in agroforestry can therefore be rewarded based on annual C accumulation. In Costa Rica, PES forest conservation contracts (US $ 210) paid US $ 42 $year^{-1}$ ha^{-1} (http://www2.gsu.edu/~wwwcec/special/lr_ortiz_kellenberg_ext.pdf). However, this C payment method implemented in Costa Rica did not take into account the differential C accumulation rate between species. More work is needed to investigate the profitability of this C payment option in different agroforestry systems in the tropics. Also, even though agroforestry systems have a great potential to provide large contributions to whole farm GHG mitigation, less is known on the processes governing C trade-offs between agroforests and the atmosphere over time.

The rate of C accumulation is also important for determining the profitability of agroforestry, as the payback period of any agroforestry system is time-dependent, and longer than conventional agricultural systems. Identifying factors responsible for variation in C accumulation would help estimate the length of the optimal rotation period, and increase C stocks and profitability of agroforestry systems, subsequently affecting their adoptability and large-scale dissemination. On the other hand, providing new insights into the mechanisms influencing C sequestration in agroforestry systems is necessary, as it would contribute to increasing the accuracy in predicting the amounts of C sequestered over time.

20.1.2 REDD+ in Tropical Agroforestry and Land Tenure

Land tenure is very important for the implementation of REDD+, as payments are made to individuals with secure land tenure (Costenbader 2011). Indeed, REDD+ incentives target landowners and users, and in many countries tree tenure is different from land tenure (for more details, please see Chap. 17). Land tenure and rights are very complex in Africa because of interactions between traditional land rights and colonial and post-colonial land tenure and rights (Cousins and Claasens 2006; Forest Trends 2002). Confrontation of these different practices has resulted in confusions between open access systems, common, and private properties (Alden 2006). Therefore, control of land and access to forest resources is very complex in Africa. For example, some farmers, who do not own land, plant and manage trees to benefit from tree products (NTFPs, soil improvement); meanwhile C payments are attributed to land owners. In this case, land users would not have incentives to put effort into management practices that would increase tree biomass, increasing C stock. On the other hand, if access to land were restricted over time, how would C payments be allocated to owners versus users? Further, in most African countries, ownership of land is based on occupancy, use and lineage and other inborn rights (Unruh 2008), making C payment more complex.

Some examples of the complexity of differed land tenure and land rights are drawn from African countries. In the Democratic Republic of Congo, the state exercises sovereignty over lands and natural resources, whether or not these lands are occupied by local people for farming. The Constitution nonetheless recognizes individual and collective property rights. This has resulted in a lack of clarity in property rights; an issue that needs to be addressed to allow for the implementation of payments to communities based on results (DRC 2010). In Cameroon, unless one has a certificate issued by the administration proving land ownership, land belongs to the state. Therefore, a farmer can "own" a piece of land by occupancy but could not benefit from C payments as he would have no way to prove before the law that he has property rights to the land. Therefore, tenure security is an important issue to participation in REDD+ in many countries in the tropics. However, land tenure and rights take on different forms in other tropical countries.

The complexity associated with land rights is also found in Latin America and Southeast Asia. In Mexico, 53 % of land (representing 70 % of the country's total forest cover) belongs to *ejidos* (i.e., area used by community members for agriculture) and local communities (www.theredddesk.org/countries/mexico). *Ejidos* and local communities have land use rights, and community members decide on adoption of management practices and choose mechanism to guarantee equitable distribution of profits. In Brazil, the framework for land tenure includes public land, protected areas, indigenous lands (owned by the federal government but indigenous populations have usufruct rights), private lands, and military lands (www.theredddesk.org/countries/brazil). The Brazil case is very complex, as there is no specific regulation addressing rights and tenure to land. This complicated land tenure system makes land ownership difficult to establish, undermining the implementation of REDD+ in the country. However, a program currently in progress aims at facilitating private land registration (Terra Legal Programme). In Indonesia, Janudianto et al. (2011) called for recognition of tree tenure as part of the REDD+ strategy, since forest authorities have tried to control collection of products from valued trees that were formerly under customary control and regulation.

Tenure insecurity is a great threat to the implementation of REDD+ in tropical agroforestry, as it raises the questions of how C credits will be defined, owned, and regulated. Indeed, tree and land tenure, as well as land rights issues need to be resolved so as to clarify C property rights and minimize potential legal conflicts over C ownership (Richards 2010). The national REDD+ projects face the challenge of equitable redistribution of C-payments without negatively affecting local and indigenous people's rights. The REDD+ can also reinforce or augment certain customary property rights, as found to be the case in Bolivia (TNC 2009). Equitable compensation and C ownership are therefore two important issues that are often contradictory when it comes to the implementation of REDD+. For instance, in some rural areas in Cameroon, women don't own land, but they plant and manage trees that will belong to their children, resulting in the problem of gender and C payment. In such households, who will receive the C-payment or compensation? Though some authors stressed that it is important to focus on how stakeholders are compensated rather than questioning who owns the C (Richards 2010), it is clear that this solution would be unfair in the latter case, unless, for example, the C payments are given to the children later on when they become adults.

20.1.3 Economics of REDD+ and PES

The production, distribution and consumption (i.e., economics) of REDD+ ecological goods (i.e., clean air and abundant fresh water) and services (i.e., biodiversity conservation, GHG mitigation, soil and vegetation generation and renewal, groundwater recharge and decomposition of wastes) are articulated around two main issues:

- How to quantify the value of these assets?
- To whom should PES be awarded?

The quantification of the value of C reduction emission in agroforestry depends heavily on the estimation of C stocks. Though C stock estimation methods in agroforestry still need to be improved and standardized, substantial efforts have been made to estimate C emissions reduction in REDD+ projects (for a practical example, please see Box 20.1) and to reward C emission reduction through the payment of C credits.

The C rights (i.e., title to C credits) constitute an issue of debate (Karsenty et al. 2012), as REDD+ ecological goods and services are public goods in nature. Carbon rights are based on the concept that REDD+ will generate revenues exceeding the full cost of corresponding efforts or "rents" (Hepburn 2000). This concept has been challenged because its underlying beliefs and interpretations may be misleading (Peskett and Brodnig 2011; Karsenty et al. 2012). Instead, Karsenty et al. (2012) suggested that rents could be created by setting a reference emission level and by possible acceptance of rules such as being remunerated for the full stock of C. Karsenty et al. (2012) also argued that compensating for easements would be a more appropriate framework for designing incentive schemes, instead of C rights

Box 20.1 Practical example: how to estimate C emissions reduction in REDD projects

Assume a forest area of 100,000 ha with a carbon density of 150 t C ha^{-1};

Baseline scenario: 1% of annual deforestation = 100,000 × 0.01 = 1000 ha $year^{-1}$

Forest carbon stock in t CO_2 = carbon density in t C ha^{-1} × 3.67 t CO_2

3.67 is the factor between C and CO_2, i, e., the molecular weight of CO_2 (44 g) over the molecular weight of C (12 g): 44/12 = 3,67

Annual CO_2 emissions = deforested area x forest carbon stock (CO_2 conversion) = 1000 ha $year^{-1}$ × 150 t C ha^{-1} × 3.67 t CO_2 = 550,500 t CO_2 $year^{-1}$

REDD++ project scenario

0.7% of annual deforestation = 100,000 ha × 0.007 = 700 ha $year^{-1}$

Annual CO_2 emission = 700 ha $year^{-1}$ × 150 t C ha^{-1} × 3.67 t CO_2 = 385,350 t CO_2 $year^{-1}$

frameworks that consist of sharing the benefits of C stock (i.e., a human production) and the sale of these benefits, because C rights are also interpreted as the right to benefit from the sale of C credit.

Carbon emission trading consists in buying C (CO_2 calculated in tonnes CO_2 eq) offsets, the money from which will contribute to fund projects that reduce GHG emissions. Two types of carbon markets exist: voluntary markets (Kyoto's framework) and regulated markets (non-Kyoto's initiatives). Regulated C markets refer to national or regulated entities' markets whose emissions are below their quotas, and which can sell excess emission credits to nations or entities that exceed their quotas (e.g., Regional Greenhouse Gas Initiative (RIGGI) and Western Regional Climate Initiative; Kossoy 2008). In voluntary markets, companies engage in voluntary programs in which they can sell or buy emission credits to meet voluntary targets (e.g., Verified Carbon Standard (VCS), Plan Vivo, European Union Emissions Trading System (EU ETS)). It is noteworthy that one of the world's bigger C emitters, China, is soon to launch pilot C trading schemes on a voluntary market basis in Shenzen, Beijing, Shanghai, Tianjin and Chongqin (www.bbc.co.uk/news/busines-22931899 accessed on June 18th, 2013). China is committed to reducing its overall C emissions per unit of GDP to 40% below the 2005 levels by 2020 (http://www.salon.com/2013/06/17/china_struggles_to_meet_carbon_emission_targets_partner/).

The EU ETS is the world's biggest C emission trading market (Table 20.1). The EU ETS, which was launched in 2005, accounted for over three-quarters of international C trading in 2010 (www.ec.europa/clima/policies/ets/index_en.htm); the sectors covered by the EU ETS include CO_2, N_2O and perfluorocarbons (PFCs). The EU ETS covers EU member states (27 on the June 17th, 2013) and works by putting a limit on overall emissions from high-emitting industries (www.ec.europa/clima/policies/ets/index_en.htm). This system allowed more than 17,000 t CO_2 eq reduction of average emissions per installation during the 2005–2010 period

Table 20.1 Transaction volumes and values, Global Carbon Market, 2009 and 2010

	Volume ($MtCO_2eq$)		Value (US $ million)	
Markets	2009	2010	2009	2010
Voluntary OTC	55	128	354	414
CCX	41	2	50	0.2
Other exchanges	2	2	12	10
Total voluntary markets	*98*	*131*	*415*	*424*
EU ETS	5,510	5,529	105,746	106,024
Primary CDM	135	94	2,858	1,325
Secondary CDM	889	1,005	15,719	15,904
Kyoto [AAU]	135	19	1,429	265
RGCI	768	45	1,890	436
Total regulated markets	*7,437*	*6,692*	*127,642*	*123,954*
Total global markets	*7,535*	*6,823*	*128,057*	*124,378*

AAU assigned amount unit, *CCX* Chicago climate exchange, *CDM* clean development mechanisms, *EU ETS* European union emissions trading system, *RGCI* regional greenhouse gas initiative (RIGCI), *Voluntary OTC* voluntary over-the-counter offset market

(www.ec.europa/clima/policies/ets/index_en.htm). It is noteworthy that the EU ETS covered more than 11,000 power stations and manufacturing plants in the EU on June 2013 (www.ec.europa/clima/policies/ets/index_en.htm).

The CDM market was unevenly shared worldwide in 2007, with 73 % in China, 6 % in India, 5 % in the rest of Asia, 6 % in Brazil and 5 % in Africa (Kossoy 2008).

Bibliography

Achten WMJ, Verchot L, Singh VP, Aerts R, Muys B (2007) *Jatropha* biodiesel fueling sustainability? Biofuels Bioprod Bioref 1:283–291. doi:10.1002/bbb

Achten WJM, Almeida J, Fobelets V, Bolle E, Mathijs E, Singh VP, Tewari DN, Verchot LV, Muys B (2010a) Life cycle assessment of *Jatropha* biodiesel as transportation fuel in rural India. Appl Energy 87:3652–3660

Achten WMJ, Nielsen LR, Aerts R, Lengkeek AG, Kjaer ED, Trabucco A, Hansen JK, Maes WH, Graudal L, Akinnifesi FK, Muys B (2010b) Towards domestication of *Jatropha curcas*. Biofuels 1(1):91–107

Aerst R, Chapin FS II (2000) The mineral nutrition of wild plants revisited: a re-evaluation of processes and patterns. Adv Ecol Res 30:1–67

Albrecht A, Kandji ST (2003) Carbon sequestration in tropical agroforestry systems. Agric Ecosyst Environ 99(1–3):15–27

Alden WL (2006) Land rights reform and governance in Africa: how to make it work in the 21st Century? UNDP, New York

Azam MM, Waris A, Nahar NM (2005) Prospects and potential of fatty acid methyl esters of some non-traditional seed oils for use as biodiesels in India. Biomass Energ 29:293–302

Bond I, Chambwera M, Jones B, Chundama M, Nhantumbo I (2010) REDD+ in dryland forests: issues and prospects for pro-poor REDD in the miombo wooldands of southern Africa, Natural Resource Issues No. 21. IIED, London

Canadell JG, Le Quéré C, Raupach MR, Field CB, Buitenhuis ET, Ciais P, Conway TJ, Gillett BP, Houghton RA, Marland G (2007) Contributions to accelerating atmospheric CO_2 growth from economic activity, carbon intensity and efficiency of natural sinks. PNAS 104(47):18866–18870

Chapagain BP, Yehoshua Y, Wiesman Z (2009) Desert date (*Balanites aegyptiaca*) as an arid lands sustainable bioresource for biodiesel. Bioresour Technol 100:1221–1226

Cole RJ (2010) Social and environmental impacts of payments for environmental servicesfor agroforestry on small-scale farms in southern Costa Rica. Int J Sustain Dev World Ecol 17(3):208–216

Costenbader J (2011) REDD+ benefit sharing: a comparative assessment of three national policy approaches, Forest Carbon Partnership/UN-REDD Programme

Cousins B, Claasens A (2006) More than simply—socially embedded: recognizing the distinctiveness of African land rights. Keynote address at the international symposium on 'at the frontier of land issues: social embeddedness of rights and public policy'. Montpellier, 17–19 May 2006

Dhakal B (2009) Carbon liability, market price risk and social impact of reducing emission from deforestation and forest degradation (REDD) programme. J For Livelihood 8(1):67–77

Diaz D, Hamilton K, Johnson E (2011) State of the Forest Carbon Markets 2011. From Canopy to Currency. Forest Trend, p 93

DRC (Democratic Republic of Congo) (2010). Final Draft: Forest Carbon Partnership Facility (FCPF), Readiness Preparation Proposal (R-PP). http://www.forestcarbonpartnership.org/fcp/node/65. Accessed 15 Aug 2011

Flugge F, Abadi A (2006) Farming carbon: an economic analysis of agroforestry for carbon sequestration and dryland salinity reduction in Western Australia. Agrofor Syst 68:181–192

Forest Trends (2002) Strategies for strengthening community property rights over forests: lessons and opportunities for practitioners. Forest Trends, Washington

Galford GL, Melillo JM, Kicklighter DW, Cronin TW, Cerri CEP, Mustard JF, Cerri CC (2010) Greenhouse gas emissions from alternative futures of deforestation aand agricultural management in the southern Amazon. PNAS 107:19649–19654

Häger A (2012) The effects of management and plant diversity on carbon storage in coffee agroforestry systems in Costa Rica. Agrofor Syst 86:159–174

Hepburn S (2009) Carbon rights as new property: the benefits of statutory verification. Sydney Law Rev 31:239–271

IPCC (2007) Glossary J-P. In (book section): Annex I. In: Climate Change 2007: Report of the Intergovernmental Panel on Climate Change (B. Metz et al. Eds.). Cambridge University Press, Cambridge, U.K., and New York, N.Y., U.S.A

Janudianto J, Mulyoutami E, Joshi L, Wardell DA, van Noordwijk M (2011) Recognizing traditional tree tenure as part of conservation and REDD strategy: feasibility study for a buffer zone between a wildlife reserve and the Lamandau river in Indonesia's REDD Pilot Province. ASB Policy Brief No. 22. ASB Partnership for the Tropical Forest Margins, Nairobi

Karsenty A, Vogel A, Castell F (2012) "Carbon rights", REDD+ and payments for environmental services. Environmental Science and Policy. http://dx.doi.org/10.1016/j.envsci.2012.08.03

Kim DG (2012) Estimation of net gain soil carbon in a nitrogen-fixing tree and intercropping system in sub-Saharan Africa: results from re-examining a study. Agrofor Syst 86:175–184

Kossoy A (2008) State and trends of the Carbon market. The World Bank. www.latincarbon.com/2008/docs/presentation/Days/AlexandreKossoy.pdf

Kort J, Turnock R (1999) Carbon reservoir and biomass in Canadian prairie shelterbelts. Agroforestry Systems 44:175–186.Leakey RRB (2001) Win:win anduse strategies for Africa: 2. Capturing economic and environmental benefits with multistrata agroforests. Int For Rev 3:11–18

Lal R (2008) Carbon sequestration. Phil Trans R Soc B 363(1492):815–830. (27 February 2008)

Leakey RRB (2001) Win:win landuse strategies for Africa: 2. Capturing economic and environmental benefits with multistrata agroforests. Int For Rev 3: 11–18

Mather R (2010) Supporting REDD Implementation in Vietnam, Laos and Cambodia through the Design of a REDD-compliant Benefit Distribution System, Final report

Montagnini F, Nair PKR (2004) Carbon sequestration: an underexploited environmental benefit of agroforestry systems. Agrofor Syst 61:281–295
Nair PKR (2011) Methodological challenges in estimating sequestration potential of agroforestry systems. Adv Agrofor 8(1):3–16
Nair PKR (2012) Carbon sequestration studies in agroforestry systems: a reality check. Agrofor Syst 86:243–253
Nair PKR, Mohan Kumar B, Nair VD (2009a) Agroforestry as a strategy for carbon sequestration. J Plant Nutr Soil Sc 172:10–23
Nair PKR, Nair VD, Mohan Kumar B, Haile SG (2009b) Soil carbon sequestration in tropical agroforestry systems: a feasibility appraisal. Environ Sci Policy 12:1099–1111
Nair PKR, Nair VD, Mohan Kumar B, Showalter JM (2010) Carbon sequestration in agroforestry systems. Adv Agron 108:237–307. doi:10.1016/S0065-2113(10)08005-3
Nelson KC, de Jong BHJ (2003) Making global initiatives local realities: carbon mitigation projects in Chiapas, Mexico. Global Environ Change 13:19–30
Paul KI, Polglase PJ, Nyakuengama JG, Khanna PK (2002) Change in soil following afforestation. For Ecol Manag 168(1–3):241–257
Peskett L, Brodnig G (2011) Carbon rights in REDD+: exploring the implications for poor and vulnerable people. World Bank and REDD-net
Richards M (2010) The REDD Opportunities Scoping Exercise (ROSE): a tool for prioritizing sub-national REDD+ activities—case studies from Ghana, Tanzania, and Uganda. Forest Trends, Washington
Scherr SJ, McNeely JA (2007) Ecoagriculture strategies for poverty reduction and biodiversity conservation. Paper presented at the International Workshop on 'Reconciling rural poverty and resource conservation: identifying relationships and remedies.' May 2–3, 2003, Cornell University, Ithaca, NY, USA
Schmitt-Harsh M, Evans TP, Castellanos E, Randolph JC (2012) Carbon stocks in coffee agroforests and mixed dry tropical forests in the western highlands of Guatemala. Agrofor Syst 86:141–157
Schoeneberger MM (2009) Agroforestry: working trees for sequestering carbon on agricultural lands. Agrofor Syst 75:27–37
Seeberg-Elverfeldt C, Schwarze S, Zeller M (2009) Carbon finance options for samllholders' agroforestry in Indonesia. Int J Commons 3(1):108–130
Shively GE, Zelek CA, Midmore DJ, Nissen TM (2004) Carbon sequestration in a tropical landscape: an economic model to measure itsincrement cost. Agrofor Syst 60:189–197
Takimoto A, Nair PKR, Alavalapati JRR (2008) Socioeconomic potential of carbon sequestration through agroforestry in the West African Sahel. Mitig Adapt Strateg Glob Change 13:745–761
Tamang B, Andreu MG, Staudhammer CL, Rockwood DL, Jose S (2012) Equations for estimating aboveground biomass of cadaghi (*Corymbia torelliana*) trees in farms windbreaks. Agrofor Syst 86:255–266
TNC (The Nature Conservancy) (2009) Noel Kempff Mercado climate action project: a case study in reducing emissions from deforestation and degradation. TNC, Washington
TNC (The Nature Conservancy) (2010) Reducing Emissions from Deforestation and Degradation (REDD): a casebook of on-the-ground experience, conservation international and wildlife conservation society. Arlington
Udawatta RP, Jose S (2011) Carbon sequestration potential of agroforestry practices in temperate North America. In: Kumar BM, Nair PKR (eds) Carbon sequestration potential of agroforestry systems: opportunities and challenges. Adv Agrofor 8. doi:10.1007/978-94-007-1630-8_2 Springer Science+Business Media BV
Unruh JD (2008) Carbon sequestration in Africa: the land tenure problem. Global Environ Change 18:700–707
Zhou X, Brandle JR, Schoeneberger MM, Awada T (2007) Developing above-ground woody biomass equations for open-grown, multi-stemmed tree species: Shelterbelts-grown Russian olive. Ecol Model 202(3–4):311–323

Chapter 21
Agroforestry Modeling

Abstract Agroforestry is a complex system that integrates biophysical, environmental, socio-economic and cultural factors. Agroforestry is also complex both functionally and economically. Therefore, sound decision-making is of high importance in agroforestry, whether it is related to species selection, components (i.e., trees and crops; trees and livestock; trees, crops and livestock) selection and integration, or maximization of benefits gained from applying agroforestry techniques. For these reasons, decision-making in agroforestry requires databases and tools for species selection, and expert systems (i.e., computer program that solves problems that normally require the abilities of human expert) to forecast different outcome scenarios. Many databases for species selection in agroforestry have been compiled and can be found on the ICRAF website. Several modeling tools exist for agroforestry research.

21.1 Introduction

Agroforestry systems are structurally, functionally and economically much more complex than monocropping systems (for more details, please see Chap. 3). In addition, agroforestry integrates biophysical, socio-economic, environmental and cultural aspects. For these reasons, proper decision-making is an important aspect of agroforestry because it must take into account the specificities of the site (climate, soil type, topographic properties, etc.), the dietary habits of farmers, the compatibility between the cultivated species (as there may be positive or neutral interactions; for more details, please see Chap. 7), and the economic interaction between the components of the system. Efforts have been made to provide agroforestry practitioners with decision-support tools to facilitate decision-making. These decision-support tools can be found by using web search engines.

Computer programs also assist agroforestry practitioners in decision-making. These computer programs, often called Expert Systems, use reasoning techniques, knowledge and facts to solve complex problems. Agroforestry computer programs were first designed to support practitioners interested in maximizing benefits gained from applying appropriate land-use management techniques. These computer programs were developed in the 1980's. The next generation of agroforestry computer programs were designed to support agroforesters interested maximizing

A. Atangana et al., *Tropical Agroforestry*, DOI 10.1007/978-94-007-7723-1_21,

benefits gained in tree/germplasm selection and introduction on farms. These programs gained momentum with the advent of agroforestry tree domestication in the 1990's. As climate change concern is now a major concern, agroforestry computer programs focus on the design of techniques that would maximize environmental benefits (carbon sequestration, erosion control and so on) gained from agroforestry management. This chapter briefly discusses some computing tools, databases and toolkits used in agroforestry.

21.1.1 Species Choice in Agroforestry

Databases useful for the selection of agroforestry species exist. Examples include Agroforestree (AFT) (http://www.worldagroforestrycentre.org/sea/Products/AFDbases/af/index.asp) and the Multipurpose Tree (MPTs) (http://www.ciesin.org/IC/icraf/mptsdata.html). These databases help in the selection of a species or a combination of species for an agroforestry system, based on the specific region, services and functions expected. The databases and toolkits suggested by ICRAF for tree selection (http://www.worldagroforestry.org/our_products/databases) are:

- The Agroforestree Database (http://www.worldagroforestry.org/resources/databases/agroforestree);
- The Useful Tree Species for Africa (http://www.Worldagroforestry.org/our_products/databases/useful-tree-species-africa)
- The Tree Seed Suppliers Directory (http://www.worldagroforestry.org/our_products/databases/tssd)
- The Botanical Nomenclature Database (http://www.worldagroforestry.org/Sites-old/TreeDBS/slides.asp)
- The Tree Diversity Analysis (http://www.worldagroforestry.org/resources/databases/tree-diversity-analysis)
- The Molecular Markers for Tropical Trees (http://www.worldagroforestry.org/our_products/databases)

21.1.2 Computing Tools Used in Agroforestry

Agroforestry is a very complex system and benefits are typically obtained after a longer period of time than for agricultural systems. Any investment in agroforestry requires a great deal of information in order to facilitate the development, acceptance, adoption and management of the system (Ellis et al. 2004). As reported by Ellis et al. (2004), "the decision-making process involved in agroforestry research, development and application is composed of several components: the person or group making the decision, the problem, the approach or method to solve the problem, and the decision". For this reason, computer-based decision support tools like

Table 21.1 Major categories of computer-based decision support technologies (Updated from Ellis et al. 2004)

Category	Description
Databases	Organizes and facilitates the management and querying of large quantities of data and information
Geographic Information Systems (GIS)	Brings in a geographic or spatial component to a database; manages, manipulates and analyzes spatial data
Computer-Based Models	Mathematical computer models that represent real world processes and predict outcomes based on input scenarios
Knowledge-Based or Expert Systems (KBS)	Adopts "Artificial Intelligence" to organize, manipulate and obtain solutions using knowledge in the form of qualitative statements, expert rules (i.e. rules of thumb), and a computer language representation system for storing and manipulating knowledge
Hybrid Systems	Integrates two or more of the above computer-based technologies (e.g., GIS, KBS and Models) for more versatile, efficient and comprehensive decision support tools (DSTs)

databases, geographic information systems (GIS), computer-based models, expert systems and hybrid systems (Table 21.1) have been developed to facilitate decision-making in agroforestry (Garcia-de Ceca and Gebremedhin 1991; Ellis et al. 2004). Databases include a database management component. GIS allows spatial representation of information due to the association of map features and inclusion of geographically referenced data values in a database. Knowledge-Based Systems (KBS) or expert systems are part of the broad field of artificial intelligence (AI). They involve the creation of computer programs that attempt to mimic human intelligence or reasoning, to "learn" new information and tasks, and to draw useful conclusions about the world around us (Patterson 1990; Ellis et al. 2004). Computer-based models involve the translation of data and information into a mathematical form that represents a real world process or system and can be used to forecast the outcomes of different scenarios. Hybrid systems are comprised of a combination of the modeling tools listed below. A non-exhaustive list of computer-based decision tools is provided in Table 21.2.

Garcia de Ceca and Gebremedhin (1991) developed a support system for planning small-scale agroforestry based systems. The program generates a set of feasible culture combination alternatives based on specified economic constraints, including the availability of land, labor and monetary resources. The variables the model uses are the area of land allocated for each culture and the amount of hired labor needed for each alternative in each season.

Table 21.2 Some computer-based decision support tools used in agroforestry (Updated from Ellis et al. 2004)

Decision Support Tool (DST)	Type	Description	Reference
Agroforestry system suitability in Africa	GIS	Spatial analysis using climate, soil, land-use and other spatial data alongside plant species to determine species and agroforestry suitability	Booth et al. (1989); Booth et al. (1990); Unruh and Lefebvre (1995)
Agroforestry system suitability in Ecuador	GIS	Spatial analysis to determine suitable areas of *Annona cherimola* agroforestry systems in southern Ecuador	Bydekerke et al. (1998)
Agroforestry system assessment in Nebraska	GIS	Spatial suitability assessment for willow and forest farming agroforestry systems in a Nebraska watershed	Bentrup and Leininger (2002)
Agroforestry Parklands in Burkina Faso	GIS	Spatial analysis of dynamics of agroforestry parklands and species distribution due to human impacts	Bernard and Depommier (1997)
Historical transformation of agroforestry landscape in Canada	GIS	Spatial analysis of census and geomorphologic data to explore dynamics of agroforestry in 19th century Canadian landscape	Paquette and Domon (1997)
Field-level spatial analysis of temperate agroforestry system	GIS	Spatial analysis using ground penetrating radar (GPR) to evaluate root biomass and distribution and soil nutrient crop-tree interactions in temperate alley cropping	Joseet al. (2001)
AME (Agroforestry Modeling Environment)	Modeling tool	Object-oriented tool to graphically visualize, construct, integrate and exchange agroforestry models	Muetzelfeldt and Taylor (1997)
RothC	Model	Model for the turnover of organic carbon in non-waterlogged topsoils	Coleman and Jenkinson (1999, 2008)
CENTURY model	Model	General model which simulates carbon, nitrogen and nutrient dynamics	Metherell et al. (1993)
Hi-SAFE microclimate module concept	Model	3D process-based biophysical model of the SAFE project. It includes the main tree functions with regard to major resources (carbon, water, nitrogen) and responses to the major climate variables (light, air temperature and humidity)	Dupraz et al. (2004)

Table 21.2 (continued)

Decision Support Tool (DST)	Type	Description	Reference
CO2FIX	Model	Carbon bookkeeping model that consists of six modules: Biomass module Soil module Products module Bioenergy module Financial module Carbon accounting module	Masera et al. (2003); Schelhaas et al. (2004)
SCUAF	Model	Model to estimate soil changes under agriculture, agroforestry and forestry	Young et al. (1998)
Carbon management tool (COMET 2)	Tool	Tool that estimates carbon sequestration and net greenhouse gas emissions from soils and biomass for US farms and ranches	USDA NRCS (2003)
EUROSEM	Model	Dynamic distributed model that simulates sediment transport, erosion and deposition	Albaladejo Montoro et al. (2003, 2010)
WBECON (WindBreaks ECONomics)	Computer program	WBECON calculates the economics of shelterbelts taking into account various factors such as shelterbelt species and characteristics, shelterbelt design, soil and climate factors, crop rotation, shelterbelt costs, crop costs and crop prices	Kort and Brandle (1991)
HyPAR	Model	Biophysical model combining crop and forest models and integrating climate, hydrology, light interception, water and nutrient competition, and carbon allocation processes in agroforestry systems	Mobbs et al. (2001)
HyCAS	Model	Biophysical model for agroforestry systems with cassava simulating competition for light, water and nutrient including phosphorus cycles	Matthews and Lawson (1997)
WaNulCAS	Model	Biophysical model of tree-crop interactions based on above and belowground resource capture and competition of water, nutrients and light under different management scenarios in agroforestry systems	Van Noordwijk and Lusiana (1999); World Agroforestry Centre (2003b)

Table 21.2 (continued)

Decision Support Tool (DST)	Type	Description	Reference
SCUAF (Soil Changes Under Agroforestry)	Model	Nutrient cycling model predicts changes in soil conditions under different agroforestry systems based on parameters of biophysical environment, land-use and management, plant growth, and plant-soil processes	Young and Muraya (1990); Vermeulent et al. (1993); Menz et al. (1997); Macadong et al. (1998); Nelson et al. (1997)
FALLOW (Forest, Agroforest, Low-value Landscape or Wasteland?)	Model and GIS	Model to evaluate impacts of shifting cultivation and fallow rotations at a landscape-scale, evaluating transitions in soil fertility, crop productivity, biodiversity and carbon stocks	Van Noordwijk (2002); World Agroforestry Centre (2003c)
BEAM (Bio-Economic Agroforestry Model)	Model	Bioeconomic model to assess physical and financial performance of agroforestry systems based on tree and crop biometric and economic models	Willis (1993); Willis and Thomas (1997)
AEM (Agroforestry Estate Model)	Model	Economic model to evaluate agroforestry in combination with other farm activities assessing effects of tree production and physical and financial resources on-farm	Middlemiss and Knowles (1996)
DESSAP (Agroforestry Planning Model)	Model	Multi-objective linear programming to assess feasible agroforestry alternatives based on land, labor and cash constraints	Garcia de Ceca and Gebremedhin (1991)
Tradeoff analysis software	Software	Software that integrates GIS-based data, biophysical production models, econometric production models and environmental models to assess the economic feasibility of carbon sequestration	Antle et al. (2007)
AKT (Agroforestry Knowledge Toolkit)	Knowledge Based System (KBS)	KBS to store, manipulate and analyze a variety of information and knowledge acquired on agroforestry systems	Walker et al. (1995)
AES (Agroforestry Expert System)	KBS	KBS and heuristic knowledge or expert "rules of thumb" to determine optimal species and spacing for alley cropping systems in the tropics	Walkertin et al. (1990)
AGFADOPT (Agroforestry Adoption Evaluation Tool)	Decision Tree KBS	KBS based on decision trees and used to assess adoption of agroforestry based on economic and social factors faced by small-scale farmers	Robotham (1998)

Table 21.2 (continued)

Decision Support Tool (DST)	Type	Description	Reference
Agroforestry Planning Tool in China	Hybrid GIS, Models and KBS	Hybrid DST integrating GIS data, regression models plus expert knowledge to assess bio-physical, social and economic suitability of *Paulownia* inter-cropping agroforestry systems	Liu et al. (1999)
PLANTGRO (Plantation and Agroforestry Species Selection Tool)	Hybrid GIS/KBS	Plantation and agroforestry species selection tool that integrates GIS and expert system on plant growth	Booth (1996); Hackett and Vanclay (2003)
SEADSS (Southeastern Agroforestry Decision Support System)	Hybrid data-base/GIS/KBS	Landscape and site-scale agro-forestry planning and species selection DST for landowners and extension agents of Southeast US that integrates GIS, tree and shrub database and expert knowledge	Ellis et al. (2003)
Conservation Buffer Planning Tools for Western Corn Belt Region, USA	Hybrid GIS/Models/Visualization	Suite of GIS, economic models and visualization tools for landowners and resource managers to evaluate agroforestry strategies in Midwest Corn Belt Region of USA	Ellis et al. (2003)

Bibliography

Dupraz C, Vincent G, Lecomte I, Bussière F, Sinoquet H (2004) Above-ground modules in Hi-SAFE; Deliverable D.4.1 (SAFE European contract QLK5-CT-2001-00560) silvoarable agroforestry for Europe (SAFE). INRA Montpellier, France

Ellis EA, Bentrup G, Schoeneberger MM (2004) Computer-based tools for decision support in agroforestry: current state and future needs. Agrofor Syst 61:401–421

Garcia-de Ceca JL, Gebremedhin KG (1991) A decision support system for planning agroforestry systems. For Ecol Manag 45:199–206

Parton WJ, Anderson DW, Cole CV, Stewart JWB (1983) Simulation of soil organic matter formation and mineralization in semiarid agroecosystems. In: Lowrance RR, Todd RL, Asmussen LE, Leonard LE (eds) Nutrient cycling in agricultural ecosystems. The Univ. of Georgia, College of Agriculture Experiment Stations, Special Publ. No. 23. Athens, Georgia

Patterson DW (1990) Introduction to artificial intelligence and expert systems. Prentice-Hall, Englewood Cliffs, p 448

Index

A. Atangana et al., *Tropical Agroforestry,* DOI 10.1007/978-94-007-7723-1,

CPSIA information can be obtained at www.ICGtesting.com
Printed in the USA
BVOW11*0430181215

430585BV00001B/3/P

9 789400 777224